TECHNOLOGY AND THE ECONOMY

The Key Relationships

ORGANISATION FOR ECONOMIC CO-OPERATION AND DEVELOPMENT

ORGANISATION FOR ECONOMIC CO-OPERATION AND DEVELOPMENT

Pursuant to Article 1 of the Convention signed in Paris on 14th December 1960, and which came into force on 30th September 1961, the Organisation for Economic Co-operation and Development (OECD) shall promote policies designed:

— to achieve the highest sustainable economic growth and employment and a rising standard of living in Member countries, while maintaining financial stability, and thus to contribute to the development of the world economy;

— to contribute to sound economic expansion in Member as well as non-member countries in the process of economic development; and

— to contribute to the expansion of world trade on a multilateral, non-discriminatory basis in accordance with international obligations.

The original Member countries of the OECD are Austria, Belgium, Canada, Denmark, France, Germany, Greece, Iceland, Ireland, Italy, Luxembourg, the Netherlands, Norway, Portugal, Spain, Sweden, Switzerland, Turkey, the United Kingdom and the United States. The following countries became Members subsequently through accession at the dates indicated hereafter: Japan (28th April 1964), Finland (28th January 1969), Australia (7th June 1971) and New Zealand (29th May 1973). The Commission of the European Communities takes part in the work of the OECD (Article 13 of the OECD Convention). Yugoslavia has a special status at OECD (agreement of 28th October 1961).

Publié en français sous le titre :

LA TECHNOLOGIE ET L'ÉCONOMIE
Les relations déterminantes

FOREWORD

This publication contains the background report concluding the Technology/Economy Programme (TEP) launched by the OECD in 1988. This report, along with the recommendations made by a High Level Group of Experts, served as the basis for a declaration concerning technology and the economy adopted by the OECD Council when it met at Ministerial Level on 4th and 5th June 1991.

The TEP addresses the concerns of OECD Member countries in a period of major change, for a better understanding of the interactions between technological development, the economy and society. An integrated and comprehensive approach of this sort is indispensable to informed policy-oriented decision-making.

The background report is based on two sources of information and analysis: firstly, work carried out within the Secretariat, particularly at the request of Committees involved in the TEP; secondly, the rich harvest of information, analysis and suggestions arising out of the ten conferences organised in the course of 1989 and 1990 at OECD Headquarters and in several Member countries.

These conferences led to an assessment of the existing work and compared alternative approaches for each of the major TEP themes. Each conference was designed and prepared with close co-operation between Member countries and the Secretariat and benefited from the active involvement of industrialists, policy-makers and academic experts.

This approach allowed for the early identification and examination of new themes and for them to be brought rapidly to the attention of the Organisation and Member countries. This was the case, in particular, for the analysis of globalisation, the importance of intangible investment, the role of networks in technological innovation, the new approaches to competitiveness and growth and the serious and urgent problems facing developing countries due to the present requisites of technological development.

This background report could not, without excessive length, do justice to the enormous quantity of facts and ideas which were shared in the TEP process by the hundreds of people who contributed to this programme. Only the principal analytical results are presented here, on which numerous studies, both in the OECD and elsewhere, are now based.

I would like to thank all those who contributed to the success of TEP, whether in Member countries or in the Secretariat. They have moved forward our understanding of the mechanisms which drive the world economy of today and which will determine that of tomorrow.

This report is based on contributions from a number of staff members of the Directorate for Science, Technology and Industry as well as the Economics and Statistics Department, the Directorate for Social Affairs, Manpower and Education, the Environment Directorate, the Development Centre and the Development Co-operation Directorate.

François Chesnais was responsible for the overall co-ordination and final preparation of the report.

Jean Claude Paye
Secretary-General of the OECD

ALSO AVAILABLE

Science, Technology, Industry Review

No. 8 – April 1991 (1991)
(90 90 01 1) ISBN 92-64-13471-9
Per issue   FF110   £13.50   US$23.50   DM45

No. 7 – July 1990 (1990)
(90 89 02 1) ISBN 92-64-13402-6
Per issue   FF110   £13.50   US$23.50   DM45

Technology and Productivity . The Challenge for Economic Policy (1991)
(92 91 01 1) ISBN 92-64-13549-9   FF560   £78.00   US$129.00 DM231

The Technology/Economy Programme (TEP) . International Conference Cycle (1991)
(92 91 03 1) ISBN 92-64-13489-1   FF150   £20.00   US$36.00   DM58

CONTENTS

DETAILED TABLE OF CONTENTS

Chapter 9

Chapter 10

Chapter 11

LIST OF BOXES

LIST OF FIGURES

LIST OF TABLES

Throughout this report Germany refers to the Federal Republic of Germany prior to German re-unification.

All references to dollars ($) are to US dollars, unless otherwise indicated.

The following symbols have been used in the tables:

    ..   Not available
    –   Nil or negligeable
    /   Break in series
    .   Decimal point
    ppp Purchasing power parities

General Introduction
Objectives, main themes and overall conclusions

This report discusses some of the key relationships which link technology and the economy today, in the setting of the national and international challenges faced by OECD countries.

The report covers a very broad area and addresses a range of issues which pertain directly to the ties between technology, human resources, patterns of industrial organisation, economic growth and international competitiveness. It also examines globalisation, the impact of recent technological changes on the situation of developing countries and questions related to technology and the environment. While they have been conceived as part of a whole, chapters can also be read and consulted separately.

The breadth of the area covered reflects the multifaceted nature of the relationships between technology and the wide variety of current economic activities. The analysis is, nevertheless, guided by several main themes which are all related to the central concern that was at the origin of this report.

The need for an integrated approach

The 1988 decision of the Council to launch the TEP programme was aimed at moving closer towards an integrated overall approach to technological, economic and social issues. The intention, in particular, was that this should lead ultimately to a better understanding and mastery of the relationship between technology and economic growth.

The need for a more integrated approach in the OECD to the problems of technology, productivity, economic growth, international economic relations, the situation of developing countries and environmentally sound long-term sustainable development is not new. As early as 1980, the Committee for Science and Technology Policy sought to respond to this need in its report *Technical Change and Economic Policy* (OECD, 1980), and this concern was expressed by OECD Ministers responsible for Science and Technology in their 1981 declaration (see OECD, 1981c, pp. 185-186).

However, it was undoubtedly the 1988 Sundqvist Report, *New Technologies in the 1990s: A Socio-Economic Strategy,* that marked the turning point for the OECD's work on issues at the interface of the economy and technology. Indeed, the Council's decision to launch the TEP originated in the discussions which followed the presentation and publication of the Sundqvist report, particularly its conclusions on "the interdependence of technical, economic and social change" and the fact that "technological change is, in its development and application, fundamentally a social process, not an event, and should be viewed not in static, but in dynamic terms" (OECD, 1988b, p. 11).

Technological change as an endogenous economic process

This report follows on from the approach of the Sundqvist Report and takes up the analysis of technological change as endogenous to the economy and society. New technologies do not originate outside the economic system and subsequently penetrate it. Rather the mechanisms and processes examined here show that technologies are invariably conceived, developed and diffused by means of long and costly investment. They respond to individual commercial demands and collective needs; they are developed and diffused under multiple economic constraints. Even as a starting hypothesis or an analytical basis, technological change cannot be treated as an exogenous factor in medium or long-term growth.

In particular, the TEP has given the OECD an opportunity to examine closely a number of factors that provide a much more precise view of the endogenous nature of technological change. Following the qualitative changes in the knowledge intensity required for the production of goods and services over the last two decades, these factors relate to the role of tangible but also of intangible investment (see below). They also include the phenomena referred to by economists in terms of externalities, interactive and cumulative processes and increasing returns to scale similarly. The forms of organisation and management of firms, the co-operation agreements established between them, the level and efficiency of investments in human resources, and the quality of networks of public communication and infrastructure, are now seen as central to the relationship between technology and growth. None of these elements is exogenous to the deployment of factors of production: each implies an investment and has a cost; each has a role in the overall relations which give rise to positive externalities and economies of scale.

While it has not yet proved possible to construct a formal economic growth model that includes these elements, this report is a contribution to encourage work along such lines. A better understanding of relationships between technology and growth, whether for the purposes of analysis or economic policy, presupposes first of all the integration of recent advances in the theory of innovation.

Innovation as an interactive process

The concept of innovation has recently changed dramatically as the focus has shifted from the single-act philosophy of innovation to the complex social mechanisms that underlie new production processes and the production of new products. At the same time, the previous reference point, the linear model of science and technology, has been supplanted by the "interactive" models of innovation. These models now emphasise the central role of industrial design, the feedback effects between downstream (market-related) and upstream (technology-related) phases of innovation and the many interactions of science, technology and innovative activities both within firms and in co-operative agreements among them.

The interactive character of the innovation process calls for organisational structures and mechanisms to ensure the appropriate interactions and feedback inside corporations as well as among the various institutions that make up national systems of innovation. For both analysis and policy, this model underscores the importance of co-operation between firms and institutions, and, thus, the role played by links and networks involving different organisations. The growth of inter-firm alliances represents a major change in the area of innovation. These emphasize, in particular, the increasing symbiosis between science and technology, the pervasive nature of some key contemporary technologies; and the synergy and even fusion of some technologies.

The interactive nature of the innovation process further calls for a more correct understanding of the relationships between science and technology. University and other long-term scientific research laborato-

ries, and the public funding authorities who support their work, are still extremely important actors in national innovation systems. The difference between science, which produces general, fundamental and abstract forms of knowledge, and technology, which is specific and practical, also requires the development of a number of "transfer sciences" situated at the interface between basic knowledge and the solution of concrete problems arising from economic and social needs. University-industry co-operation plays a major role in the symbiotic process between science and technology.

Long-term research

In the twentieth century, new technology systems have invariably emerged in conjunction with major scientific advances. Even the renewal of the most standard technologies is quicker and more efficient if these technologies are connected to scientific disciplines which have reached a predictive stage, but whose development is continuing. Long-term research thus has considerable economic implications.

To increase the stock of totally public scientific knowledge, accessible to all the actors in the innovation process, it is necessary to finance the activities of certain institutions dedicated to long-term research, such as universities. The notion that such research has become a factor in competitiveness is reflected in the efforts of some large firms. However, only a small number of firms are in a position to undertake the high-level of R&D investment required. Moreover, the knowledge acquired in this manner raises questions of intellectual property rights and access for the whole research community.

Universities and other higher education and research institutions provide the qualified personnel required by an economy based on specialised professional knowledge, and they guarantee to applied research and technology development the upstream long-term research that they require. Yet public expenditure on education and long-term research in the higher education sector came, in some OECD countries, under increasing pressure through the 1970s and 1980s. This is an unfortunate trend that unless rapidly reversed will have detrimental effects on growth and welfare in the long run.

Support for long-term research does not imply equal support for all disciplines. New ones are born which require nurturing, while others as they grow older may be prone to rather unproductive forms of scientific perfectionism. Periodic assessments of resource allocation between disciplines are required. Policies for resource allocation should likewise give priority to interdisciplinary research and to disciplines that play a natural interface role. This is the case for the engineering and other transfer sciences as well as for research related to pervasive generic technologies.

The cumulative nature of technology and the learning process

The concept of "cumulativeness" in relation to technology involves several important complementary dimensions of the innovation, creation and diffusion process. The first is that key learning processes are part of the full development and use of new technology. This is true for firms initiating innovation as well, of course, as for firms that adopt the new technology at later points in the development *cum* diffusion process. These learning processes include learning-by-doing, i.e. increasing the efficiency of production operations, learning-by-using, i.e. increasing efficiency by the use of complex systems, and learning-by-interacting, involving users and producers in an interaction now seen as driving product innovations.

The accumulation of skills, experience and technical know-how, whether at the level of firms or of countries, takes time and is a process pertinent to long-run economic development. The heritage of technology and human capital can only increase through a snowball effect, through gradual accretion, from the local, national and international environment, to an initial kernel of knowledge.

Since innovation reflects learning, and hence accumulated experience and know-how, as much as it does novelty, and since learning arises partially out of routine activities, innovation will be firmly rooted in the prevailing economic structure. The countries, firms and institutions which have been able to exploit the opportunities over many decades and create a base for technological accumulation are the best placed to adapt to the transitions and transformations of structural change.

Technological change renders equipment obsolete and eliminates the demand for given kinds of worker skills, it leads to the scrapping of plant and to some forms of technological unemployment. But, it cannot *per se* destroy institutions or firms, nor the knowledge which they have built up through technological accumulation and institutional learning. The foundations on which technological learning has taken place can only be destroyed as a result of bad corporate or institutional management, insufficiently considered budgetary cuts, mergers and acquisitions carried out without regard for accumulated learning within the organisations concerned.

Technology diffusion and organisational change

The interactive factors listed above have resulted in the breaking down of the earlier distinctions between the creation of technologies and their diffusion. This is particularly relevant to the so-called "generic" technologies. It has long been recognised that, in the area of science, the assimilation of new knowledge and its creation are part of a single process. Today the same holds true for technology: the range of complementary technological knowledge and skills required for an even larger part of innovation means that firms will need to absorb, create and exchange (transfer) knowledge as part of a single broad process.

An important dimension of technology diffusion concerns the absorptive capacity of firms and research organisations. Absorptive capacity refers to the ability of firms to learn and use the technology developed elsewhere through a process that involves substantial investments, particularly of an intangible nature. In this respect, R&D has a dual role: in addition to developing new products, it enhances the capability of firms to learn, to anticipate and follow future developments. This again implies the creation of formal or informal networks of industries and firms, within which participants are at the same time supplying and acquiring new technology.

Technology diffusion occurs widely through components, machinery, equipment (such as computers or computer assisted-equipment), which are produced by very high technology sectors. Technology transfer to other sectors is only accompanied by a productivity transfer if firms accept often profound changes in organisation. This has become more important still with the contemporary wave of new technologies, in particular the information technologies. Today, organisational changes imply, among other things, continuous training, acquisition and use of software, adaptation of work organisation, labour relations and management structures, market exploration and the formation and development of technological links with other firms and with suppliers and customers.

Human resources

The issue of human resources is tackled in this report from two angles. The first is that of scientists and engineers whose work is directly related to technology. In a large number of OECD countries there are indications of a growing scarcity of scientific personnel and engineers. Furthermore, science and engineering at all levels of technical competence do not seem to be attracting a sufficient proportion of students to meet the needs of the future. Unless attitudes change, this may become a serious impediment to technological development and growth. Given possible world-wide shortages and the trend towards international sourcing, countries both within and outside the OECD could also be faced with increased rates of migration of highly skilled labour. Such a development would create new tensions among and within nations.

An adequate supply of appropriately educated and trained personnel is likewise one of the critical factors in creating and diffusing technology. For the foreseeable future, dividends will accrue to the capacity of the work-force to absorb new learning. A dynamic economy requires life-long education and retraining. Learning is increasingly becoming a life-long occupation as the nature of the demand for skills changes ever more frequently. Given the large and complex effort required to increase the level and homogeneity of human resources, firms must link their efforts to those of national and local authorities. Certainly some progress has been achieved, but at the same time most OECD countries are seeing the emergence of many pockets of illiteracy.

At the heart of these two problems is the notion of "professional" or "technical" personnel. Both intuition and experience suggest that the role of this personnel is essential. Even more than the number of scientists and engineers, the absence of sufficient numbers of qualified technical personnel may make the difference between the industrialised economies and the developing countries.

The central role of tangible and intangible investment

Broadly speaking, investment is of two types: the traditional material or tangible investment in physical assets, such as machinery, plant and buildings; and the range of outlays on human and financial resources that are covered by the umbrella of intangible investment. In addition to investments in technology (R&D expenditures or the purchase of its results), it covers investments in training, outlays for a growing number of diversified business services, expenditures on market exploration and the acquisition and exploitation of software. Private intangible investments will often have a close interface with certain categories of public investment such as education and R&D.

There is a clear complementary relationship between the more traditional investment in physical assets and intangible investment. Intangible investments play a crucial and increasingly important role in reaping the potential benefits, in terms of greater output and productivity, associated with technical change. Investments in marketing or in the management of scientific and technological information, for example, are an important element in the interaction of demand and supply for goods and services. They develop the necessary information structures that help identify market signals and transform them into instructions for the deployment of productive resources. Alternatively, these information structures are used in order to stimulate the response of consumers to innovative efforts by firms.

Investment of both a material and an intangible nature thus has a central mediating role in the creation and diffusion of innovations and in transforming new technologies into economic growth. It determines both the physical and innovative capacity of firms and thereby has a direct impact on the speed of technological change and the pace at which firms replace production processes. It tends to be particularly effective as the link between knowledge production and economic growth in macro-economic and structural environments characterised by high savings rates, low cost of capital and a long-term perspective on the part of enterprises and capital markets. In this sense, investment lies behind many of the processes and issues examined here.

Technology and growth

The relationship between technology and growth is a central concern of OECD Member countries. In the second half of the 1980s discussions, as in the OECD International Seminar (OECD, 1991c), took as their starting point the contradiction between the apparently rapid acceleration in technical progress since the end of the 1970s and the absence of any significant measurable impact on the growth of total factor productivity. In analytical frameworks which viewed technology and its creation as an exogenous variable, the growth of this indicator has always been considered as a proxy for technical change.

Despite the inherent difficulties in measuring the effects of technical progress on factor productivity in a context of steady growth in the service sector and the blurring of its frontiers with those of the manufacturing sector, the analysis here suggests that the causes of the observed contradiction lie in a mismatch between earlier forms of corporate organisation, as well as those of the public sector, and the characteristics of the new technologies, notably information technology. The slow rate of productivity growth could also be due to deficiencies of countries and firms in the training and management of human resources.

At the analytical level, the materials and concepts examined during the TEP lead to a concept of growth that has undeniable affinities with the "new growth theories" although diverging on some important points. Previous growth models generally assumed that investment opportunities are created by scientific advance or by inventive activity at a pace that is independent of investment. Another hypothesis was that investment opportunities exhibit diminishing returns and are thus gradually exhausted as a consequence of undertaking investment. The "new" theories suggest instead that the very process of undertaking investment creates and reveals further opportunities for both investment and growth.

The "new" theories are based on the partial public good features of innovation and in particular on its cumulative nature and the possibility for collective appropriation: unlike normal economic goods that are exhausted in use, the use of a new design or production

18

process by the innovator does not preclude others from using it. New knowledge or technology therefore benefits more than just the originating firm; other firms or industries improve their productivity by building on, and adding to, the cumulative stock of knowledge. This is also expressed in the concept of "social return" (or "social profit") to innovation activities which goes far beyond the private return to the innovator. Modelling of this approach has begun, although the initial results cannot yet be considered completely satisfactory. Given the novelty and potential importance of this research, more theoretical and empirical work should be carried out in this direction.

The increasing reference in growth theory to the existence of increasing returns to technology and other forms of intangible investment reflects concepts close to those mentioned above. However, on certain key analytical points the presentation in this report distances itself from the new growth theories. Three of these points deserve mention.

First, nothing in this report supports the idea, contained in the new models, of exponential or explosive growth. On the contrary, both the very long gestation time for some forms of investment (in particular, private intangible investments and public investments in the training of human capital) and the vulnerability of the technological accumulation process to some forms of industrial restructuring and financial operations introduce many built-in breaks and obstacles that may work against optimal interaction between technology and the economy. The long-term result is rather to set ceilings to growth, except in historically exceptional situations (e.g. 1950-70).

Second, this report emphasises the role played by non-market co-ordination between private agents in the creation of externalities and interactive mechanisms. This may take forms such as inter-firm co-operation agreement or the enhancing institutional environments found in such or such a country. Moreover, once the key role of externalities and public goods in economic growth is recognised, a place for non-market institutions and policies appears. In developing countries, such policies may be the only way to create important externalities and to trigger off the cumulative "virtuous" processes that might reduce the divergence in growth paths.

Finally, the report gives some importance to externalities resulting from links between producers and users or consumers. The hypothesis that growth will only be possible with the diffusion of new approaches to consumer demand emerged quite strongly from the TEP analyses. Firms must not only satisfy tastes that are more and more refined and personalised but also meet society's collective needs, expressed through a wide range of democratic and associative mechanisms. Hence, the rapid development of new forms of interaction between producers and customers. The latter include firms which are currently multiplying R&D

partnership agreements with their suppliers. In fact, many studies show that interaction with demanding and informed consumers is an essential factor in both growth and industrial competitiveness.

The different dimensions of competitiveness

The overall approach and the findings of the TEP also have implications for the understanding of competitiveness. While satisfactory macro-economic conditions and appropriate national policies are a prerequisite for competitiveness, many key factors now concern the micro-economic level and the capacity of firms to organise for technological change. Further decisive processes occur at a regional or local (e.g. site) level where dynamic, interactive and cumulative processes take place between firms and their "environment".

The importance attributed to micro-economic factors in competitiveness is reinforced by the analysis contained in this report. Efficient competition depends on the quality of products (as much as on their intrinsic novelty, which most competitors can copy easily), superior process technology and organisation of production, speed of delivery and quality of after-sales service. All these aspects are necessarily, to a large extent, an expression of the quality of the management of the firm, whose success also depends on its capacity to participate in co-operation networks and maintain efficient trade with suppliers and users of its products.

Numerous studies have also shown, however, that the performance of firms depends not only on their own qualities, but also on the structural characteristics of the sectors and countries in which they operate. The concepts of "structural competitiveness" and of the "diamond of national competitive advantage" now offer two ways of bridging the gap between macro-economic policies, corporate performance and the main findings of TEP with respect to the interactive nature of the innovation process, the systemic features of technology, the learning processes associated with innovation, the role of inter-firm co-operation in its many forms, the vital importance of human capital, and the significance of organisational and institutional factors in innovation.

The TEP has also witnessed a renewed interest in the funding of industrial R&D and other costly innovation-related investments. In considering the issues, the focus has shifted: formerly, stress was laid on the measures taken by each country to ensure public R&D funding on favourable terms for enterprises; now greater emphasis is placed on the organisation of capital markets and banking policies, and on the way that these factors influence enterprise investment decisions. Recent studies have pointed to significant differences in the micro-structure of capital markets in different OECD countries. The role played by national banking systems in the financing of R&D and intangible investment is now considered an important factor in the shaping of structural competitiveness in each country.

Virtuous circles of interaction and feedback – with respect to the accumulation of physical capital, the development and improvement of human capital, technological accumulation and the competitive performance of firms – are generally more easily created in certain sectors or groups of industries, or, in locations which present specific advantages. National competitive advantages will stem from the number and nature of such industries and sites. But the strength of these advantages depends on the extent to which industries and sites have interactive cumulative feedback relationships among themselves and do not remain isolated "islands of prosperity" within countries otherwise marked by the stagnation or falling back of whole areas and regions. The building of such relationships for the innovation system and throughout the national economic structure is a challenge now faced by governments as a result of globalisation.

Technology and globalisation

Globalisation refers to a set of emerging conditions in which value and wealth are produced and distributed within world-wide networks. Large multinational firms operating within highly concentrated supply structures are at the hub of these conditions. During the 1980s, two main factors accelerated changes in earlier patterns of internationalisation and led to globalisation. The first is deregulation and the increasing globalisation of finance. The second is the role played by the new technologies, which have acted both as an enabling factor and a pressure towards yet further globalisation.

Globalisation is marked by a new ranking of the factors creating interdependencies. International investment, rather than trade, dominates internationalisation and thus determines the structures that govern production and trade in goods and services. Intra-firm trade flows are becoming ever more important. International investment is clearly strengthened by the globalisation of banking and financial institutions which facilitates mergers and international acquisitions. New forms of inter-firm agreements bearing on technology have developed alongside the traditional means of international technology transfer – licensing and trade in patents – and they have often become the most important way for firms and countries to gain access to new knowledge and key technologies. Finally, new types of multinational enterprises with network-like forms of organisation are emerging.

The transition from internationalisation to globalisation has been accompanied by an increase in industrial concentration. In a growing number of industries in manufacturing and services, this implies that the prevailing form of supply structure is world oligopoly. This new situation raises problems relating to the measure of international concentration and possibly its control in the face of the potential dangers of weakly contestable markets and international cartelisation. Further issues requiring special attention have been identified. The first is the marginalisation of developing countries and the risks of selective exclusion from globalised information networks. The second is the question of government support to firms engaged in global competition and the need for new rules and codes of behaviour. The third relates to the setting of standards and norms in the new context of world concentration and globalisation.

Globalisation is also modifying the basis of competitiveness. It tends to reinforce the cumulative character of innovation-based competitive advantages for large firms, but it may weaken the resource base and organisational cohesion of domestic systems of innovation. In the future, the competitiveness of large network firms will increasingly stem from their ability to make the best use of R&D and human capital resources located in several countries. This move towards "techno-globalism" – a term used in the TEP meetings (see OECD, 1991h) – could prompt reactions of "techno-nationalism". Careful comparisons of the various national innovation systems, leading to a better understanding of similarities and differences, can help in the evaluation of possible conflicts between systems and, if necessary, the elaboration of appropriate rules.

The special situation of developing countries

It was stated above that technology, *per se,* cannot and does not destroy institutions or firms nor the knowledge which they have built up as a result of institutional learning. For this statement to hold, the conditions that allow technological accumulation and learning must really have occurred. This is the case for most OECD countries, but it is certainly not the case for developing countries. In the absence of effective technological learning and the establishment of the material and social conditions for its use, some countries will see technical change principally or even exclusively in the form of substitution effects and loss of markets in international trade – in the face of new norms for the productivity of labour and other factors of competitiveness.

What separates these countries from OECD countries is a huge gap in their capacity to use technological change as a motor for growth, structural transformation and modernisation. They have a cruel lack of the institutional structures and human capital which would allow them to absorb, reproduce, adapt and improve imported technologies alien to their traditional know-how. They often even lack the relatively simple qualifications required for operation, assembly and repair. In fact, since the early 1980s, the main indicators of the volume of international technology flows – foreign direct investment, capital goods imports, payments for licences and know-how, and official technical assistance – show an unprecedented shrinkage of technology flows to developing countries, with only two exceptions: China and the East Asian Newly Industrialising Economies.

Yet the technologies of today, in particular the new information technologies, contain a huge and still largely untapped potential for creating new technology combinations that could be based on local know-how. At present many developing countries seem unable to make the changes necessary for productivity growth through the new technologies. The present assessment is that despite the new technologies, and indeed on account of their particular requirements in terms of resources and skills, the gap separating rich and poor nations is likely to increase.

The overall assessment for developing countries in terms of growth rates and participation in world trade and trade balances is in line with predictions of international economic models coming from, or close to, the new models of endogenous growth. The many sources of externalities and the interactive mechanisms that determine competitiveness and growth lead directly to the possibility of cumulative processes, both positive and negative, resulting in very different growth paths for different countries.

Technological trajectories, social assessment and environmental management

The understanding that the development and application of technological change is fundamentally a social process has been heightened by increased analytical attention to the way in which technologies can be seen to follow trajectories. The path of technological advance is by no means determined. The use of technologies and the selection of potential applications will depend on a wide range of economic factors (e.g. relative prices, income distribution), social values and arbitrations on the part of the main actors involved. The concept of "increasing returns to adoption" expresses the fact that technologies may not be selected on account of their superior efficiency but rather, may become efficient because, and after, they have been chosen.

This vision of technological change should lead to a better grounding in technology assessment and to better understanding of the potential for contemporary societies to modify developments. Experience shows that technological progress is not only a matter of innovation and diffusion. It is also a matter of social acceptance.

The interactions between modern technology and the natural environment represent an area where public concern is very strong and the issue of social acceptance particularly sensitive. Over the greater part of this century, the choice of technologies and the shaping of technological trajectories have been dictated almost exclusively by considerations of labour and capital productivity, cost and competitiveness. However, there has recently been a brutal recognition that respect for the environment should be added to the determining factors in technology development. After a phase of adaptation, even conflict involving environmentalists and firms, one can now see emerging a new approach to the relationship between technology, growth and the environment, in which environmental concerns are not simply a constraint on economic growth but offer possibilities for choosing new paradigms and trajectories leading to new growth paths.

Consequences of the TEP for the development of new and improved indicators

Understanding the relationships between science and technology and economic and welfare performance is largely conditioned by the availability of relevant statistics and indicators. Special attention was paid during the TEP to assessing the data needs, both to profit from the new analytical tools and conceptual frameworks outlined above and to help in their further improvement. The most important result of the exercise was to bring together users and producers of statistics and indicators to discuss the findings of the TEP and to encourage them to look beyond their own specialisation, to clarify the conceptual frameworks underpinning their current work and to participate in building a consensus on desirable yet feasible developments in the medium term.

The main general area for improvement lies in work aimed at integrating statistics previously collected, analysed and published separately (especially science and technology statistics, industrial statistics and education and employment statistics). New indicators should be developed for innovation and its diffusion and for intangible investment and its components. More attention needs to be paid to data with an international dimension in order to contribute to a better analysis of globalisation. There is a need to collect better data on human resources, especially data on training and the supply and demand of scientists and engineers. More controversial, given the rules governing the official collection of statistics, is the need for indicators on firms, especially MNEs. The indicators of long-term research should be improved, particularly in the higher education sector.

By way of a conclusion

The TEP describes a complex system in which numerous public and private actors interact. The information that can be drawn from this report as a basis for action will depend on the actor's situation and responsibilities.

For the *industrialist,* whether in the manufacturing or service sector, the principal argument of the TEP is that the ability of the firm to take advantage of the technology that it has produced or imported will depend on the organisational and relational framework into which the technology is integrated. This means that the central role of human resources must be recognised and that the firm must devote resources to their development, i.e. make investment in training. Further, the report implies that attention be paid to co-operative relationships, not only among firms and institutions but also among all those within firms who participate in the technological creation, production and diffusion process.

For *government,* to think in terms of a complex system is to understand that while the firm is the main link in technology-economy relationships, its intrinsic economic and technological efficiency also depends on the institutional characteristics of the national innovation system as well as the network of international relationships within which it operates.

For *ministers responsible for innovation,* attention will be paid to the evaluation and improvement of the national innovation system in terms of its ability to create, diffuse and take advantage of technological change. The creative as well as the adaptive capacity of a national system depends on the number and quality of its components, but also on the nature of the relationships established among client firms, suppliers, markets, public agencies, financial and banking institutions, education and training institutions at different levels (both technical and university), university or industrial research centres, infrastructures, etc. National systems of innovation involve a set of networks linked in such a way that the creation and diffusion of technology and its transformation into commercial products depend as much on the vitality of the whole set of relationships as on the individual performance of any given element of the system.

This approach is not only of interest for government officials responsible for technological policy, education and training, and research. It also concerns those responsible for *economic policy* in all its forms. As other reports (in particular, the Sundqvist Report) have done, this study underscores the fact that investment, both physical and intangible, is the necessary channel for all forms of innovation and technological change. Thus these can only develop in countries and regions where the rate of investment is high and the climate favourable. This obviously depends on the banking system, the attitude of financial markets to the long term, and the full range of fiscal and monetary measures. For this reason, it is important that governments evaluate the impact of their macro-economic decisions on enterprise innovation and engage in long-term investment in education, research and infrastructures which prepares for the future.

Recognition of the systemic aspects of the relationships of technology, competitiveness and growth, together with the importance of social, institutional and cultural factors in a country's or region's ability to profit from technological change, gives rise to novel issues concerning the organisation of *international relations.* The possibility of more acute competition among firms in association with friction between national systems with distinctive institutions and economic performances, is a hypothesis which follows directly from the TEP conclusions. The OECD has begun to look at the inherent risks of this eventuality and at possible solutions (OECD, 1991h).

In the face of the combined effects of globalisation and cumulative growth mechanisms highlighted by the TEP, developing countries are particularly vulnerable. Many developing countries already face significant risk of losing touch with technological progress in the industrialised regions of the world. If this trend is to be halted and reversed, special attention must be paid to facilitating the access of *developing countries* to new product and process technologies, but also to assisting them in establishing the appropriate physical, educational and managerial infrastructures for assimilating and harnessing technological change.

Chapter 1

TECHNOLOGICAL INNOVATION: SOME DEFINITIONS AND BUILDING BLOCKS

ABSTRACT

The understanding of technological innovation has recently changed dramatically. Interactive models, differing significantly from the earlier linear approach, now emphasise the central role of industrial design, the feedback effects between downstream (market-related) and upstream (technology-related) phases of innovation and the numerous interactions between science, technology and innovation-related activities within and among firms.

Innovation is now deeply embedded in firms throughout OECD, and during the 1980s, business financed and business performed R&D increased strongly. University laboratories, government research centres, and non-profit organisations can contribute significantly and sometimes decisively to scientific and even technological breakthroughs, but they are generally only marginally responsible for commercial innovation.

The shortening of the period separating scientific breakthroughs from their first market application and the close interface between basic science and production-related technology now represent permanent features of the innovation system. However, the development phase remains very lengthy and R&D and other start-up costs are rising rapidly.

Proper relationships between science and technology are increasingly crucial for successful innovation. University and other long-term scientific research laboratories, and hence the public funding authorities who support their work, are still extremely important actors in national innovation systems. The difference between science, which produces general, fundamental and abstract forms of knowledge, and technology, which is specific and practical, also calls for the support of a number of "transfer sciences" situated at the interface between basic knowledge and the solution of concrete problems arising from economic and social needs.

Technological change involves important cumulative learning processes for users and producers. These processes include learning-by-doing, learning-by-using, and learning-by-interacting. Research institutions and firms form the foundation of these learning processes. Technologies are now seen to follow "trajectories" and to be subject to complex processes of selection. The use and application of technologies depend on a wide range of economic factors (relative prices, income distribution), social values and arbitrations on the part of the main actors. The related notion of "increasing rates of adoption" expresses the fact that technologies may not be selected on account of their superior efficiency, but rather become efficient because they have been chosen.

INTRODUCTION

This chapter examines the characteristics of the innovation process: its nature, its sources and some of the factors which shape its development. It aims to introduce the reader to concepts that are central to understanding the process of technological change and to emphasise the notion of process. These concepts will recur in later discussions of technology diffusion, issues of organisation, human resources and development, competitiveness and globalisation.

Section 1 draws attention to the interactive nature of innovation and to the numerous and costly investments entailed. Particular stress is laid on showing how the characteristics of innovative activities go beyond "linear" models of technological change. Section 2 turns to the production of scientific and technical knowledge to discuss the central position occupied by firms in the development of new technologies. After looking at science and technology as the bodies of knowledge on which innovation draws, it examines the interactions and linkages between them and argues that these sources must be continuously replenished. Section 3 addresses continuities and discontinuities in technology. It points in particular to the cumulative features of scientific and technological knowledge. The notions of "technological trajectories" and of "increasing rates of adoption" are then discussed in order to present a more complete picture of the development and selection of technology.

1. INNOVATION AS A PROCESS

To quote the first OECD study on the subject, in 1971, technological innovation must "be defined as the first application of science and technology in a new way, with commercial success" (OECD, 1971, p. 11). The definition focuses on the products and production processes which simultaneously embody some degree of novelty and receive market recognition. It implies that often in capitalist economies "certain kinds of R&D that would have high social value simply are not done" (Nelson, 1988, p. 313), thus creating conditions of market failure calling for government action. Today, R&D related to environmental issues is an obvious example (see Chapter 9). Use of the term innovation in this first sense has led to the development of a number of classifications, in particular the recent distinction among i) incremental innovations, ii) radical innovation, iii) new technology systems and iv) pervasive generic technologies (Freeman, 1987b). These terms have been used in earlier OECD reports and will be found elsewhere in this one (OECD, 1988b).

The definition centred on "first application", while useful as a starting point for analysis, has its limitations. It can be interpreted as lending support to the view of innovations as representing well-defined, homogeneous things that enter the economy at a precise point in time. In reality, most important innovations go through drastic changes over their lifetimes. Subsequent improvements can be vastly more important, economically, than the original invention[1]. In the case of totally new products or processes, many knowledge-related inputs and numerous feedbacks and trials enter the development and production process. Thus, the accent has shifted from the single act philosophy of technological innovation to the social process underlying economically oriented technical novelty.

It is now considered important "by using terms such as 'the process of innovation' or 'innovation activities' to indicate that traditional separations between discovery, invention, innovation and diffusion may be of limited relevance in this perspective" (Lundvall, 1988, p. 350). Throughout this report, whenever possible, the term "innovation" will be used in this sense.

An interactive model of the innovation process

For almost three decades, thinking about science and technology was dominated by a linear research-to-marketing model. In this model, the development, production and marketing of new technologies followed a

Figure 1. An interactive model of the innovation process
The chain-linked model

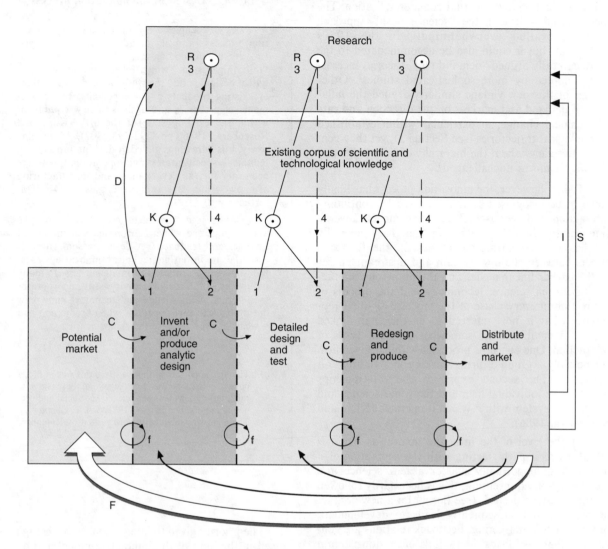

Individual firms and productivite sectors at a more aggregate level

Symbols used on arrows in lower boxes:

C = Central - chain - of - innovation
f = feedback loops
F = particulary important feetback

Vertical links:

K-R: Links through knowledge to research and return paths. If problem solved at node K, link 3 to R not activated. Return from research (link 4) is problematic - therefore dashed line.
D: Direct link to and from research from problems in invention and design.
I: Contribution of manufacturing sector scientific research by instruments, machines tools, and procedures of technology.
S: Financial support of research by firms in sciences underlying product area to gain information directly and by monitoring outside work. The information obtained may apply anywhere along the chain.

Source: Adapted with minor changes from Kline and Rosenberg (1986).

well-defined time sequence that originated in research activities, involved a product development phase, then led to production and eventual commercialisation. The model fitted well with the "science push" approach which prevailed overwhelmingly in the 1950s and 1960s, but it could also be accommodated to the more carefully argued "demand pull" theories increasingly adopted by more sophisticated studies[2]. Other studies of this new vintage similarly argued the influence of demand and markets on the direction and rate of technical change, in particular within established technological trajectories (see Section 3), yet they generally left views about the hierarchy of knowledge and the way it moves unchallenged[3].

Today, however, the innovation process has finally come to be recognised as characterised by continuing interaction and feedback. Interactive models diverge significantly from the linear approach. They generally emphasise the central role of design, the feedback effects between the downstream and upstream phases of the earlier linear model and the numerous interactions between science, technology and the process of innovation in every phase of the process. Figure 1 represents what is now referred to as the chain-linked model. It combines two somewhat different types of interaction. One concerns processes within a given firm (or possibly a group of firms working in a tightly-knit network). The second expresses the relationships between the individual firm and the wider science and technology system within which it operates (Kline and Rosenberg, 1986).

At the level of the firm, the innovation chain is visualised as a path starting with the perception of a new market opportunity and/or a new science and technology-based invention; this is necessarily followed by the "analytic design" (see Box 1) for a new product or process, and subsequently leads to development, production and marketing. Feedback relations are generated: short feedback loops link each downstream phase in the central chain with the phase immediately preceding it and longer feedback loops link perceived market demand and product users with phases upstream. Problems thrown up by the processes of designing and testing new products and new processes often spawn research in engineering disciplines but also in science.

In-house R&D by large firms and the establishment of corporate industrial laboratories represented the first and, over many decades, an exclusive and very successful way of organising the linkages in the lower block, especially when the relevant technologies were somewhat idiosyncratic and tacit. The centralised R&D function was associated with vertical and horizontal integration within hierarchical corporate structures. Integration facilitated better information flow between the R&D laboratory and those who implemented the new technology and also served to limit information leaks (Nelson, 1986, p. 10).

Box 1. **Design in the innovation process**

Design occupies a central place in the innovation process, and it includes several concepts. Broadly understood, it may convey the sense of "initiating design", which reflects invention; "analytical design"; and "the study of new combinations of existing products and components, re-arrangements of processes and designs of new equipment within the existing state of the art" (Kline and Rosenberg, 1986, pp. 292-293, 302). More narrowly, it is "drawings aimed at defining procedures, technical specifications and operational features necessary for the development and manufacturing of new products and processes" (OECDa, forthcoming).

Design activity is not a lower level or routine activity, but one which can originate a number of linkages and feedbacks. A design in some form is essential to initiating technical innovations, and redesigns are essential to ultimate success. In many industries, this design activity still incorporates tacit forms of knowledge and technical know-how dating back to earlier periods when production had a weak or even no base in science at all[a].

---

a) The first important author to stress this was in fact a pure scientist who was nonetheless highly sensitive to the interrelations between science and social and economic factors. See Bernal (1971, Vol. 1, Chapter 1). The point was subquently documented by Rosenberg (1982).

The priority given to integration is now being reassessed in the light of the interactive model and of new approaches to the organisation of the firm. The organisation of research within many large US and European firms and its relationships to engineering and production have come under scrutiny (Bienaymé, 1988). In particular, economists and social scientists in the United States have found considerable disconnection from the problems of production in many corporate laboratories as well as insufficient regard for design. Japanese firms appear to exhibit a much higher degree of recognition of the interactive character of the innovation process and consequently search for better ways of creating appropriate feedback relationships (see Chapter 4). When complementary skills external to a given firm are required, such relationships may entail inter-firm co-operation[4]. This is discussed in Chapter 3.

The second set of relationships visualised in Figure 1 link the innovation process embedded in firms and industries with the scientific and technical knowledge base (K) and with research (R). In an industry-focused, interactive approach to innovation, a useful analytical distinction can be made between the two different uses of science and technology by firms, namely i) the use of the available knowledge about physical and biological processes, and ii) the work undertaken to correct and add to that knowledge. Generally, innovation takes place with the help of available knowledge. When corporate engineers confront a problem in technical innovation, they will call first on known science and technology, most often in serial stages. Only when those sources of information prove inadequate does a need arise for research.

This analysis of the role of industrial R&D in the innovation process applies directly to large firms. Firms below a certain size cannot bear the cost of an R&D team. The critical size has been calculated to be of the order of one thousand employees in low technology industries, and 100 employees for high technology, using simple indicators such as the share of turnover devoted to R&D activities, and the average cost of an industrial researcher (Mordchelles-Regnier *et al.*, 1987), thus excluding most SMEs. However in the case of such firms, technological renewal does not need to be continual. While the average product-life has fallen, typically shelf-life is still a number of years, and consequently incremental innovation within the normal process of production can ensure competitiveness.

A number of solutions help ensure the technological "leaps" necessary for all firms. A large firm can find it advantageous to transfer to an SME, belonging to its network of sub-contractors, the necessary technology through partnership agreements for collaboration between its laboratories and the engineers or technicians of the SME. In other cases, the SME sub-

contracts its R&D effort to a contract research organisation (CRO). The relative importance of this arrangement is growing rapidly (there are currently 50 such organisations in Europe employing 25 000 people) and merits particular attention. Kline and Rosenberg show that R&D effort is only successful where a very strong interaction between all the active elements in the firm exists, which excludes research being done externally. Successful CROs actively collaborate with their client SMEs in the analysis of market prospects and in the evaluation of their know-how and "technological past";

Box 2. From description to prediction in basic science

"Knowledge in the physical and biological sciences tends to move through recognisable major stages. In the earliest stage the work in a science is descriptive; in the next stage the work becomes taxonomic; then the work passes to formation of generalising rules and hypotheses and finally, in some sciences, to the construction of predictive models. A science in the predictive stage, such as mechanics or classical electromagnetism today, is usable immediately by anyone skilled in the art for purposes of analytic design and invention. A science still in a descriptive or taxonomic phase is far less reliable for the latter purposes, despite the fact that it may still be very important in guiding the innovative work" (Kline and Rosenberg, 1986, p. 295).

Box 3. Two dimensions of the science-technology interface

There are crucial portions of high technology industries where attempts to advance the technological frontier are painstakingly slow and expensive because of the limited guidance available from science. For example, the development of new alloys with specific combinations of properties proceeds very slowly because there is still no good theoretical basis for predicting the behaviour of new combinations of materials, although materials science may now be approaching the point of developing models with predictive powers. Many problems connected with improved fuel efficiency are severely constrained by the limited scientific understanding of something as basic as the nature of the combustion process. The development of synthetic fuels has been seriously hampered in recent years by scientific ignorance of the relationship of the molecular structure of coal, which is known, to its physical and chemical properties. The requirements of computer architecture remain badly in need of an improved scientific underpinning. The design of aircraft and steam turbines are both hampered by the lack of a good theory of turbulence.

On the other hand, scientific breakthroughs are only the first step in a very long sequence of knowledge accumulation. The first lasers were developed around 1960 and have expanded into a remarkably diverse range of uses in the past thirty years. But, from the point of view of the historian of science, the basic science underlying the laser was formulated by Einstein as long ago as 1916.

The present-day world-wide search for products in which to embody the recently acquired scientific knowledge of high temperature superconductivity may only yield results, permitting large-scale commercial exploitation of this knowledge decades from today, just as the great breakthroughs in molecular biology of the 1950s are only now beginning to find an embodiment in the products of the new biotechnology industry (Rosenberg, 1991, pp. 3-4 and 17).

they work throughout the contract hand-in-hand with the SME personnel. Whether the result of this process is a prototype or a turn-key process, it will have been developed just as if there were an internal R&D team in place. This process is therefore different from the usual industry-university agreement (expect perhaps some Japanese universities that operate on the same basis).

The business enterprise sector brings some financial support to research and to the ongoing creation of new knowledge, either directly or through non-profit foundations (D). The rationale for this support is the fact that industrial production and the engineering disciplines that go with it require "predictive science" (see Box 2) in order to solve technological problems. When the state of the appropriate predictive science is unsatisfactory and key components of basic knowledge are lacking, engineers in industry will seek ways around the technical problems they meet and fall back on older knowledge. As Kline and Rosenberg have shown, "today there still remain crucial portions of high technology industries in which attempts to advance the state of the art are painstakingly slow and expensive because of the limited guidance available from science. (...) The results of this lack of predictive science are very high costs in development, long lead times (e.g. for the combustion space in new models of jet engines), and a strong and reasonable conservatism on the part of designers" (Kline and Rosenberg, 1986, p. 296), (see Box 3).

The length and costs of the innovation process

Earlier studies on innovation tended to lay considerable stress on the spectacular shortening of the period separating scientific breakthroughs from their first application in the form of marketed products or production processes. This dimension is now taken for granted: the extremely close interface between basic science and production-related technology in fields such as biotechnology or information technology (e.g. super-conductivity and information storage capacity on chips) confirms that, at present at least, speed continues to be a characteristic of the contemporary innovation system. Today, emphasis is placed rather on the length of the development phase and the high costs of innovation.

The different phases of the innovation process all entail outlays for investment (see Box 4). These outlays include "classic" capital investments and R&D expenditure, but also a wide range of outlays for intangible investments (see Chapter 5). They are often very costly, and time plays an important role. In the field of electronics, for example, the design and development of reliable, high-capacity memory chips have drastically raised the stakes for commercial survival. In the

---

Box 4. The pattern of investment in innovation

During the innovation phase, outlays will concern principally investment in technology (R&D, licences, design and engineering) with some fixed assets needed to produce sufficient quantities of the goods for market introduction, some market testing and market exploration, and possibly some exploratory "enabling" intangible investment (worker training and production reorganisation). In the market expansion phase, stress is placed on fixed capital investment (process innovation) accompanied by enabling intangible investment (particularly worker-training and outlays related to the reorganisation of production) and marketing investment associated with large-scale production. Once markets for the new product have peaked and competitiveness is based on price, there will be strong pressure to introduce new processes for rationalisation and reconstruction. This requires further enabling intangible investments in training and reorganisation, as well as investments in physical capital.

Box 5. Lead time and rising costs in the semiconductor industry

In the semiconductor industry, technology progresses so rapidly that each generation lasts only about four years, and the useful life of capital equipment rarely exceeds five. Typically, each generation of R&D must begin at least five years before commercialisation and must be co-ordinated with development of new capital equipment and materials, whose technologies are themselves becoming extremely complex and capital-intensive. Any semiconductor maker aspiring to hold or gain market share must therefore make very large investments. Current technology requires, on average, $200 million to $1 billion for each generation of process development, $250 million to $400 million for each factor, and $10 million to $100 million for each major device design. These costs are expected to double again by the late 1990s, when the world semiconductor market will exceed $100 billion (Ferguson, 1990).

Figure 2.   **Annual spending on motor vehicle R&D by the automobile
industries in the United States, Japan and Europe**
1967-88

Spending (millions of 1988 dollars)

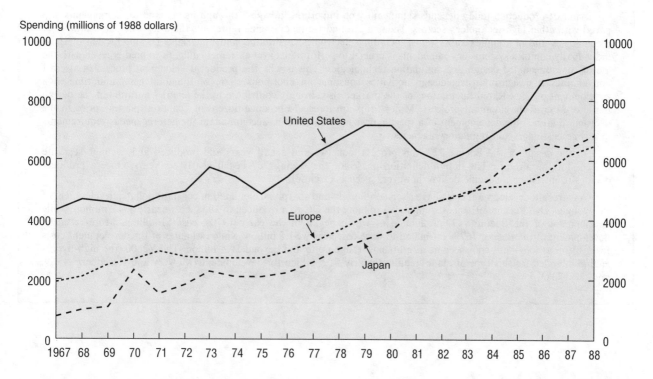

semiconductor industry, since the mid-1970s, every technology generation has seen more than a doubling in the scale of R&D and capital investment required to compete (see Box 5). Similar trends are apparent in other "new" areas, such as liquid-crystal displays, software, and telecommunications.

The length of the development phase and the rising costs of R&D are also an issue in older industries, such as the automobile and aircraft industries. The average development time of an automobile embodying a reasonably significant degree of technical novelty is about five years in the United States, a little less in Europe (60 and 57 months respectively) and a little under four years in Japan (46 months). For the same industry, flexible manufacturing systems (FMS) took two and a half to three years to be developed from conception to operation in the United States and about one and a half years in Japan (Womack *et al.*, 1990, Chapter 5). Lead times in the design and development

of new models are now recognised to be a central feature of competition between producers, but reduction in design and development periods has been accompanied to date by rising costs in R&D (Figure 2).

Design and development periods are longer still for large civil aircraft, and, in contrast to automobiles, they continue to increase. The design and development phase of the Boeing 727 lasted two and a half years; it increased to nearly six years for the Boeing 767. Accompanying the longer development time are dramatically increasing R&D costs (see Box 6).

The scale of the investments required for innovation in a growing number of industries is a factor in processes of industrial concentration and globalisation (see Chapter 10). Inter-firm networks and technological alliances (see Chapter 3) furnish a means of partly offsetting, through inter-firm co-operation, the scale of R&D costs.

2. THE PRODUCTION OF SCIENTIFIC AND TECHNICAL KNOWLEDGE

The innovation process involves the use, application and transformation of scientific and technical knowledge in the solution of practical problems. During the last thirty years, innovation has become deeply embedded in firms throughout the OECD area. While government research centres, university laboratories and non-profit organisations can contribute significantly and sometimes decisively (as in biotechnology) to scientific and even technological breakthrough, they are generally only marginally responsible for commercial innovation. The success of such innovation is nonetheless crucially determined by the underlying long-term basic research and its institutional set-up and by "transfer sciences", which provide a bridge between basic research and industrial innovation.

The central role played by firms: industrial R&D

R&D expenditure by firms as a percentage of value added has grown regularly in all countries. The role which industrial innovation has come to play in competitiveness and growth is reflected in the scale of outlays for R&D performed by the business-enterprise sector (see Table 1).

Since the beginning of the 1970s, the swing from public to private financing has effected a major structural change in the financing of R&D. In some OECD countries, business financed R&D was already predominant in the 1960s and 1970s; it is even more so today. In others – including the largest R&D spender, the United States – business financed R&D was relatively weak at that time in comparison with government financed R&D, which it gained on and overtook in the 1980s. Today, as shown in Table 2, about half the recorded outlays on industrial R&D are financed by business enterprises in most countries, and in some, the percentage is considerably higher. Overall, the shift to the business sector continues to be significant.

The trend towards industry-financed R&D is one reason why the slowdown in R&D expenditure observed in the mid-1970s proved to be of limited duration. The drop in government financed projects was quite rapidly offset by increased outlays by the business enterprise sector. In spite of short-term fluctuations, R&D expenditures as a percentage of GDP increased in virtually all OECD countries, particularly after 1979, even if differences among countries remain large. In most cases, industry has been the main motor

Table 1. **Expenditure on R&D performed in the business enterprise sector**

	% of GDP			% of GERD*		
	1975	1981	1989	1975	1981	1989
Australia	0.23[1]	0.25	0.52	23.2[1]	25.0	41.5[3]
Austria	0.47	0.65	..	50.8	55.8	..
Belgium	0.84	/0.96	/1.18[3]	64.2	..	/73.6[3]
Canada	0.41	0.60	/0.74	37.3	49.6	55.7
Denmark	0.45	0.54	0.84	44.2	49.7	55.0
Finland	0.48	0.65	1.11	52.1	54.7	60.8
France	1.06	1.16	1.40	59.6	58.9	60.3
Germany	1.41	/1.70	2.10	63.0	/70.2	73.0
Greece	..	0.05	0.10	..	22.5	/22.3
Iceland	0.03	0.06	0.12[4]	2.9	9.6	16.1[4]
Ireland	0.25	0.32	0.49[3]	30.6	43.6	56.8[3]
Italy	0.47	0.49	0.74	55.7	56.4	57.1
Japan	1.14	1.41	2.12	56.6	60.7	69.7
Japan (adj.)				62.7	66.0	74.3
Netherlands	1.08	1.00	1.32	53.6	53.3	/60.0[3]
New Zealand	0.20	0.22	0.27	21.7	21.7	28.6
Norway	0.64	0.68	/1.16	48.0	52.9	/60.8
Portugal	0.06[1]	0.09[2]	0.12[3]	21.1[1]	28.6[2]	24.6[3]
Spain	0.20	0.19	0.45	56.6	/45.5	56.3
Sweden	1.17	1.46	1.83	65.3	63.7	/66.2
Switzerland	1.84	1.70	/2.14	76.7	74.2	74.8
Turkey	..	..	0.01	..	..	8.4
United Kingdom	1.27	1.49	/1.37	58.4	61.8	/66.6[3]
United States	1.53	1.72	1.98	65.9	70.3	70.2
Yugoslavia	..	0.43	0.51	..	56.4	55.9

Notes: * Gross Domestic Expenditure on R&D.
/ Break in series.
1. 1976.
2. 1980.
3. 1988.
4. 1987.
Source: OECD, STIID database.

Table 2. **Funding by industry and by government**

	Percentage of GERD financed by			
	Industry		Government	
	1985	1989	1985	1989
Austria	49.1	52.5	48.1	44.9
Belgium[1]	66.5	71.6	31.6	26.7
Canada	40.8	41.5	47.5	44.9
Denmark	48.9	46.8	46.5	45.5
France	41.4	43.9	52.9	48.1
Germany	61.8	65.1	36.7	32.8
Greece	25.6	19.2	74.4	69.1
Ireland[1]	46.0	51.3	45.8	38.4
Italy	44.6	46.4	51.7	49.5
Japan	68.9	72.3	21.0	18.6
Netherlands[1]	51.7	53.4	44.2	42.7
Norway	51.6	45.6	45.3	50.8
Spain[1]	47.2	47.5	47.7	48.8
United Kingdom	46.0	50.4	43.4	36.5
United States	50.0	49.6	48.3	48.3
Yugoslavia	59.6	53.6	38.4	42.9

1. 1988.
Source: OECD, STIID database.

of this growth. The embeddedness of innovative activities in firms and industries within OECD countries is further reflected in the role played by the business enterprise sector as an employer of qualified scientific and technical personnel, in R&D, in production and at different levels of management. The percentage of researchers employed in industry is somewhat lower than the figure for R&D expenditure. This is due principally to the many teacher-researchers working in the higher education sector and to the probable overestimation of the number really engaged in R&D. Country to country differences may reflect institutional and cultural differences or may simply reflect differences in reporting R&D statistics.

Furthermore, in many industries (both R&D-intensive and non-intensive) only a fraction of the technological effort of firms is carried out in dedicated R&D facilities and counted as R&D. Reported R&D expenditures are thus only a proxy of innovation-related outlays by firms, a sort of "tip of the iceberg". Case study research on machine building and transport equipment industries reveals that R&D data underestimate activity aimed at improving technology in industries which produce products to order (like certain kinds of machinery), where, despite its overriding importance, design is often not counted as R&D[5]. Small mechanical and electrical engineering firms often do not have separate R&D facilities; they do much inventive work which may or may not be caught in the R&D statistics. Process engineering, which is important in many industries, also tends not to be fully counted as R&D (Sahal, 1981). In order to complete the data provided by formal R&D reporting, a growing number of countries have begun to carry out innovation surveys (see Chapter 5).

In different industries, innovation stems from very different sources within science and engineering; it also occupies a place of varying importance within the overall set of factors which affect profitability and competitiveness. In some industries, firms will make strong demands on technology provided by other industrial sectors; others rely heavily on their own R&D (Pavitt, 1984). These inter-industry differences are reflected in the widely divergent outlays made by firms for R&D and thus in strong variations in industry-specific R&D intensities (see Table 3). The data on the average R&D intensity of manufacturing industry by countries shown in Table 4 reflect differences in industrial specialisation as well as different moments in the build-up of R&D capacities within industry.

Considerable technological effort is needed to stay up with and employ new technology created upstream and to respond to changing user demands. In industries where technology is advancing rapidly, firms often invest considerable resources in monitoring the innovations introduced by their competitors and advances in basic and applied scientific and technical knowledge. Generally, every large corporate laboratory develops

Table 3. Intensity of R&D expenditure in the OECD area

Weighting of 11 countries[1] – R&D expenditure/output ratio

1970		1980	
HIGH		**HIGH**	
1. Aerospace	25.6	1. Aerospace	22.7
2. Office machines, computers	13.4	2. Office machines, computers	17.5
3. Electronics & components	8.4	3. Electronics & components	10.4
4. Drugs	6.4	4. Drugs	8.7
5. Scientific instruments	4.5	5. Scientific instruments	4.8
6. Electrical machinery	4.5	6. Electrical machinery	4.4
MEDIUM		**MEDIUM**	
7. Chemicals	3.0	7. Automobiles	2.7
8. Automobiles	2.5	8. Chemicals	2.3
9. Other manufacturing industries	1.6	9. Other manufacturing industries	1.8
10. Petroleum refineries	1.2	10. Non-electrical machinery	1.6
11. Non-electrical machinery	1.1	11. Rubber, plastics	1.2
12. Rubber, plastics	1.1	12. Non-ferrous metals	1.0
LOW		**LOW**	
13. Non-ferrous metals	0.8	13. Stone, clay, glass	0.9
14. Stone, clay, glass	0.7	14. Food, beverages, tobacco	0.8
15. Shipbuilding	0.7	15. Shipbuilding	0.6
16. Ferrous metals	0.5	16. Petrol refineries	0.6
17. Fabricated metal products	0.3	17. Ferrous metals	0.6
18. Wood, cork, furniture	0.2	18. Fabricated metal products	0.4
19. Food, beverages, tobacco	0.2	19. Paper, printing	0.3
20. Textiles, footwear, leather	0.2	20. Wood, cork, furniture	0.3
21. Paper, printing	0.1	21. Textiles, footwear, leather	0.2

1. Australia, Belgium, Canada, France, Germany, Italy, Japan, Netherlands, Sweden, United Kingdom and United States.
Source: OECD.

Table 4. R&D expenditure in manufacturing as a percentage of value added

	1971	1975	1979	1981	1983	1986
United States	5.7	5.4	4.9	5.9	6.9	7.6
Japan	2.8	3.4	3.7	4.5	4.8	5.6
Germany	3.4	3.7	4.5	4.3	4.8	5.2[1]
France	3.5	3.6	3.7	4.2	4.5	5.0[1]
United Kingdom	4.5	4.3	4.3	4.9	4.7	4.9[1]
Italy	1.5	1.5	1.5	1.9	3.0	3.2[2]
Netherlands	4.1	4.5	4.6	5.3	6.2	6.5
Belgium	2.1	2.9	3.2	3.7	4.1	4.4[1]
Denmark	..	..	2.3	2.6	3.0	3.2[1]
Ireland	..	0.1	0.1	0.3*	..	..
Greece	..	..	..	0.2	..	0.4
Portugal	..	..	0.2	0.4	..	..
Spain	..	..	..	0.6	0.8	1.2[1]
Canada	2.1	1.7	1.9	2.7	..	..
Australia	0.4*	0.6	0.8	0.8	..	1.4[1]
New Zealand	0.6	0.7	0.7	0.6*	0.6	..
Austria	1.2	1.5	1.9*	2.1	2.3*	2.9[1]
Finland	1.8	1.7	2.0	2.4	2.9	3.2[1]
Norway	1.6	2.0	2.4	2.6	2.9	3.7
Sweden	3.6	3.9	5.4	6.3	6.3	7.0
Switzerland	..	..	..	6.5*	..	7.4*

* Secretariat estimates.
1. 1985.
2. 1984.
Source: OECD (1989).

its own in-house "watch post" and absorption mechanisms. These are particularly important for industries which rely heavily on intra-industry transfers of technology[6]. This work may or may not be counted as R&D, but independently of accounting, the corporate R&D laboratory must also bring scientific and technical knowledge into the firm. In addition, their network linkages with university laboratories (see Chapter 3) have often been likened to "windows of opportunity" on developments in science and technology. These observations lead to a new perception of the relationships (discussed in Chapter 2) between the production of scientific and technical knowledge and its absorption and diffusion.

Newly industrialised economies (NIEs) and developing countries illustrate some further aspects of these relationships. Industry and country case studies have long established that for them, innovation may largely involve learning to produce products or employ technologies that have long been in use in the industrialised economies. Economists specialised in technical change have come to understand that considerable technological effort may be required in such learning (see Chapter 2). In the case of NIEs (see Chapter 12), reverse engineering – learning to produce by taking apart products and processes to find out how they work – may involve considerable intellectual effort. While not generally counted as such, reverse engineering is very much like development and could be considered as R&D. As firms and countries catch up, such work increasingly aims at production and begins to be counted as R&D.

The funding of long-term basic research and its institutional foundations

From World War II until the mid-1960s, basic research was supported on a massive scale by governments. The underpinning for this support lay in the many militarily decisive scientific discoveries and inventions of the preceding decades, which found their application during the war, whether for destruction (e.g. the atom bomb or long-range missiles), the saving of lives (e.g. penicillin), or the organisation of operations (operations research and computing). Subsequently basic research received much weaker financial and political support from governments.

Governments have observed industry's growing interest in university research and its readiness to finance some parts of it for its own needs and under its own conditions. Faced by budgetary constraints, many governments have tended to reduce their support to science, in particular to the fundamental research component of higher education sector budgets, with the expectation that this reduction in support could be compensated by industry. Universities have been encouraged to seek industry support in whatever form

available, including short-term R&D contracts or trouble-shooting for firms which do not possess adequate in-house technical facilities. The latest OECD *Science and Technology Indicators Report*[7] found that:

i) In 1975-85, higher education R&D (HERD) grew at an average annual rate of no greater than 3.5 per cent (as compared with 5 per cent for gross expenditure on R&D – GERD) and 6 per cent for industrial R&D. Thus, the share of the higher education sector in the total OECD R&D effort has declined. The most recent figures published in 1990 show that this trend has persisted.

ii) Closer analysis showed that during the 1980s, these figures would have been lower had it not been for the fact that in the United States (which accounts for about 43 per cent of total OECD outlays for higher education R&D), the universities had received sizeable R&D contracts from the Space Defense Initiative Agency of the Department of Defense (DOD).

iii) Gradual changes have taken place in the structure of university R&D funding, with the share of "general funds" decreasing in favour of direct government funding, often "mission-oriented" towards programmes with an economic, social or regional slant.

As a result, in many countries, R&D in the higher education sector as a percentage of gross domestic expenditure on R&D has fallen, sometimes quite significantly (Figure 3). Among the major economies, the United States is a partial exception to this situation and among the other economies, R&D in the higher education sector has recently risen in Canada.

This shift in government policy may reflect misunderstanding about the nature and function of long-term or basic science and the conditions under which it is successfully produced. In contemporary conditions, the generation and utilisation of relevant areas of scientific knowledge is an integral part of, and often a necessary condition for, the development of a new technological paradigm. During the early phases, scientific advance plays a major role, and the relationship between science and practical technical activity is extremely intimate. Thus, since the last decades of the 19th century, beginning with electricity and chemical inventions, changes in the basic "technological paradigms" lying at the heart of industrial production and communication systems have occurred increasingly in conjunction with major scientific advances. This was demonstrated in the early phases of micro-electronics: the discovery of certain quantum mechanics properties of semiconductors yielded a Nobel Prize for physics and co-incided with the technological development of the first micro-electronics device.

Since the late 1970s, this type of intimacy has occurred again on a large, even spectacular, scale in genetic engineering and biotechnology. A

Figure 3. **R&D in the higher education sector as a percentage of gross domestic expenditure on R&D**

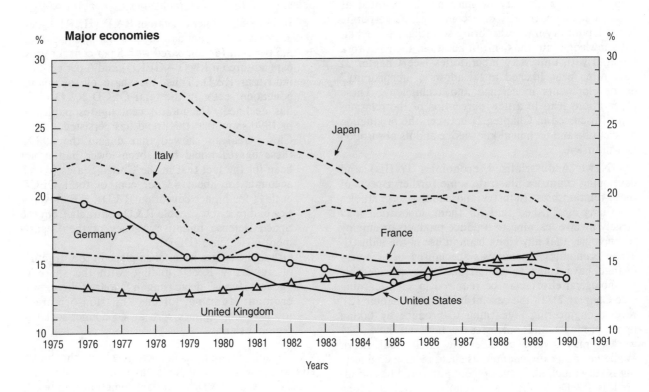

Major economies

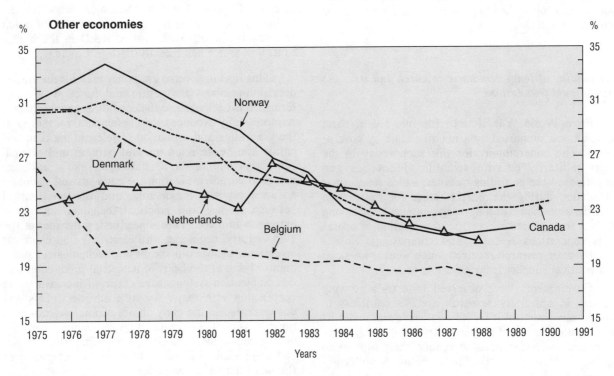

Other economies

Source: OECD, STIID Database, June 1991.

scientometric study carried out in the mid-1980s, and published under the significant title "Is Technology Becoming Science?" found that "the science being relied upon in patented technology, at least in biotechnology type patents, is quite recent. It is, in fact, just about as recent as the technology (patents) referenced in those patents. Furthermore, if one allows for an extra year or two to prosecute a patent, the cited papers are just about as recent as the papers cited by papers. These biotechnology patents are not using the old, codified science found in texts and reference books: rather they are using current science just about as quickly as it emerges from the research labs (...)" (Narin and Noma, 1985, p. 3). For this reason, university laboratories do and should receive industrial or public R&D contracts for development research or technical activities. However, these contracts, and the funds they imply, must come in addition to the general funds for fundamental research and not be a substitute for them. They do not and cannot replace the funds allocated to increase the pool of predictive science required by all participants in the economic process, first and foremost by firms.

Fundamental research of the quality and on the scale which can lead to major scientific advances takes place in relatively few firms[8]. It calls for high thresholds of R&D investment and a corporate research environment conducive to developing and discussing ideas freely with other research workers. Knowledge developed within firms also raises proprietary issues. For such reasons, the advance towards reliable and totally public "predictive knowledge", available to all the actors in the innovation process and in particular to small firms which cannot afford large R&D laboratories, must also continue to take place within the institutions specially devised for the production of fundamental, general, vetted and public knowledge.

The main institutions created by Western society to meet these purposes have been its universities, its learned societies and academies. These institutions have also provided ethical foundations for science, with professional rules, attitudes and codes of behaviour governing the disclosure, critical inspection and transmission of new knowledge[9]. However, because of weakened government support and its increasingly close relationship with industry, science, considered as an independent institution with a code of professional rules, has come under great and ever increasing pressure. In particular, the norm of rapid and total disclosure of new knowledge has been subjected to extraordinary strains. Great financial awards can be earned by keeping certain vital scientific knowledge secret and by moving with it to the business enterprise sector, which can reward the production of knowledge through participation in the flow of profits which result from it.

Concerns about necessary priorities in research and the most appropriate mechanisms for determining R&D priorities are legitimate, both on the part of finance ministries and of science and technology policy makers. A recent OECD study of the institutions and mechanisms of priority setting shows recognition of this issue. It notes a number of common trends among OECD countries, despite a high degree of institutional variety (OECD, 1991f). Suggestions for improving the mechanisms employed almost invariably involve closer co-operation among scientists (who use peer assessment of projects) and policy makers (who apply economic and social criteria for priority setting). This can help avoid shifting onto the research system the responsibility for the weakening of basic research. However, the central problem lies outside the sphere of priority setting within science.

It is that of wider political and social priorities as reflected in the salary levels and the status of non-industrial research personnel. Yet it is essential to maintain an adequate supply of scientific and technical personnel (see Chapter 6). There are two related dimensions. The first is the need to guarantee to the whole economy through the higher education sector a sufficient supply of qualified personnel for both production-related tasks and different varieties of research. The second is the need to ensure for the research system *stricto sensu,* and the higher education sector in particular, the conditions for satisfying the personnel requirements necessary to maintain and expand the knowledge base. This entails paying university researchers and their staff salaries somewhat closer to the levels in industrial R&D than is the case in most OECD countries. This will allow new, freely available knowledge at the highest level of quality and sophistication to continue to be produced.

If the remuneration of work in universities aimed at producing knowledge which either opens up new research vistas or strengthens the "predictive" nature of science in vital areas continues to drop in comparison with rewards offered by industrial R&D, some future outcomes are predictable (Dasgupta and David, 1988). If the research base outside industry weakens seriously, engineers will seek ways around technology-related production problems, but development costs will rise; very large firms with very sophisticated laboratories will try to produce a part of the knowledge for themselves, but secrecy will preside over this work and bring with it increased duplication of R&D. Firms will make proprietary knowledge available only on their terms. This is the basis on which much tacit or semi-codified knowledge is already transmitted among firms (see Chapter 3), but there would be very serious consequences if economic and social conditions extended these terms to all knowledge production[10].

Transfer sciences

Once industrial society's need for basic science has been re-asserted, attention must be given to the disciplines that can bridge the gap between the type of knowledge produced by basic science and the type of knowledge needed by firms or government agencies in day-to-day activities. The industrial R&D laboratory and industry-specific research associations are geared to bridging the gap from the side of industry, but the bridging process must also be organised from the side of the university laboratories and the government research centres. This effort may be complementary to industry measures or a necessary first step when firms are not yet ready to undertake the necessary bridging activities themselves. Historically, for example, governments were active in agriculture, where they supported the development of agronomy, the training of agronomists and the establishment of agricultural extension services. As a result, agriculture has had a long and particularly successful record in creating appropriate "transfer science"[11].

The notion of transfer sciences approaches the "bridging" issue in a systematic way and involves a distinction between two groups of sciences: "pure sciences" and "transfer sciences". Characteristics of "pure sciences" include their fundamental activity (the exploration of the boundaries of knowledge without concern for the practical implications of the findings); their rules and codes of behaviour regarding knowledge disclosure and recognition; their location (in universities or public laboratories closely related to universities); their funding (mainly from government sources); and their priority with respect to training (the training of graduates for joining laboratories dealing with basic science). The subjects tackled by these disciplines essentially belong to the realm of the physical and biological sciences, and the scientists involved constitute closeknit communities, at national and international levels.

"Transfer sciences" (which include the various branches of engineering) share with the pure sciences a concern for predictive science, but otherwise they have rather different characteristics: their activity is driven principally by the urge to solve problems arising from social and economic activities; their research centres are located in technical universities, engineering

Figure 4.   **The place of transfer sciences**

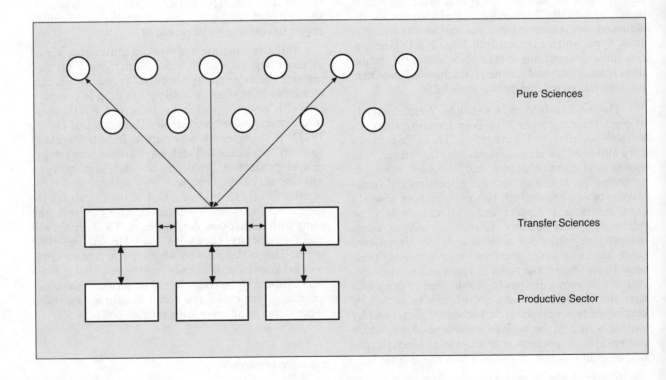

Source: OECD, see text.

schools, sectoral government R&D establishments, and industry; a large part of their funding comes from industry; their graduates are normally employed by industry. They tackle subjects broadly relating to artificially made objects and phenomena, and the communities of scientists active in research in these areas are very close to the professions most concerned by application of their results. It would be wrong, as in the case of engineering (which must of course be classified among the transfer sciences), to see them simply as "applied science" just "downstream" of fundamental science. Their bridging formation and function do not imply that they are not fields with their own internal organising principles.

Transfer sciences play an essential role in providing an interface between the world of "pure science" and the world of industry (Figure 4). They investigate concrete problems arising in all fields of human endeavour. Seen as "fields" or "disciplines", transfer sciences may straddle the normal borders separating "science" from "technology" (see Box 7); the same organisational structures (be they firms or universities) may in some cases generate technological innovations and scientific knowledge. Their boundaries are not always clear-cut, they are often multidisciplinary (e.g. materials science) and their analytical development largely reflects social and economic needs. Their functions include those of any scientific discipline (namely creation, transmission and organisation of certain types of knowledge) together with the aim of undertaking or improving technical projects. From the standpoint of the sociology of science, they involve "hybrid communities" (Blume, 1990) composed of people with the same background and initial training, some of whom are located in or close to "pure science" and others in industry[12]. Because of their increasing importance and their complexity, one goal of science policy should be to create and sustain a network of relations and links between the different types of research-producing and research-using institutions (see Chapter 3).

3. CONTINUITIES AND DISCONTINUITIES IN TECHNOLOGY

The two traditional means of describing the main determinant of technological change are the demand-pull and the technology-push hypotheses. The former signals the centrality of market forces in shaping the directions of technology, with changes in demand, costs, prices and profit opportunities altering firms' incentives and bringing forth technical advance. The latter attributes the main factor underlying innovation to autonomous advances in pure science and technological expertise and more generally in the body of knowledge[13]. Both approaches encounter difficulties[14].

A more promising synthetic approach recognises that the innovative process has some rules of its own which cannot be described as simple and flexible reactions to changes in market conditions or the scientific base of the industry. The fuller definition of an economic environment that determines technological advance and within which technology and public policy interact takes into account changes in market demand, prices and profitability, as well as the role of institutions and other systemic factors. The "cumulativeness" of technology, the notion of "technological trajectories" and of "increasing rates of adoption" are all central to this broader concept.

The cumulative features of scientific and technical knowledge

During the last ten years, much stress has been laid on new technologies' potential for radical change. In other work, the accent is placed almost equally strongly on the ubiquitous, often incremental and necessarily cumulative features of technology. During TEP emphasis has tended to be placed on this latter aspect as well as on the relationship of technology to certain important structural transformations and organisational change discussed below (e.g. in Chapter 4 on corporate organisation and Chapter 10 on globalisation). Emphasis on the cumulative dimensions of technology serves to stress the very important learning processes involved in the full development and use of new technology, both by the firms initiating innovations and by those firms which approach the new technology at later points in the development cum diffusion process (see Box 8). These learning processes include learning-by-doing, e.g. increasing the efficiency of production operations; learning-by-using, e.g. increasing the efficiency of the use of complex systems; learning-by-interacting, involving users and producers in an interaction resulting in product innovations; and even "learning-by-learning", whereby the capacity of firms to absorb innovations developed elsewhere depends on their learning experience, which in turn is enhanced by R&D and other related intangible investments (Cohen and Levinthal, 1989).

Firms, institutions and countries which have had sufficient opportunity to build up skills and create a cumulative learning base will have better ability to adapt during periods. Technological change destroys transitional equipment and the demand for specific skills; per se, it cannot and does not destroy institutions or firms nor the knowledge built up as a result of institutional learning[15]. The destruction of the foundations on which technological learning has taken place can occur, but it will only come about as a result of radical budgetary cuts, bad management, or mergers and acquisitions carried out without regard for accumulated learning within organisations.

Box 8. The "snowball effect" as a metaphor for the growth of knowledge

The growth of the stock of knowledge and technology can be likened to the formation of a snowball. For a snowball to move and grow, a number of factors are important:

- It needs an initial critical mass: a knowledge/technology base is essential to the virtuous circle of technology and growth; where this is absent or insufficient, governments can help create it and get the snowball rolling.
- It has to find snow in its path: an environment rich in publicly accessible knowledge i.e. an environment where spillovers, created in an environment of formal and informal networks, are abundant. The role of policy is to ensure the public availability of such knowledge.
- It has to be able to gather snow and thus speed: the absorptive capacity of firms is crucial to turning a conducive environment into actual economic benefits; policy here can help develop the capacity of firms to use externally available knowledge.

To the extent that prior technological accumulation and a long build-up of skills has taken place through learning processes internal to firms or immediately related to their activity, existing firms will have fairly good opportunities to restructure, adapt and survive in the face of marked technological change. This is the case both when the learning base exists in isolation (as for small mechanical and electrical engineering firms) or in combination with size, finance and a certain degree of market power (for larger ones). Studies on multinational enterprises (MNEs) now emphasise a fair degree of stability within the group of world leaders which can be largely attributed to technological accumulation (Cantwell, 1989). Cumulativeness offers an explanation of the extent to which "reverse causation" (Nelson, 1981b) (e.g. the strengthening rather than the weakening of concentration in the face of technical change) has occurred during the last few decades.

For these statements to hold, however, technological accumulation and institutional learning must have had time to take place. Such is the case in all but a few OECD countries, but it is not the case for developing countries, where lack of "cumulativeness" must be included among the factors shaping their present situation. The newly industrialised economies and their firms lie at the borderline with respect to technological accumulation, learning processes and institutional strength. This explains why factors other than technological capability (in particular social and political factors) have been decisive in determining the greatly differing performance of NIEs (e.g. the contrast between Korea and Brazil) during the 1980s and into the 1990s (see Chapter 12).

The notions of technological paradigms and trajectories

Two further substantial advances recently made in research on the economics of innovation and technical change throw light on how technologies develop and how a given technology may be selected by economic and social mechanisms and survive to the detriment of another. The first is based on research related to the idea of technological paradigms and trajectories, while the second is connected with the development of the theory of increasing returns to adoption.

The term "technological trajectories" expresses in analytical terms the basic cumulative and evolutionary features which mark the developments and changes experienced by technologies as they are diffused and employed in production and services (see Box 9)[16]. It also expresses the idea that, following paradigmatic scientific and technological breakthroughs, one will find an initially unmapped technological potential to be exploited and expanded. Experts in the theory and/or history of technological innovation refer to this potential in a variety of different ways: "generic technical principles", "technological guide-posts", "technological systems", "dominant designs" and increasingly, today, "technological paradigms".

Box 9. **The steps in a technological trajectory**

- The exploitation of an initial impulse from a new technological paradigm emerging from the techno-scientific breeding ground, resulting in a new industrial development path which will not necessarily be based at this stage on substantial R&D.
- A cumulative process implying that as evolution proceeds the range of potential choices will become narrower. This has the effect of setting the technological trajectory along a main stem growing at a gradually steadier rate.
- The multiplication of differences and diversification of the applications of the original paradigm, based on R&D which tends to expand through offshoots from the main stem in the form of subtechnologies generating a variety of products.
- The onset of a saturation phase where renewal and productivity gains, matching increasing efficiency in the exploitation of the pool of available knowledge (this efficiency is difficult to measure), require ever increasing creativity for more and more limited advances. At this stage, these advances consist of combinations of technical functions (e.g. personal stereo) or improved services to users.
- Possible revival, provided the basic technology can recover its potential for development in ways that counteract the tendency towards depletion. One reason why revival is possible relates to the fact that trajectories are not independent of one another and feed each other while competing. For example, the development of new materials revived the steel industry by raising and shifting the performance threshold. Revival may also be caused by scientific progress (e.g. transistors for computers), or an impulse from the market (e.g. the return to the production of sailing ships for leisure).

Like a scientific paradigm (Kuhn, 1962), a technological paradigm (Dosi, 1982 and 1984b) embodies a definition of the relevant problem and a pattern of enquiry. It defines the needs to be fulfilled and the scientific principles and material technology to be used. It represents a "pattern" of selected principles which prescribes the directions to be pursued. Technological paradigms have powerful heuristic features; the efforts and technological imagination of engineers and of the organisations in which they work are thus channelled in precise directions, so that they are "blind" to other technological possibilities. The definition of a technological paradigm relates to the generic task to which it applies (e.g. amplifying and switching electrical signals), to the material technology it selects (e.g. semiconductors such as silicon), to the physical/chemical properties it exploits (e.g. the "transistor effect" and "field effect" of semiconductor materials), to the technological and economic dimensions and trade-offs it focuses on (e.g. density of circuits, speed, noise immunity, dispersion, frequency range, unit costs).

Technological trajectories correspond to specific processes. In the light of studies carried out in the 1980s, the concept of "trajectory" cannot be confined to the initial understanding of a "natural trajectory" of technologies shaped mainly by increasing mechanisation and exploitation of latent economies of scale. In each industry or sector, technological evolution follows a specific pattern which depends on the initial technological area. Industries differ significantly in the extent to which they can exploit the prevailing general natural trajectories, and these differences influence the rise and fall of different industries and technologies (Nelson and Winter, 1977).

Competition between technologies and increasing returns to adoption

Although the notions just discussed offer some explanation of how technological growth proceeds, they do not throw light on the selection and stabilisation of technologies. In other words, they do not explain why a given technology is selected where another is rejected, and succeeds where another fails. This issue is now being examined at two levels: that of the part played by economic, institutional and social factors in shaping the overall selection mechanisms for technology (see Box 10), and that of a narrower set of elements leading to increasing returns to adoption.

Box 10. Factors in the selection of technology

The first and crucial role of economic, institutional and social factors is to select from possible paths at each level, from research to production-related technological efforts. This selection is based on some rather obvious and broad criteria such as feasibility, marketability and profitability. Thus the choice technological paradigms is wide and it is impossible to rank them *ex ante* given the intrinsic uncertainty of their outcomes, in terms of both technological and economic success.

Other more specific variables are also likely to come into play, such as:

   i)   the economic interests of the organisations involved in R&D in the new technological areas;
   ii) their technological history, the fields of their expertise, etc.;
   iii) the institutional variables *stricto sensu* such as the procurement policies of public agencies (as in arms, space, energy, etc.).

All these factors are likely to channel technological development in defined directions. In particular, the role often played in the establishment of a particular trajectory by government priorities and levels of funding must be stressed. An obvious example is electronics, especially in the fields of semiconductors and computers during the two decades of the post-war period. Military and space programmes operated as a powerful focusing mechanism for defining technological targets, while providing financial support for R&D and guaranteeing public procurement.

Another powerful selection criterion in capitalist economies is likely to be the cost-saving capability of the new technology and, in particular, its labour-saving potential: this is obviously consistent with Nelson and Winter's suggestion of "natural trajectories" towards mechanisation and exploitation of economies of scale (1977, pp. 36-77). Certainly, in societies where industrial and income distribution conflict are structural features, substitution of machines for labour will be a powerful determinant in the search for new technologies.

More generally, patterns of industrial and social conflict are likely to operate both as negative criteria (which developments to exclude) and as positive criteria (which technologies to select) in the selection process of new technological paradigms. In this respect, it might be possible to define a long-run relationship between patterns of social development and the technological paradigms selected. (One quite clear example is the association between industrial relations at the turn of the last century and the selection and development of "taylorist" patterns of technical change in mechanical engineering.) (Dosi, 1984b, pp. 84-87.)

The conventional explanation for the non-adoption or death of a technology is that it was intrinsically inferior to its rival. However, if technology is regarded as an "end-point", intrinsic inferiority (or superiority) can only be established when the two technologies have reached an equivalent level of development. Therefore, any explanation concerning technological selection and competition mechanisms must first ascertain whether the processes involved have indeed reached their end-point, and secondly, whether the ousted technology was not eliminated during, or even at the start of, a process leading to its abortion, so that *a posteriori* it appears to be the one that has been defeated. To explore this issue further, it is necessary to turn to the concept of "increasing returns to adoption" and the related notions of "network externalities" and "lock-in".

The basic idea behind recent research on the theory of "increasing rates of adoption" is that, in many cases, a technology is not chosen because it is efficient, but becomes efficient because it has been chosen. Many of the characteristics which make one of two rival technologies superior to the other are in fact acquired during its development and diffusion. This new approach to the analysis of the selection of innovations is pursued through study of the mechanisms by which a sequence of choices, made by the adopters of the technology, determines the superiority or inferiority of each of the two technologies concerned. "What makes competition between technologies interesting is that usually technologies become attractive – more developed, more widespread, more useful – the more they are adopted" (Arthur, 1988, p. 590).

The main source of increasing rates of adoption is learning through practice: the more a technology spreads, the more is learnt about its use, the more it improves and the more it is likely to be adopted by subsequent users. Accordingly, the outcome of the competition between two technologies is governed mainly by the selection strategies of the first users, who focus technological change on a specific technology, making it more attractive than its rival. The latter lacks the learning benefits of adoption in the gravitational orbit of the "winning" technology, from which it will be increasingly difficult to escape. Diffusion processes are influenced by such "network externalities" (see Chapter 2).

Network externalities intervene likewise by modifying the economics of the technology as it spreads. Even without significant improvements or material changes, there is a "lock-in" effect (see Box 11). As more and more users enter a technology's orbit, it becomes increasingly attractive to subsequent users. Furthermore, a highly successful new technology from the functional standpoint creates and expands its own market. It can therefore reap economies and offer users correspondingly better performances, in particular in terms of price and convenience. There was no demand for the telephone or the computer before they were developed. As the number of subscribers and users expand, it becomes more expedient to join the system. The analytical and policy-related approach to standard setting in industries marked by strong globalisation and very large investment outlays such as telecommunications and telematics has recently been modified by recent work on returns to adoption and lock-in (see Chapter 10, Section 3).

Box 11.   Stages in lock-in and increasing returns to adoption

On the basis of the present state of understanding (for an overall discussion, see Foray, 1990*b*), five "stylised facts" appear to characterise competition between technologies, "lock-in" (and lock-out) and increasing rates of adoption:

- As competition proceeds, i.e. during the selection sequence, a "lock-in" situation arises at a given juncture, i.e. the process locks into one of the technologies concerned.
- The technology adopted at this stage is not necessarily the better one (the one that would ensure the best performance when the learning procedures are over).
- There may be very little opportunity for public authorities to intervene, for instance to prevent the inferior technology from taking over, especially if the role of the State as a technology adopter is disregarded.
- Even where the winner is not the better technology, it is extremely difficult to dislodge or convert it, owing, among other things, to lack of information, lack of co-ordination between users and lack of technological interrelationships.
- Because the leading technology is practically unbeatable, competition between different trajectories proceeds under a system which tends to establish complementarities or relaunch saturated trajectories rather than undertake radical conversions or substitutions. This is fully apparent in the interactions between new and old materials, for instance.

This kind of irreversible selection process may well develop on an international level. It can contribute to industrial barriers to entry and affect the competitiveness of firms and even whole industries. Comparative studies on the smelting industry in Germany and France have shown that in Germany, some technologies had the advantage of a broader and more dynamic market. They became more widely and more firmly established and put up stronger resistance to subsequent market fluctuations, owing to the accumulated advantages arising from the experience of users and resultant continuous improvement of processes (Foray, 1989). The implications of these aspects of innovation for international competitiveness, international technology transfers and the situation of newly industrialised countries and other developing countries will be touched on in Chapters 10 and 12.

4. SOME POLICY CONCLUSIONS

Although the aim of this chapter has been to give an overall view of the workings of the innovation process, some policy conclusions are apparent; they should be read in conjunction with those of the next two chapters.

Industrial R&D

Industrial R&D is not the only innovation-related activity that firms undertake, and a simple measure of the resources involved does not suffice to measure innovative effort. Nonetheless, measuring the resources devoted to R&D by industry is still the least controversial and most popular tool used by governments to monitor industrial innovation. The best known of such indicators measures the relationship between industrial expenditure on R&D and gross domestic product (GDP) (see Table 1).

This type of indicator reflects two phenomena: first, the propensity of firms in a given country to spend more or less on R&D than the OECD average for firms in the same industry and second, the industrial structure of that country, both with respect to industrial profiles (with fewer high technology industries average R&D intensity will be low) and size profiles (large firms take better advantage of scale economies and do proportionately more R&D than small ones). Despite its shortcomings (e.g. technology acquired through the purchase of sophisticated machinery is not reflected in the numbers), governments will continue to watch this indicator closely. Industrial R&D expenditures below 1 per cent suggest a weak national system of innovation. Where they fall below 0.2 per cent, an exceptional and concerted effort is required from both firms and governments.

Since public efforts cannot act as a substitute for those of the firms themselves, government action in this area can generally only be a limited one. However, environments conducive to innovation can be created, and governments have at their disposal a battery of policy instruments. They can encourage the diffusion of generic technologies and support strategic technologies (see Chapter 2). They can encourage firms and public laboratories to interact and create networks. They can encourage long-term research in large firms, develop contract R&D firms and technical centres and encourage the industrial development of high technology industries through foreign investment and international inter-firm collaborative agreements. Well focused studies are now required to evaluate the effectiveness of existing policies in the new context. Already, many countries that are catching up (smaller economies for all activities or some others for specific industries) consider such programmes to be an important component of their economic policy.

Long-term research

It is essential, even from an exclusively economic viewpoint, for all countries to support long-term, and in particular basic, research. The development of basic research is by nature an international process. Scientists were among the first to give their activities the global character towards which business sector activities are currently moving. Policy measures should give priority to promoting international co-operation among laboratories, the mobility of research personnel, the building and exploitation in common of equipment for big science and the international co-ordination of programmes that study global concerns such as climatology and oceanography.

The public good character of science also implies that each country should contribute to its development according to its economic and financial capacity. This is something more than simple "burden-sharing"; it is to the clear advantage of every country to maintain substantial capabilities for basic research on its territory. Basic research is part of the adoption and diffusion mechanism. The interactive character of innovation and the increasingly important links between science and technology give an advantage to firms that can effectively communicate with neighbouring universities and high quality public laboratories.

These considerations are not new. Nevertheless, they require strong reaffirmation, because the level of resources for basic research has generally tended to decrease in comparable terms and because the work conditions of researchers have also tended to deteriorate. Governments should also appreciate the fact that even when the public sector budget for research experiences low or zero growth, the claims on it increase, due to the multiplication of disciplines and the increasing complexity of projects.

Support for basic research does not imply that all disciplines should be developed at the same rate. Exercises in technology assessment and forecasting should be better developed and co-ordinated. Although very difficult, attempts should be made to evaluate the impact of each discipline on the development of others. New disciplines are born that require nurturing; others, as they grow older, are prone to rather unproductive forms of scientific perfectionism. Finally, a periodic assessment of the allocation of resources among disciplines should be instituted. Resource allocation policies should support the development of disciplines that play a natural interface role, since innovation increasingly depends on cross-fertilisation of ideas from different disciplines.

Institutions, equilibria and interactions

Innovation has an interactive character; it operates as a system, so that its success depends not only on the quality of its various elements, but also on the synergy among them. Understanding the process of innovation as well as the relationship between technology and public policy requires an understanding both of the sources and basic determinants of the rate and direction of innovation and of the ways policy can affect them.

The extreme complexity of the situation explains in part the difficulty that innovation policies often encounter when different actors (firms, governments, the academic sector) pursue partial or not necessarily converging strategies. The institutional organisation of research may be decisive, as different institutional forms – universities, national research councils, institutes and centres – may encourage or, on the contrary, hamper exchanges and even mergers between disciplines, the birth of new disciplines or the re-channelling of research into new fields.

It is therefore imperative to establish active mechanisms for institutional co-ordination at national, regional and local levels. The institutions whose role it is to promote interaction – such as "network laboratories", technopoles, scientific associations, technology assessment agencies and science and technology museums – must improve and multiply.

Some areas requiring particular policy attention can be identified. They include disciplines that play a natural interface role, such as interdisciplinary sciences; pervasive generic technologies, which, while first developed in a limited number of industrial sectors, are subsequently used by virtually all industries, whose organisation and methods of production they often completely transform; and transfer sciences.

Transfer sciences (including engineering) play an essential role in establishing productive relations between science and industry. As such, they should constitute an especially important target for government policies designed to promote technological development. Such sciences are constantly evolving under the pressure of economic and social needs and as new knowledge is produced by the natural sciences and mathematics. Events of the last few decades show the importance of launching programmes for stimulating and supporting the emergence of a new transfer field early. For instance, in many OECD countries government action has facilitated and speeded up research in data processing, new materials, biotechnology, etc. There are fields such as the environment where the need for new transfer disciplines seems abundantly clear, although "ecological engineering" is slow to emerge.

In these three areas, governments and public organisations should play a determined and positive role, since these interface disciplines have intrinsic difficulty in obtaining support. Firms are, above all, interested in the short-term and in technologies that are specific to them; universities have a tendency to support the "purest", and thus to them the most "noble", areas of research; scientific communities often have a tendency towards isolation from each other.

Technological development is only sustainable in the long-term with the help of balanced efforts by the public and the private sector. If public R&D, and in particular university research, are at a mediocre level, the long-term prospects of OECD economies may be endangered: the environment of firms deteriorates, investors are discouraged and the industrial fabric starts to weaken. But if public effort is not followed up by private firms, the scientists trained in the universities will emigrate, and the research performed by public laboratories will be stifled or its results will be taken up and developed by firms abroad.

NOTES

1. These processes have been studied by historians of technology or by economists interested in historical processes. For a review of the evidence, see, for example, Rosenberg (1976).

2. See Schmookler (1966). In examining cross-sectional data for a large number of industries in the years before and after World War II, he found a very high correlation between capital good inventions for an industry and the volume of sales of capital goods to that industry. He concluded that demand considerations, through their influence upon the size of the market for particular classes of invention, are the decisive determinant of the allocation of inventive effort. Far from being the exogenous variable postulated by a large part of economic theory and almost all economic modelling, innovation outlays as measured by patenting can be endogenised and treated like any other form of investment. See Scott (1989) for a full discussion of this.

3. For a critique, see Mowery and Rosenberg (1979, pp. 102-53).

4. For further data and references and a full discussion, see Roos (1991).

5. This point (as well as others discussed in this sub-section) is made in Nelson and Rosenberg (1990).

6. A study of the innovation-related activities and investments of large food-processing firms, in particular Unilever and Nestlé, including the organisation of "watch-posts" and the role of engineering departments and formal R&D facilities, can be found in OECD (1979*b*).

7. See the section on Higher Education R&D, in OECD (1989*e*).

8. Bell Telephone Laboratories have provided several spectacular examples. In the late 1920s and early 1930s, Jansky's pioneering research on the sources of static in the new transatlantic radiotelephone service was motivated by the need to reduce or eliminate background noise and improve the quality of calls. But his discovery of this "star noise", as he labelled it, marked the birth of radio astronomy. More recently, fundamental research in the same area conducted by Penzias and Wilson in work aimed at improving the quality of satellite transmission led to the discovery of cosmic background radiation, which is now taken as confirmation of the "Big Bang" theory of the formation of the universe. For a further discussion and more examples, see Mowery and Rosenberg (1989, pp. 12-13).

9. The role played by disclosure in ensuring the availability and reliability of scientific knowledge is important and may be underestimated. See Dasgupta and David (1988, pp. 19 and 29): "Disclosure increases the expected span of application in the search for new knowledge. To put it in other words, disclosure raises the social value of new discoveries and inventions by lowering the chance that they will reside with persons and groups who lack the resources required to exploit them. Second, disclosure enables peer groups to screen and evaluate the new finding. The result is a new finding containing a smaller margin of error."

10. Dasgupta and David (1988) make the point eloquently: "The existing pool of knowledge is an essential input in the production of new knowledge. That is why technology today draws so heavily upon the infrastructure provided by science. If, to take an extreme example, science were to close down, each enterprise in technology would, roughly speaking, have to rely on its private knowledge pool. This would dampen technological progress enormously, as firms in science-based industries in particular would then for the most part be conducting duplicative research. The public-good-producing aspect of science is of course recognised to be of considerable importance to the technology community, which is why one on occasion sees groups of technological enterprises spontaneously joining to support activities organised under the cultural rules of science" (pp. 52-53). By contrast the public-good-producing attributes of science and its social role have come to be dangerously overlooked by some governments. In the words of Dasgupta and David, "The argument is given that if there is some useful research to be done it will be performed as R&D, by organisations working in technology, and that it will be done there by cheaper means and without recourse to the public purse. This, quite simply, betrays a staggering lack of understanding of the socio-economics of science and technology. Modern economic growth under such conditions of 'privatisation of science' might still continue to be grounded in the exploitation of scientific and technological knowledge, but it would lose the sustained character that has been taken by many to distinguish it fundamentally from the process of economic change in earlier epochs" (p. 58).

11. The term "transfer sciences" was coined and developed at the French Centre National de la Recherche Scientifique (CNRS). It has been used at the CNRS in a policy-making perspective from the mid-1970s. See the recent history of the CNRS published for the 50th Anniversary: Picard (1990, pp. 262-263).

12. Social sciences are a special case: they are both fundamental theoretical disciplines and transfer fields. "Social engineering" has not developed as a recognised and institutionalised discipline, in spite of numerous examples of

applied psychology, sociology and economics. The contributions of social sciences promote the development of technical projects (e.g. architects use them as much as "exact" sciences) but they also generate their own objects in the form of explanations concerning the operation of society. Here again, the relationship between a discipline and an allied profession is not unequivocal. Sociologists and economists no doubt apply the concepts derived from sociology and economics but also acquire tools taken from mathematics, life sciences or psychology. Progress in solving social problems requires multidisciplinary research in the same way as technical problems.

13. For a detailed discussion of the two hypotheses, see Kamien and Schwartz (1982).

14. The demand-pull view implies a reactive and mechanical view of change in productive techniques, with an economic environment defined too narrowly in terms of markets represented by changes in prices, costs and profitability, and in terms of "users' needs". It has difficulty in accounting for the discontinuities and non-market elements in the innovation process that mark its unique nature. It can therefore be described as setting out the necessary, but not the sufficient conditions for innovation (Mowery and Rosenberg, 1979). The difficulties of the technology-push view are opposite but in a sense complementary. Instead of introducing the role of institutions and the interaction of market and non-market factors in the innovation process, technology-push approaches ignore economic factors and provide a linear, unidirectional account of the science-technology-production relationship. The growth of the scientific base is taken to be exogenously determined, and no attempt is made to link it with the evolution of technologies and with the pattern of the industry, market or the institutions that support it.

15. See Johnson (forthcoming).

16. For a recent comprehensive presentation of several parallel, slightly differing approaches to the notion of technological trajectories, see Dosi (1988, p. 1128 ff). Also Gaffard and Zuscovitch (1988, p. 628) and Gaffard (1990a).

Chapter 2

TECHNOLOGY DIFFUSION

ABSTRACT

   This chapter analyses the mechanisms of technology diffusion and discusses the need for making diffusion policy a major element of overall technology policy. The analysis is based on a distinction between disembodied technology diffusion and equipment-embodied technology diffusion. Disembodied technology diffusion is character-ised by research spillovers and by the absorptive capacity of firms. The term "research spillover" refers to knowledge developed by one firm becoming potentially available to others. Absorptive capacity refers to the ability of firms to learn to use technology developed elsewhere through a process that involves substantial investments, particularly of an intangible nature. R&D thus both develops new products and enhances the capability of firms to learn to anticipate and follow future developments. Formal or informal networks of industries and firms, within which participants are at the same time suppliers and demanders of new technology, facilitate this process.

   Equipment-embodied diffusion recalls the more traditional pattern, in which a few industries act as suppliers of new technology through their sales of technologically-intensive goods to downstream industries. Certain key technol-ogies [mostly related to information technology (IT)] play a central role. International differences, market structure, the expectations of investors, characteristics of innovations in modern production systems, skills availability and the organisation of firms all influence this process. Market structure in the supplying industries determines prices of technologically-intensive inputs and thus the degree to which benefits due to the original innovation are captured by the user industries or consumers. The real or perceived pressures of competition and the relative advantages of using competing technologies affect the pace of investment and adoption of new technologies.

   The economic performance of the bulk of manufacturing and service industries that lie outside new technology sectors depends to a large extent on adopting ideas and products developed elsewhere. Since society benefits from the R&D efforts of firms, public policies should provide innovators with an environment which stimulates innovative activity while allowing maximum use of their products. A stable macro-economic environment that encourages investment in the creation and adoption of new technologies is an important prerequisite. More important, however, are micro-economic policies that incite firms to share information, develop absorptive capacity and increase rates of adoption of new technologies, either directly (through subsidies, financial schemes, etc.) or indirectly (through alteration of the institutional and regulatory environment).

INTRODUCTION

Chapter 1 examined the nature and characteristics of the process of technological innovation, stressed the dynamic and interactive nature of the process and explored its patterns and determinants. This chapter moves to the process of technology diffusion (see Box 12) and the application of an innovation after its initial development. The approach taken stresses how closely interlinked the two processes are. Rather than thinking of innovation (the supply of technology) and diffusion (the demand for technology) as two separate activities, it is more accurate to think of the creation of new technology and of its adoption and management as two aspects of a single process[1].

Many of the measures for promoting technology diffusion depend on increasing intangible investments, notably investments in R&D, training and human capital, as distinct from investments in physical capital. Even though intangible investments have not traditionally been viewed in many countries as directly connected to science and technology or even industrial policy, such investments are critical for capturing the economic potential of new technology, and there is increasing evidence that they determine economic returns on investment in new equipment. The efficiency of these investments depends in turn on the infrastructural environment and on technology-specific factors

---

Box 12. Conceptual dimensions of technology diffusion

The notion of technology diffusion must be taken today to "include adoption by other users as well as more extensive use by the original innovator. More generally it encompasses all those actions at the level of the firm or organisation taken to exploit the economic benefits of the innovation" (OECD, 1988a, p. 49).

Thus, diffusion cannot be reduced to the introduction of new machinery onto the factory floor or into the office or to the adoption by firms of new intermediate goods. It must include the other active and vital steps taken by firms to adapt technology to their needs and thus increase the economic efficiency with which new technology is utilised. These steps include the reorganisation of factory work and materials flows (such as just-in-time production programming) and improved management practices on the shop floor, in production development and in marketing (see Chapter 4). More generally, the notion of technology diffusion must also include the process whereby knowledge and technical expertise spread through the economy.

Conceptual work on technology diffusion has advanced hand in hand with the changes in the way policy makers regard the process. As a result, a number of useful related notions are emerging. These include the need to distinguish between the cost of adoption and the cost of purchasing and the importance of analysing diffusion as a competitive selection process between old and new technologies. The recognition that *every act of adoption involves certain transformations and is thus an act of incremental innovation in itself* is central to these new views.

Underlying these issues is a fundamental conceptual distinction between two types of technology diffusion. It is the distinction between *disembodied* and *equipment-embodied* technology diffusion. The former is the process whereby technology and know-how spread through channels other than embodiment in machinery. It originates in the *externalities* that characterise the innovation process and the *research spillovers* that occur when the firm developing a new idea or process cannot fully appropriate the results of its innovation. Equipment-embodied diffusion on the other hand is the process whereby innovations spread in the economy through the purchase of technologically-intensive machinery, components and other equipment.

relating to the systemic and dynamic characteristics of innovation. Chapter 5 will examine available data on intangible investment by the business enterprise sector, and an analysis of education and training in relation to new technology will be presented in Chapter 7.

This chapter focuses attention on the importance of technology diffusion, analyses its mechanisms and discusses the need for making it a major element of overall policy. Section 1 discusses disembodied tech-nology diffusion, and Section 2 discusses the more traditional equipment-embodied technology diffusion. Section 3 analyses the economic implications of the diffusion process, in particular the transfer of produc-tivity from innovating to user industries and the rela-tionship between diffusion and the incentives to inno-vate. Finally, Section 4 draws conclusions on the role of governments and sketches out elements of a diffu-sion policy[2].

1. THE SPREAD OF KNOWLEDGE: DISEMBODIED TECHNOLOGY DIFFUSION

This section examines disembodied technology diffusion: its actors and mechanisms of transmission, its main determinants (knowledge spillovers and the absorptive capacity of firms) and the factors that underlie the efficiency and speed of the process, as they relate to the type of technological knowledge involved and to some systemic characteristics of innovation.

Actors and mechanisms

The diffusion of disembodied technology may be organised, as when firms sell the rights to a patent or license an innovation. More often, however, it is just a consequence of the firm's innovative activities, as when the knowledge that it develops becomes available to other firms. In either case, its channel of transmission is mainly through research personnel. Scientists and engineers who have participated in the elaboration of new ideas or techniques within a particular firm or laboratory initially communicate this knowledge to other parts of the organisation they belong to (the production or marketing departments, subsidiaries, etc.); eventually the know-how, "rules" or expertise involved leaks out and becomes part of the public domain.

The transmission of knowledge has many chan-nels. It is the result of reverse engineering by a firm on its rivals' products; it is contained in descriptions of new products or processes to be found in publications, catalogues or patent applications; it may be dissemi-nated in conferences or seminars; it can be part of the human capital that research personnel take with them when changing jobs; or it can be the by-product of mergers and acquisitions, joint ventures or other forms of inter-firm co-operation[3]. These different channels of transmission generate different diffusion patterns with distinct effects on productivity, competitiveness, and the incentives that firms have to invest in innovation in the first place.

Early studies explain diffusion patterns by focus-ing mainly on the way that information spreads, the influence of expected profitability and the size of firms. They argue that following the introduction of a signifi-cant innovation, its spread among potential adopters is governed by the spread of information about its bene-fits. In turn, the speed of adoption is a function of the characteristics of the innovation (principally its per-ceived profitability or the length of the pay-back period) and of the potential adopters (mainly the size of firms and the cost of the innovation relative to the firm's assets). The rate of adoption increases as more firms come in contact with the growing number who have already purchased the new technology. As the market approaches saturation, the rate decreases[4].

Empirical studies have investigated how quickly new knowledge developed by one firm comes into the public domain. In a study surveying 100 US firms in thirteen manufacturing industries, Mansfield (1985) showed that information concerning development deci-sions for a major new product or process were in the hands of at least some of the firms' rivals within 12 to 18 months, on average, after the decision had been made. If the creation and commercialisation of a major new product or process takes three years or more, there is a better than even chance that the decision will leak out well before completion. Furthermore, information about the detailed technical content of innovations (both process and product) reached most rivals within a year after the development of the innovation (see Table 5). Corroborating studies have also shown that the traditional means of protection against this type of knowledge spillover are not very successful: the mobil-ity of personnel and practices of "reverse engineering" compromise efforts at secrecy, while patents tend to operate as a means of communication among firms (see Rogers, 1982; and Levin, et al., 1987).

Despite the emphasis placed on the central role of the spread of information in shaping diffusion patterns, Mansfield's study points out that the fact that informa-

Table 5. **Percentage distribution of firms, 10 industries, United States**

By average number of months after development before the nature and operation of a new product or process are reported to be known to the firm's rivals

	Products				
	Less than 6	6 to 12	12 to 18	18 and more	Total[1]
Chemicals	18	36	9	36	100
Pharmaceuticals	57	14	29	0	100
Petroleum	22	33	22	22	100
Primary metals	40	20	0	40	100
Electrical equipment	38	50	12	0	100
Machinery	31	31	31	8	100
Transportation equipment	25	50	0	25	100
Instruments	50	38	12	0	100
Stone, clay and glass	40	60	0	0	100
Other[2]	31	15	15	38	100
Average	35	35	13	17	

	Processes				
Chemicals	0	0	10	90	100
Pharmaceuticals	0	33	0	67	100
Petroleum	10	50	10	30	100
Primary metals	40	40	0	20	100
Electrical equipment	14	14	57	14	100
Machinery	10	20	30	40	100
Transportation equipment	0	67	0	33	100
Instruments	33	33	33	0	100
Stone, clay and glass	0	20	20	60	100
Other[2]	27	0	36	36	100
Average	13	28	20	39	

Note: The data cover a sample of 100 US firms in the industries mentioned in the table spending over $1 million (or 1 per cent of sales, if sales were at least $85 million) on R&D in 1981.
1. Because of rounding, figures sometimes do not sum to total.
2. Fabricated products, food, rubber and paper are included in the "other" category.
Source: Mansfield (1985, p. 220).

tion leaks out relatively quickly does not imply equally rapid imitation. It often takes time to invent from a patent, to develop prototypes, to alter equipment, and to engage in the manufacturing activities required to introduce an imitative product or process. Thus, a more complete appreciation of the mechanisms involved must take into account "learning by using" effects – the participation of user firms in the development of an innovation and the costs that they incur in learning how to adapt it to their needs (Rosenberg, 1982). As an innovation diffuses, more information about its technical and economic properties becomes available; as this information spreads, the innovation diffuses more easily. It has been shown that observed diffusion paths reflect changes in the innovation and adoption environment; the process is a distinct one and different from learning within a static situation (David, 1987; Davies, 1979; and Gold, 1989).

In the light of findings that emphasise the efforts that firms must make in order to adopt technology developed elsewhere, two central notions help explain both the pattern and the determinants of this type of technology diffusion. The first is that of *research spillovers*. These reflect the appropriability characteristics of particular technologies and explain how new knowledge or technology developed by one firm becomes potentially available to other firms or industries, domestically or abroad. The second notion is that of *absorptive capacity*. This is how firms learn to use technology developed elsewhere; it involves substantial investments, particularly of an intangible nature.

Research spillovers

Research spillovers have been defined to include "any original, valuable knowledge generated in the research process which becomes publicly accessible, whether it be knowledge fully characterising an innovation, or knowledge of a more intermediate sort" (Cohen and Levinthal, 1989, p. 571). They have been labelled disembodied or knowledge spillovers to emphasise that they do not necessarily relate to knowledge embodied in machinery or equipment. Most definitions imply that a firm or industry may use the knowledge developed elsewhere without providing compensation. Such knowledge spillovers occur when some of the inventive activity is not fully appropriable: the R&D efforts by firms generate externalities which affect the decisions of other firms and industries. Thus, spillovers related to knowledge arise because innovation has certain characteristics typical of public goods (see Box 13).

While knowledge and technology have some of the characteristics of public goods, they are privately provided by firms that invest in R&D and other technology-related activities. While one might consider knowledge spillovers as "leaks" or as an unjust loss of profits for the innovator, in reality they are the sine qua non condition for the development of knowledge and of the economy. It is because innovations benefit more than just the originating firm and because they become widely diffused that knowledge can develop in a rapid and cumulative manner.

This suggests potential gaps between the private and the public interest with respect to developing and using new technologies. The corollary is that the role of policy is to reconcile the opposing aims of a framework that must provide innovators with an environment which stimulates innovative activity (by restricting use of the innovation and thereby guaranteeing some gains to the innovator) while at the same time allowing maximum use of its product (by keeping its price low and thereby ensuring imitation, adoption and diffusion).

The centrality of research spillovers in the process of innovation is at the root of the formation of formal or informal "networks" (see Chapter 3). The "public good" aspects of innovation make it an activity with strong elements of "collective creation", whether codified in joint-venture arrangements between firms – "learning by interacting" – or implicit in the use of knowledge as "barter" between otherwise competing firms. The existence of knowledge spillovers thus suggests that production of knowledge by a particular firm or industry depends not only on its own research efforts but also on outside efforts or, more generally, on the knowledge pool available to it. The productivity of a firm's own research may therefore be affected by the size and nature of the knowledge pool that it can draw upon. Thus, and as is suggested by the "interactive" model of innovation (Chapter 1), innovation and diffusion are shown to be "two faces of the same coin" (Jacobs, 1990): innovation leads to diffusion which in turn influences the level of innovative activity.

Absorptive capacity

While spillovers determine disembodied diffusion flows, the actions of receiving firms and industries determine to what extent innovations developed elsewhere are actually adopted into production processes. As many authors have argued, research and development expenditures and other intangible investments enhance the firm's ability to assimilate and exploit information in the public domain. Thus, the ease of learning within an industry is directly affected by the level of expenditures on R&D and will indirectly determine the influence that spillovers have on effective flows of diffusion (Cohen and Levinthal, 1989).

This line of argument recognises a dual role for R&D: in addition to developing products – or new information – and to helping maintain a market position for the innovating firm, R&D also develops the capacity to learn to anticipate and follow future developments. This second aspect has been referred to as "learning to learn" or learning by learning, as distinct from "learning by doing". In this area, R&D helps firms identify, track and potentially take advantage of knowledge initially developed elsewhere: it develops its "absorptive" capacity. The two aspects of R&D go hand in hand: adoption of new technology presupposes the capacity for *absorption;* the latter depends in large measure on the capacity for *innovation.*

A number of authors have drawn attention to the role of learning. Nelson and Winter (1977) have argued that in order for firms to be able to use freely available knowledge they often have to invest in R&D. Rosenberg has likened performance in (basic) research to "a ticket of admission to an information network" (1990, p. 170). Tilton has asserted that one of the main reasons why firms invest in R&D in the semiconductor industry is in order to "facilitate the assimilation of new technology developed elsewhere" (1971, p. 71). These expenditures are necessary because knowledge is not simply "on the shelf"; substantial research capability is often necessary in order to understand and assimilate knowledge. This is particularly so since assimilation usually involves transformation and adaptation at the same time.

While recognition of the role of intangible investments in learning is not new, its implications for shaping the pattern of technology diffusion have not been explored until recently. The capacity to imitate and take advantage of technological advances made elsewhere may depend crucially on a firm's own R&D expenditures. In the past, imitation costs were viewed as consisting primarily of the costs of transmission of information, and they were considered small when compared to the cost of generating the innovation. Imitation costs may, however, depend crucially on the technological level achieved by a firm (its stock of accumulated R&D, its training and organisational practices, etc.). The costs are only low when the firm has already invested in the development of its absorptive capacity in the relevant field.

These conditions are also significant for the research strategies of small and medium-sized enterprises (SMEs) and for industries situated on the periphery of the "core" sectors which develop science-based generic technologies and form the heart of the innovation system in large OECD countries. In-house research capacities and research directed at adapting generic technologies are normally inadequate (see Chapter 1 on innovation). Firms without substantial means for research need to develop their absorptive capacity in order to participate in networks and take advantage of disembodied technology flows. At the same time, in order to choose, absorb and adopt technology, they will also have a greater change of success if they have some creative activity and capability of their own, even it it only design capability, coupled with some kind of monitoring and absorption function within its management. They will then be better able to take advantage of public initiatives to diffuse technological information, notably the networks of technology advisers established in some countries. However, it is also necessary to adapt external technology to the firm's existing stock of knowledge. Chapter 1 discusses ways in which this might be done, in particular the role of contract research organisations (CROs).

Some of the smaller OECD countries with weaker systems of innovation and a later start in industrialisation fall into the same category as SMEs that lack an inventive capacity of their own. They are thus, to an extent, unable to take full advantage of the process of disembodied technology diffusion and will rely mainly on equipment-embodied diffusion mechanisms in order to make use of innovations developed elsewhere.

Factors influencing the efficiency of disembodied technology diffusion

The effect of research spillovers and the absorptive capacity of firms on the speed and efficiency of disembodied technology diffusion is conditioned by a number of factors that relate to the nature of technology and its systemic characteristics. The extent to which knowledge becomes publicly accessible, the learning of firms, and hence the pattern of disembodied technology diffusion all depend on different types of knowledge and technology (basic, applied/firm-specific or generic/diffusing), as well as on systemic characteristics of innovation such as cumulativeness.

The degree to which R&D is critical to the development of a firm's absorptive capacity, and hence to the pattern and speed of disembodied technology diffusion, depends on the characteristics of outside knowledge. A number of factors may play a role: the degree to which a particular field of knowledge is cumulative; the degree to which it is targeted to the specific needs of the firm; the degree to which it is "tacit"; the pace of advance. The more complex outside knowledge is and the more generic and less targeted its nature, the more important are R&D expenditures for identifying and facilitating the exploitation of valuable knowledge. The same is true when knowledge is tacit and uncodified; substantial time and effort are needed to transform such knowledge into a set of rules or routines that can be useful to firms. The faster the pace of advance of the field, the greater the effort required to keep up with developments.

Different types of R&D with different appropriability characteristics imply differences in ease of diffusion and adoption of innovations. Nelson (1980), among others, has drawn a distinction between two types of technological knowledge. One relates to basic ("upstream") research and consists of inferences about how things work, the identification of constraints and of possible ways of overcoming them. The other relates to "operative techniques" (or development) and consists of ways to make things work that are specific to the task at hand.

The type relating to basic research possesses "public good" features to a greater extent, since this type of knowledge has a wide range of applications. It can be communicated to someone in the field without major learning costs and it could considerably limit the capabilities of those denied access. It is also here that most pre-competitive co-operative research tends to be concentrated and that formal and informal networks tend to form. In contrast, the type relating to "operative techniques" or development usually possesses public good properties only to a limited extent. The range of "technique" applicability is narrow, learning entails high costs because it needs to be adapted to the specific needs of the user, and denial of access may not eliminate chances for success in developing new processes or products if the basic technology has been assimilated.

This suggests that specificity and localisation are determining characteristics of most innovations, both in terms of the innovating firm's ability to appropriate relevant knowledge and in terms of the pattern of adoption that results[5]. The cost of acquiring technological knowledge in a form usable for a firm's own production may thus be high. Evidence suggests that the resources spent on reverse engineering or spying are in fact substantial (Levin, 1988).

An additional category of "generic" or "diffusing" technologies (e.g. information technology, certain materials and equipment) has, as a main characteristic, wide applicability in almost all industries and an ability to transform the economic environment drastically. Generic technology underlies production in the same sense that basic knowledge does; however, while in basic research the body of knowledge concerned is specific to certain technological areas, generic technologies underlie innovations and production in industries that rely on a variety of technologies. Generic research also shares a characteristic with "operative techniques": it often requires substantive investments on the part of firms in order to be adapted to their use. Since generic technologies are developed in certain industries and then diffused widely, investments by users do not build on their own accumulated "technological capital"; the knowledge required is external to the firm. In such cases, large public support programmes can help develop the capabilities for adoption (for example, the IT-related public training programmes available in many countries).

In all three types of technology, certain attributes of innovation dilute its public good nature to some degree and make substantive absorption investments a prerequisite for diffusion. These attributes are best understood in the light of the notions of technological trajectory and the *cumulative* nature of technological change (see Chapter 1) at firm, country or world level. These notions reflect the importance of learning processes. Technological advance is not a random process: its direction is defined by the "topography" of advances. Most innovations today build on previous technology and incorporate many of the features of the displaced products and processes; the probability of success in innovating is thus a function of the level of achieved results (Stiglitz, 1987). In this situation, spillovers and disembodied diffusion lead to high levels of intangible investment by adopters. The demand for new technology (the diffusion process) acts as a spur for the supply side (the innovation process): the two are complements, rather than substitutes (OECD, 1988*a*).

2. EQUIPMENT-EMBODIED TECHNOLOGY DIFFUSION

This section discusses once again the main actors and mechanisms and examines a different concept of "spillover", relating to market structure as the main determinant of the extent to which economic benefits flow across the economy. The discussion then turns to firm-specific issues (investment, strategy and timing in adoption decisions) and to issues relating to the environment within which firms operate and to the systemic characteristics of innovation. The section closes with a look at some issues that arise in the diffusion of consumer products.

Actors and mechanisms

The traditional interpretation of the process of technology diffusion describes the spread and introduction into production processes of machinery, equipment and components incorporating new technology. Equipment-embodied diffusion recalls this pattern, in which a few industries act as *suppliers* of new technology. They sell technologically intensive intermediate and capital goods (both manufacturing and non-manufacturing) to downstream industries, consumers and the government. All these buyers reflect the demand of users for technologically intensive machinery, equipment and components. The supplying industries are mainly in the R&D intensive manufacturing sector and include electrical industrial machinery, parts for electronic appliances and communications equipment, drugs and medicines, and the scientific and chemicals industries. They receive relatively little inflow of embodied R&D from other industries and primarily use their own technology to improve their productivity.

Evidence on the patterns of the flow of technology across industries comes from a number of studies using different methodologies: innovation surveys, studies of patent data, studies of data on purchases of technologically intensive intermediate and investment goods.

Some of the most comprehensive data assembled on the production and use of technology derive from a UK study of 4 378 "significant technical innovations" identified by industrial experts working in the 28 industries concerned in the period from 1945 to 1983. On the basis of interviews with experts and questionnaires sent to innovating firms, significant innovations that have been successfully commercialised are identified and allocated to an innovating industry and to a first user industry[6].

The study reveals that five "core" manufacturing sectors – chemicals, machinery, mechanical engineering, instruments and electronics – account for about 65 per cent of all innovations. A further six "secondary" sectors – metals, electrical engineering, shipbuilding/offshore engineering, vehicles, building materials, and rubber and plastic goods – account roughly for another 23 per cent. The five remaining manufacturing sectors account for the rest.

Innovations produced in the core sectors are particularly pervasive, being used in a large number of other sectors. Out of 26 user sectors, innovations originating in chemicals and electronics found applications in 18, mechanical engineering in 25, and instruments in 26. Table 6 gives the broad composition of user sectors of innovations originating in the core sectors. It shows a much higher proportion of use outside manufacturing for chemicals, instruments and electronics innovations than for mechanical engineering and machinery innovations. The non-manufacturing proportion of use is highest in electronics. Unlike the other core sectors, it also has a relatively high proportion of use intra-sectorally, principally for the application of electronic components and computing in electronic products.

The pattern of technology diffusion across industries has also been examined with the help of data on purchases of technologically intensive intermediate and investment goods. A study carried out at the US Department of Commerce, focusing on inter-sectoral and international technology flows between the United States, Canada and Japan, reinforced the widespread perception of the central role of certain key technologies (mostly IT-related) but also showed substantial international differences in the adoption of technologically sophisticated machinery and equipment (Davis, 1988).

The study emphasised that indirect technology inputs embodied in upstream intermediate and capital inputs account for a large share of total technology embodied in total output (for export or domestic use). Moreover, the relative importance of direct (innovation-based) to indirect (diffusion-based) technology inputs differs widely among countries. In Japan, for example, (domestic) indirect technology inputs in 1984 averaged over three-quarters of the direct inputs, while in the United States they were around 50 per cent. This suggests that Japanese industries are both more dependent on technology from key indirect technology sources and more able to diffuse technology across industrial sectors. Canada, meanwhile, is shown to be

Table 6. **The sectors of use of "core sector" innovations, 1945-83**

Percentage of total use

	Own use	Other manufacturing	Non-manufacturing	First three sectors of use	
Mechanical engineering and machinery	14.2	58.1	27.7	Textiles	(19.8)
				Mining	(11.5)
				Electrical engineering	(10.2)
Chemicals	24.9	32.1	43.0	Health	(24.9)
				Textiles	(13.1)
				Agriculture	(6.3)
Instruments	9.9	47.9	42.2	Textiles	(9.9)
				R&D	(9.7)
				Health	(8.4)
Electronics	37.4	11.7	50.9	Defence	(10.1)
				Business equip.	(8.4)
				R&D	(7.6)
Total sample	30.5	34.0	35.5		

Source: SPRU (1984).

largely dependent on the domestic diffusion of technology embodied in imports of intermediate and capital goods. The study thus pointed to the importance of technology diffusion for shaping competitiveness. It also underscored the need to develop more meaningful measures of *intensity of technology* that incorporate both innovation (R&D performed within an industry) and diffusion (R&D performed in other industries domestically or abroad and embodied in machinery and equipment) in order to make informed policy choices[7].

Market structure and a different concept of spillover

The mechanism by which equipment-embodied technology developed in one industry spreads to other industries involves a second type of spillover[8]. It relates to the prices that user industries pay for their R&D-intensive inputs or that consumers pay for R&D-intensive products. The issue is whether the prices at which R&D-intensive inputs are purchased by users accurately reflect changes in the user value, or marginal productivity, of the commodities[9]. If products are sold to user industries or consumers at prices below their true social cost, the fruits of R&D efforts can be partially captured by the buying industry. R&D conducted in supplier industries will therefore have positive side effects in the user industries or in terms of additional benefit to consumers.

The key to the size and effect of this type of spillover lies in the structure of the market and the extent of oligopolistic conditions and barriers to entry in the supplier industries[10]. Competition determines the prices at which R&D-intensive inputs are sold across industries and thus how much of the benefits due to the original innovation are captured by users. If the innovation supplier industries are concentrated and exhibit strong oligopolistic tendencies, they will be able to charge high prices for their technology and thus capture most of its social benefit. Competitive pressures in the innovation supplier industries, on the other hand, will force down prices so that the price at which inputs are purchased by user industries will not fully reflect their increased downstream value. In this case, most of the benefit will be passed on to users (Mohnen, 1989).

The role played by competition and the nature of supply structures in determining the speed of technology diffusion has long been recognised. It extends well beyond "spillovers" to affect decisions regarding the scrapping of existing capital and the extent of material and intangible investment to replace and expand existing productive capacity. Thus, the real or perceived pressures of competition, or lack of it, are major explanations of the rate of investment in and adoption of new technologies[11].

To the extent that the expectations of future profits associated with a certain degree of monopolistic power is a necessary prerequisite for innovation, the positive effects of these spillovers are maximised in an environment where the "profit carrot" or the competitive threat (leading to innovation) and the results of competition (leading to low prices and diffusion of the benefits of technology) are balanced. Furthermore, the realisation of large, economy-wide gains from the type of spillovers discussed here rests on the public good aspect of technology. The technology embodied in a particular good does not depreciate with use; it retains its superior properties until it is replaced by a better technology. Thus the (relative) loss to the innovating industry associated with the loss of some of the potential profit stream is transformed to a benefit for all users.

The role of investment and timing in adoption

Even when embodied technology is available in the form of equipment, machinery or components, it will only be absorbed by user firms and industries when certain conditions are met:

i) the equipment must be purchased, which involves investment (mostly material but also of an intangible nature, involving training and reorganisation);
ii) the equipment must be compatible with technologies already in use and with the firm's technological environment; and
iii) the technology must be easy to assimilate with respect to the organisational norms in place and the types of skills and training at the firm's disposal.

These factors warrant further analysis.

The decision to adopt new technology involves cost-benefit calculations based on expected future costs and profitability of alternative technologies. Investment in new technology can take place more quickly and earlier when capacity is expanding, and investment is less reliant on replacement of existing capital. Replacement (or modernisation) investment, on the other hand, is determined by the profitability of adopting new equipment compared with that of running existing equipment. As was realised early in the development of diffusion theory (e.g. Mansfield, 1968; and Salter, 1960), adoption of new kinds of equipment incorporating new technologies can be delayed by the age of the existing capital stock and by sunk costs. The decision also involves competition between new technologies and existing ones (Metcalfe, 1990). Firms which have successfully mastered old technologies face difficulties in surmounting the skill and knowledge limitations inherent in their existing technologies and in acquiring the new skills necessary to the successful use of the new one.

The moment at which innovations occur will also affect diffusion. The rate and extent of diffusion will be affected by firms' expectations of the path and pace of future technical and market change. In the initial stages of diffusion, design may be fluid and technical or regulatory standards uncertain (see Box 14). Firms may expect that incremental innovations will improve performance, decrease costs and stabilise the fundamental technical and regulatory specifications of the original innovation. Options may be added, making it more suitable for specialised markets and increasing the range of potential users. When technical characteristics are both uncertain and changing, decision-making criteria are modified (Silverberg, 1990). This is particularly the case for radical departures from a well-developed technological trajectory.

Early adopters are risk-takers who, once they are numerous enough, create a "critical mass", a gauge by which late adopters can judge the benefits of adoption. Rapid formation of critical mass is necessary for the spread to domestic markets of innovations generated abroad, because it affords the opportunity to weigh profitability against risk. This is especially true for radical innovations like advanced communications that require a complete overhaul of skills and organisational routines. Adoption costs can be measured more accurately through evidence provided by a large number of users. The very existence of this critical mass enhances the profitability potential because it creates a widening

pool of skilled manpower, better technical assistance and the overall understanding of the new technology that comes from "learning-by-using". Thus, as the original innovation evolves and as thresholds or critical masses of diffusion are reached, the rate of adoption will change.

In addition, strategic considerations are also crucial to the rate of adoption of new (embodied or disembodied) technology. In order to secure a competitive lead, a firm may resort to leader-follower strategies. It may delay the adoption of new technology until a rival innovating firm has borne the costs associated with the learning curve. It may delay in order to pursue a "leap-frogging" strategy in an intensely competitive market[12]. Thus, delay in adoption may well be rational (Metcalfe, 1990).

Timing in adoption is also affected by the strategies of supplier firms. In reality, the diffusion cost/benefit and profitability ratios involved are twofold: one set is the adopter's, the other the supplier's. Thus, diffusion curves are in fact the outcome of two processes: one relates to the development of a market for the technology and the other to the creation of a capacity to supply that market. Moreover, it is the relative profitability of competing technologies, not their absolute profitability, which is important. During the diffusion process, this will change under the influence of changes in the supply of competing technologies and in their diffusion environment. Even in the special case where sufficient capacity already exists to meet the maximum rate of demand for an innovation, supply factors (pricing policy and rate of output decisions) can influence investment in new technologies and hence the diffusion process.

Asymmetries in investment and adoption patterns are also due to the fact that firms differ in their capacity to assess and finance new technology and to appropriate the benefits from adoption. In general, adoption will be easier (and diffusion will occur faster) if external conditions are favourable. This implies both low cost of capital and an appropriate financial structure, in terms of access to capital and the existence of investors with a long-term view. The importance of the design of fiscal policies, which can encourage or act as an obstacle to the creation of such an environment, should not be underestimated.

Systemic characteristics of technologies as factors in adoption

Factors relating to the systemic characteristics of technology also affect equipment-embodied diffusion. The interrelatedness of many technologies and the associated network externalities form two types of linkages. The first relates to production: technological networks or "clusters" are central to the production of many innovations. The second relates to use: the networks of users of many technologies are a crucial factor in both their development and pattern of adoption.

The concept of *interrelatedness* reflects the complex nature of many production technologies as systems of multiple interdependent parts. An innovation which changes one of these parts may be incompatible with the rest of the system and require costly system changes to make it fit. These "costs of interrelatedness" must be taken into account in any rational investment decision (Metcalfe, 1990). The greater the degree of interrelatedness in an existing technological system, the less likely it is that a given innovation will be compatible with it. This helps explain the increasing importance of minor incremental process innovations over the life cycle development of manifestly system technologies, for example in the automobile industry (Abernathy, 1978). Interrelatedness also helps explain "technology fusion": the creation of new technologies (with new rules and procedures) by the merging of previously distinct ones (Kodama, 1990).

Case study material on the diffusion of telecommunications equipment suggests that the rate of diffusion of new technologies across countries is affected by the strong cumulative effects of "interrelatedness". Latecomers (late-adopting countries) with lower levels of penetration – smaller amounts of the innovative capital goods that embody the new technology – may also have lower rates of diffusion. This is because it is difficult to obtain all the relevant information and skills for information-based technologies without having already reached a certain level of interrelatedness with other users.

In a more general sense, innovations rarely work in isolation. The productivity of each innovation depends on the availability of complementary technologies (Rosenberg, 1982). Building on the effects of interrelatedness and on associated complementarities among industries is essential to ensuring efficient diffusion patterns. A recent study focusing on industrial performance identified such inter-sectoral complementarities as a key factor (Amable and Mouhoud, 1990). Countries such as Germany and Japan were shown to benefit greatly from the complementarities between information technologies and medium technology industries that use IT-related innovations as inputs in production (mainly mechanical and electrical engineering industries). Italy was shown to benefit from complementarities between much lower technology industries, such as textiles and related equipment-providing industries. The existence of "islands" of high technology, on the other hand, cut off from the rest of the industrial base, retards diffusion and creates structural problems for other industries: examples offered are the United States (with aerospace) and France (with aerospace and chemicals).

The utility of many technologies (and by implication the benefits from adoption, as well as from further development) increases simply because the number of users has increased (Foray, 1990a). The effects of these "network externalities" are particularly evident in innovations such as telecommunications-based services and equipment, where the notion of costs of adoption as distinct from the mere cost of purchasing (the market price of new capital goods) has emerged as a key factor. These costs depend on the behaviour of other firms (the technology they adopt) and on the penetration of the technology into a given country or wider region. As the number of adopters increases, so does the availability of skilled manpower. Maintenance and spare parts' costs decline and collective learning improves.

Analysis of the distinctive network features of new technologies must build on an understanding of the role played in general by externalities in the adoption of capital goods that incorporate innovations. Careful examination of the pattern of adoption and diffusion of a wide variety of new technologies suggests that network externalities, or more generally "dynamic externalities", are a factor in many instances (see Antonelli, 1991b). Dynamic externalities are characterised by technical interrelatedness and vertical complementarity in upstream markets. In the case of advanced telecommunications, diffusion of digital switching and efficient transmission of data enable manufacturers to fashion efficient organisational innovations such as "just-in-time" techniques that cut down on warehousing and related costs. Dynamic externalities appear to be especially strong at the regional level, because of this tight meshing of product and process innovations and the corresponding efficiencies. This suggests that there is a "geography of externalities" or a "spatial" dimension to technology diffusion as well, built around a concentration of technologies, their producers and users, which in turn helps shape future generation and diffusion of technology.

The diffusion of consumer products

With the exception of some brief references, the diffusion of consumer products has been largely omitted from the preceding discussion. While many of the factors and obstacles outlined above also apply here, the importance of the adoption of innovations by consumers and the special characteristics of the process merit a separate discussion.

The first issue here is consumer demand for technological progress. New products and services develop and survive only to the extent that there is sufficient demand for them. Consumers provide both the stimulus for the development of new technologies and the necessary feedback for their continuous evolution and adaptation. A sophisticated group of users with versatile needs forces firms to experiment with product diversification and customisation and in the process to innovate, either incrementally or more radically. The interactive model of innovation described in Chapter 1 shows that the absorption of products in final demand provides a necessary feedback for the development and supply of innovations.

Consumers thus have an important role in the economic and social selection of technologies and in determining the direction of technical change. There are, however, some key differences between firms and consumers as users of innovations. Firms usually have well-defined needs; this allows them to focus attention on certain user-producer relationships and to influence innovative activity actively. Consumers, on the other hand, spread their attention among many different commodities and many potential user-relationships. In addition, they are much more concerned with short-term benefits. As a result, they often play only a passive and reactive role in the innovation process, and their choices may often result in the adoption of intrinsically inferior technologies[13].

A number of writers have recently examined the implications of this phenomenon and the associated risks of "lock-in" into particular technological choices (e.g. Arthur, 1988; Foray, 1990b; and David, 1987). As discussed in Chapter 1 arguments are based on the systemic characteristics of some technologies and the effects of increasing returns related to network externalities: the benefits of having a FAX machine, for example, increase with the number of users. In addition, as the community of users increases, so does the quantity and quality of complementary products. Thus, the increase in the number of owners of a particular type of video recorder leads to a decrease in cost and an increase in the variety of tapes available for that equipment. Here again, however, technological choices that persist and diffuse rapidly because of "network effects" are not necessarily the result of a conscious search and assessment of all available techniques (see Box 15).

Box 15. Standardising the typewriter keyboard

The establishment of QWERTY as the standard keyboard layout in English language typewriters and now word-processors is a classic example of "lock-in". At the beginning of the diffusion process in 1880, there were a number of designs and keyboard layouts on the market. By the end of the century, the QWERTY layout dominated, despite the fact that it is generally agreed not to be the most efficient. In detailing the QWERTY story, David (1986) attributes this to a combination of historical accident and self-enforcing mechanisms: early training of "touch typing" was by chance based on the QWERTY keyboard; an increasing number of QWERTY-based touch-typists made it more profitable for purchasers to use QWERTY-based typewriters; and as buyers preferred QWERTY-based layouts so all suppliers came to use them.

The QWERTY story points to the costs and benefits of standardisation (having a single technology dominate the market) and to a potential role for public intervention. The benefits relate to the exploitation of economies of scale and network externalities and to the additional flexibility to consumers, when, for example, they can mix and match stereo components (see Cowan, 1991). The costs arise for users who adopted the non-standard technology and are also in the resulting loss of variety. Furthermore, the market may fail to choose the right standard, thus making the costs of standardisation potentially greater. Locking in to an inferior technology may hinder the ability to pursue a more promising technological path in the future.

3. THE ECONOMIC IMPORTANCE OF TECHNOLOGY DIFFUSION

This section addresses three issues by drawing on a number of representative empirical studies. The first is the difference between private and social rates of return to innovation. The second is the transfer of productivity due to equipment-embodied diffusion flows from an innovating industry to user industries. Finally, the interaction of the demand and supply of new technology – the effect of disembodied and embodied diffusion on the incentives for technological development – is briefly examined. This allows discussion of whether innovation and diffusion should be considered as substitutes or as complementary activities.

Social and private rates of return to innovation

When deciding whether to invest in R&D or in other intangible investments with the aim of developing a new process or product, firms undertake some kind of cost-benefit calculation. In its simplest form, estimating private returns to innovation entails forming expectations concerning the future stream of profits, net of any costs for producing and marketing the innovation and net of the profits that the innovator would have earned on products displaced by the innovation. Adjustment must be made for unsuccessful R&D. Firms will undertake the necessary investments if they expect these net returns to be positive.

This simplified decision rule ignores many of the uncertainties and risks associated with such investments. It takes into account neither the dynamic aspects of technological change nor strategic considerations. For instance, it may be impossible to form expectations about a unique optimal pay-back period independent from market structure and about pricing strategies over which to calculate investments (see Silverberg, 1990). Similarly, the continual evolution of the characteristics of innovations as they are diffused complicates assessments of future benefits due to commercialisation. Lack of information about how quickly innovations will be imitated also makes the forming of expectations difficult. Finally, firms may be willing to invest in developing a new technology even if the returns appear to be negative *ex ante,* either in order to maintain a market share or in order to be able to follow future technological developments in the area[14].

Nevertheless, simple "rules of thumb" allow *ex post* calculation of private returns from specific innovations and comparison with the social returns of these innovations, which include the net benefits that accrue to users of the innovation after its original development. However, these social returns are even more difficult to calculate than private returns. A simple attempt involves obtaining a measure of social returns

by adding to the net private benefits the change in consumer surplus due to lower prices and profits by innovators minus the costs incurred by consumers or by firms other than the innovating firm. Studies have concluded that the social returns from R&D (especially in basic research) are significant and higher than private returns, since only a fraction of the benefits of inventions are captured by the inventor or the innovating firm even when patent rights exist.

In what is perhaps the best-known group of studies in the area, Mansfield and his associates have established that in the United States, the median return from investment to firms which adopted new products or processes was 56 per cent as compared with 25 per cent for the firms that developed and marketed the innovations. Furthermore, in a number of cases, returns to firms which developed the innovation were so low "that no firm with the advantage of hindsight would have invested in the innovation". Yet in these cases, the return from diffusion was so high that, from society's point of view, the investment was well worthwhile (Mansfield *et al.,* 1977, and Mansfield, 1985). This type of result is widely corroborated. A study examining the effects of the diffusion of computers in the United States, for example, estimates that between 1958 and 1972 the benefit from the adoption of mainframe computers in the financial service sector in the United States was at least five times the size of the expenditure for it in 1972 (Bresnahan, 1986).

This type of difference between private and social rates of return to innovation is rooted in the characteristics of the process of disembodied and of equipment-embodied technology diffusion discussed above. The public good nature of much innovation associated with knowledge spillovers implies that the benefit to society exceeds the net benefit to the firms that develop new technologies. The policy implications of this market failure are explored in Section 4.

Technology diffusion and productivity transfer

Studies like Mansfield's quantify the gap between private and social rates of return to innovation without explicitly identifying the inter-industry or intra-industry links through which technology diffusion takes place. A second group of studies explores the economic implications of diffusion by explicitly modelling the channels of transmission of technology across industries[15].

Studies that focus on the productivity effects of technology diffusion strongly support arguments that industry productivity often depends more on technology developed elsewhere than on own innovation. Early work by Terleckyj indicated that the rate of return on R&D embodied in goods purchased from other industries in the United States was almost twice the rate of return to own R&D; these findings were confirmed by subsequent work that also revealed strong inter-industry differences[16]. Recent OECD work focusing on data for 16 industries in six countries from about 1970 to 1983 finds that for understanding total factor productivity (TFP) growth at the industry level, it is important to analyse the industry using the new technology as well as the industry creating it. The study also finds wide discrepancies in the flow of new technology into industries, with non-manufacturing benefiting less than manufacturing. Finally, manufacturing and the chemicals and machinery industry groups, where R&D use is heavily concentrated and which contain most of the high-technology industries, show a very strong response of TFP growth to R&D diffused from other industries (as well as to their own R&D expenditures) (Englander et al., 1988).

The effect of technology diffusion on industrial productivity is not limited to the impact of purchases of technologically intensive intermediate and capital inputs. Technology developed in one industry may also affect productivity in other industries through knowledge spillovers. Such non-embodied inter-industry spillovers may affect the productivity of the user industry even if the prices at which inputs are sold across industries fully reflect their quality improvements. Goto and Suzuki (1989) examined the electronics industries in Japan in order to investigate whether innovation in these industries has also contributed to the growth of other industries that use electronics technology in their production processes. They were surprised to find that, despite the fact that the prices of electronics products have fallen significantly with quality increasing dramatically, there was no evidence linking the rate of growth of total productivity of manufacturing industries to the inflow of R&D embodied in intermediate and investment goods purchased from electronics-related industries. They concluded that the reason for this apparent paradox was that their methodology could not capture the effects on productivity due to disembodied diffusion through knowledge spillovers.

By adopting a methodology based on the *technological closeness*[17] of industries rather than on their inputs purchases from each other, Goto and Suzuki were able to show that the impact of electronics technology on other industries productivity growth was mainly achieved through the diffusion of technological knowledge, rather than through the purchase of intermediate and investment goods embodying the electronics technology. Industries whose technological positions were similar to those of electronics-related industries were able to exploit technology developed through the R&D activity of the electronics-related industries, in order to make their production processes more flexible and/or to manufacture the electronics-related products themselves[18].

Other studies have confirmed the cost-reducing and productivity-enhancing effects of both intra-sectoral and inter-sectoral knowledge spillovers. Bernstein (1989) estimated the effects of intra-industry and inter-industry spillovers in seven Canadian industries (food and beverages, pulp and paper, metal fabricating, non-electrical machinery, aircraft and parts, electrical products, and chemical products). He concluded that both types of spillover affect production costs, with inter-industry spillovers exerting greater downward pressure on costs of production than intra-industry spillovers. Levin and Reiss (1984 and 1989) concluded that the extent of spillovers is higher for processes than for products and that it varies considerably among industries, with electronics industries appearing to have significantly higher spillovers than others. The importance of different channels of spillovers for their effects on costs and productivity was also stressed.

Thus, empirical studies show that total factor productivity depends not only on the technology-related expenditures undertaken directly by industries but also on technology developed elsewhere that becomes available to user industries. The effect of diffusion on productivity thus occurs both through the purchases of technologically sophisticated machinery, equipment and components (equipment-embodied diffusion) and through the simple "borrowing" of ideas, know-how and expertise (disembodied diffusion).

Diffusion and the incentives to innovate

The empirical work which establishes that disembodied technology diffusion through knowledge spillovers has a significant effect on industrial productivity leads to the question of how this type of technology diffusion affects incentives to innovate. Does the fact that firms can take advantage of knowledge developed elsewhere diminish their own incentives to innovate? In other words, are innovation and diffusion substitute or complementary activities? An extensive literature based on surveys of the impact of disembodied knowledge spillovers and on (econometric) estimations of production and cost functions has developed in the last few years to address these questions[19].

In addressing the question of whether disembodied technology diffusion "crowds out" technological innovation, the evidence is mixed and industry-specific. Bernstein and Nadiri (1989) provide some evidence supporting the traditional view of the disincentive effects of spillovers. In a sample of four US industries (chemicals, petroleum, machinery and instruments), they observed that both the short-run and the long-run demand for R&D capital (and for physical capital) decreased in response to an increase in the intra-industry spillover. They did not find any complementarity effects between intra-industry spillovers and firms' own R&D. Rather, own and borrowed R&D appear to be substitutes, and firms tend to substitute diffusion for innovation.

Using a different methodology and a different data set, Bernstein (1989) finds that the response of firms to intra-industry spillovers depends on the nature of the industry they operate in. Firms operating in industries with relatively low propensities to invest in R&D tend to substitute their rivals' R&D for their own. In contrast, industries with relatively higher R&D propensities show a complementary relationship between own and rival R&D. Here, disembodied diffusion due to spillovers has a positive incentive effect on the demand for R&D. In a sense, innovation and diffusion tend to be complementary activities in industries where R&D expenditures represent an important competitive tool.

Levin (1988), using survey data and cluster analysis, arrives at similar conclusions. He argues that industries reporting the highest levels of spillovers (primarily electronics-related) also have average rates of innovation higher than industries relying on their own R&D. He suggests that his results support the hypothesis that spillovers are conducive to rapid technical progress but not the hypothesis that they discourage R&D investment. He further suggests that the reason for this complementarity between innovation and diffusion can be traced to some of the systemic characteristics of technological advance discussed earlier in this chapter. To the extent that technologies are cumulative, with each advance building on previous findings, there are important complementarities between the research of a given firm and the research of other firms in the same technological area. As a response to a diffusing and available technology flow, firms tend to intensify their own innovative efforts, in order to be able to absorb external knowledge and, in a more general sense, follow future developments.

4. SOME POLICY CONCLUSIONS

Diffusion phenomena can be described most simply as a propagation of technology from source of emission to reception by users. Abandonment of the linear scheme of innovation points, however, to constant interaction between sender and receiver and to the fact that technology is transformed as it is absorbed. Policy can be motivated by problems or obstacles related to *spillovers* or to *absorption*.

Policies related to the spillover phenomenon

The perception of the gap between private and social returns to innovation, due to knowledge spillovers and to spillovers related to market structure, furnishes a starting point for policy design. Policies must maintain an environment that reconciles two aims: on the one hand, the environment must be rich in incentives so that the expectation of significant *private returns* will spur the creation of new technology. On the other hand, it must have high spillover, so that firms only appropriate a fraction of the benefits of innovation and *social returns* are maximised through low-cost diffusion of technology.

In the first instance, therefore, governments must intensify their efforts to disseminate information about new processes, designs and technologies by setting up information centres, affording easy access to public data banks, etc. This information is often publicly available in principle; in practice, most of it does not reach the firms that could use it. Policies that actively bring new technologies to firms thus tend to be more successful.

In order to maintain a balance between the private and the social benefits of innovation, competition policy is important for both innovating and adopting industries. Excessively monopolistic market structures in the supplying industries allow innovators to maintain high price/cost margins for the equipment-embodied technologies that they provide to user industries. As a result, they keep a larger part of the social product associated with new technologies and can retard further innovation. The competitive threat, in addition to keeping prices low, also induces firms to continue to innovate. In consequence, a growing stock of knowledge and technology is created, on which all firms can build. Similarly, competition in adopting industries acts as an incentive for investment and thus accelerates the diffusion process. In both situations, however, policy needs to be applied with the understanding that the type of environment most conducive to the creation and diffusion of new technology combines elements of both competition and monopoly, with the monopolistic elements diminishing when there are important technological opportunities.

Intellectual property policy provides another example of a regulatory/institutional instrument that plays on the balance between the private and the social benefits from innovation. The role of patents is threefold: to increase appropriability by conferring a temporary monopoly power on the originator of an innovation; to disclose the new knowledge publicly; and to prevent its unauthorised use. Because in many industries the loss of appropriability does not act as a disincentive for expenditures for innovative activities, a case can be made for maximising public disclosure of information, while reducing the temporary monopoly conferred on the originators of inventions. One example is a relaxation of patent policy for universities and small firms engaged in government-sponsored programmes. Sectoral differences are important in this respect, and differences in the nature of technology should be reflected in the importance of patents and other such tools in restoring appropriability.

Policies related to the development of absorptive capacity

A capacity for the absorption of new technologies is the sine qua non condition for taking advantage of technological externalities. This is the strongest message of this chapter. While the development of such a capacity is primarily the responsibility of firms, policies can help. They have at least three goals: to develop the capacity of individual firms to absorb new technology, to build an environment conducive to the rapid adoption of new techniques and products and to encourage the absorption of specific technologies such as IT. They may differ depending on whether they relate to disembodied diffusion or to the adoption of technologically sophisticated machinery and equipment and of technologically advanced consumer products.

For the individual firm, policies to help build absorptive capacity involve "tuning of antennas", so that firms can more easily track, identify and adapt technology developed elsewhere. The transfer of know-how and of accumulated learning-by-doing skills, which is extremely firm-specific in nature, is important here, particularly for SMEs whose innovation capabilities will thus be reinforced. Institutional arrangements can cover information awareness programmes, search and assessment proposals or joint development schemes (which help overcome minimum economy of scale barriers). Such policies also involve assistance with research, with the training and skills improvement of personnel (through sharing of costs) and with managerial and organisational expenditures. Furthermore, experience has shown that government support is important for contract research organisations, even if by definition they depend on commercial contracts. It helps them to renew their stock of knowledge, stay at the centre of technological development and fully play their role in the wide diffusion of the most advanced technologies. A wide range of mechanisms are used, e.g. support proportional to the level of contracts won, or to their participation in large technological programmes.

Such policies affect the capacity of firms to absorb innovations developed elsewhere, whether disembodied or embodied. For assistance specifically focused on the introduction of new machinery into production processes, several countries have policies that attempt to ensure that the high cost of capital does not act as a disincentive for adoption of new technology. These include finance and tax incentives, such as accelerated depreciation allowances or special tax treatment of new equipment, that are mainly targeted for smaller firms.

In addition to assistance for developing absorptive capacity at the individual firm level, diffusion policy should ensure that the infrastructural environment is conducive to the rapid and efficient flow of technology. A number of countries (e.g. Germany, Switzerland and Sweden) place emphasis on strengthening institutional mechanisms for technology adoption (Ergas, 1987a). Education and training systems (in particular the vocational component), are strengthened, and local training centres and university programmes are set up for improving the infrastructural base for understanding and adapting to new technologies. The creation of formal and informal networks of co-operative research between firms is also promoted. Whether through industry-university links, or through industry-wide co-operative research laboratories, the aim of policy is to increase technology transfer and to help firms pool their perceptions of future technological threats and opportunities. Evidence suggests that the most successful applications are decentralised, with a wide range of end-users participating in setting the overall direction.

Encouragement to adopt specific technologies is the third dimension in a policy designed to increase the capacity of firms to absorb innovations. With new technologies characterised by technical interrelatedness and economies of system scale, there are significant network externalities associated with widespread adoption and use. Policies that focus on the adoption of certain key innovations such as the technological programmes of the European Community (ESPRIT, BRITE, etc.) can therefore have far-reaching economic benefits. More generally, policy should encourage the formation of an industrial structure that builds on the systemic aspects of modern manufacturing systems, rather than creating "islands" of innovation by selective support of certain high technology industries.

This "diffusion-oriented" approach[20] can also clarify issues raised by the increasing globalisation of technology. Governments question the support of research teams of multinational firms, on the grounds that the innovations developed with the help of national policies will, because of spillovers, rapidly diffuse and be used in other countries where the firm operates and indirectly benefit international competitors. However, this support has as its first effect to strengthen local research teams and thus to increase the national technological absorptive capacity and thus the quality of the system which produces local innovation. It is also important to ensure the spread of technologies which diffuse easily (primarily of the basic and generic type), through collaborative international arrangements, for example, among a number of competing nations.

The role of standard setting is also important, in particular for technologically sophisticated consumer products. Standards have a significant bearing on the diffusion of new technologies and products. In their absence, because costs of adoption are high, adoption will often be delayed until an industry standard for a particular technology emerges, either through market domination or through administrative guidelines or legal rulings. The efficient diffusion of a new technology can also be hampered if the market leads to a premature "lock-in" into a certain technological trajectory and the "wrong" standards (see David, 1987). The role of policy in this respect is to encourage "plurality in research" and discourage premature standardisation. This involves the support of nascent network technologies and of exploratory long-term research.

A more general policy-related point concerns the tendency of new products to be more complicated than existing ones. They therefore demand a considerable learning effort by users, who may instead prefer to follow more familiar paths. Training and informing both the public at large and specialised users in the characteristics of new technologies will lead to more informed decision-making and better participation in the elaboration of technological choices. The development of a technological information structure is essential to making educated choices, and users and consumers therefore need a large and comprehensive body of information that follows technological developments but does not serve particular interests. Hence, the necessity to make decision-making mechanisms transparent, from those relating to the choices in large technological infrastructures to those that relate to problems of health, education and the environment.

Experience shows that such efforts in "technology assessment" lead to a better understanding of the characteristics of new technologies as well as a more realistic evaluation of possible risks (OECD, 1979a). In addition, the technology assessment processes and procedures can have highly beneficial effects on diffusion: they stimulate public discussion of technological options and their implications and thus may influence users and consumers to consider the longer-term implications of their choices.

The policies discussed here are all micro-economic. They operate on incentives to share information, develop absorptive capacity and increase the rates of adoption of new technologies. They operate directly through subsidies, financial schemes, etc., or indirectly by altering the institutional and regulatory environment. In addition to these policies, however, the pace and efficiency of technology diffusion is critically influenced by macro-economic conditions. Where macro-economic policy conditions discourage (traditional and intangible) investment, they potentially slow diffusion. Thus, while by themselves insufficient for overcoming the various obstacles to diffusion, policies to improve the general economic climate will necessarily accompany more specifically micro-economic policies.

NOTES

1. The methodological approach followed in this chapter builds on the converging elements from two different traditions of theoretical and empirical work in technological change. The first tradition, which could be labelled evolutionary or neo-Schumpeterian, has always stressed the dynamic and systemic nature of the innovation and diffusion processes, as well as the close links and feedbacks between the two. It has sought to demonstrate its theoretical framework empirically with work in the history of technology and with detailed case studies. The second tradition, which can be called neo-classical, has steadily moved towards a view that stresses dynamic and systemic aspects and thus shares many common elements with the first. Its rigorous theoretical and empirical framework is helping to clarify and quantify many of the as yet unanswered questions in this area. This chapter proposes an orientation based on the reconciliation of two, until recently opposing, methodological approaches.

2. This chapter does not explicitly address the problem of international technology diffusion, although many of the factors underlying domestic diffusion of technology are equally applicable to international diffusion. Spillovers and the importance of building an absorptive capacity, for example, are crucial to the international diffusion of technology. When considering public policy, however, the issues are more complicated. The process of international technology diffusion demonstrates that national technology policies often cannot be contained within a country's borders; this implies that trade and technology policy are intertwined. These issues are discussed briefly in the final section of this chapter; Chapter 10 on globalisation and Chapter 11 on competitiveness treat them in more detail.

3. Levin *et al.* (1987) provide some information based on evidence from surveys on the importance of various channels of knowledge transmission.

4. This is usually visualised with the help of an S-shaped curve: the "S" is irregular and its precise shape different for each technology and each country. For a discussion of the S-shape literature see Jacobs (1990), p. 5. See also Ray (1984 and 1989) on differences between countries and technologies. Ergas (1987*b*) discusses how it is related to the product-cycle concept.

5. The "localised" nature of technological progress is a concept first introduced by Atkinson and Stiglitz (1969). It is related to the distinction made between basic knowledge, with wide applicability, and technical knowledge, which is much more specific and localised. The argument is elaborated in Stiglitz (1987).

6. While for the purposes of diffusion analysis the SPRU (Robson *et al.*, 1988) data are more significant and relia-

ble than patent data because they concern both patented and unpatented innovations whose use has been effectively ascertained as measures of technology flows, they still have limitations. They do not measure the continuous incremental improvements in products and processes that were not considered by industry experts as significant innovations. Nor do they measure disembodied technological know-how transferred in the form of licences, expert personnel, published and other forms of information. Furthermore, they measure only the declared first use of significant innovations, without providing a comprehensive picture of their complete application, when subsequent users can be different from, and possibly more important than, the first ones. The other serious caveat on the data relates to the 1945-83 period covered. These decades correspond to the previous technology regime and to the start of the transition from the late 1970s from that regime to the one which we are now entering.

7. The methodology employed in the Davis (1988) study sought to capture the importance of embodied-technology diffusion; however, the argument about the need for the development of tools that measure technology flows is equally applicable to disembodied technology diffusion. In a more general sense, it is not easy to identify separately in empirical work embodied and disembodied diffusion flows, especially since both types of diffusion can occur among industries and within industries. The key here is the particular links used in the analysis. Thus, the use of input purchase flows suggest that embodied diffusion is mainly – but not exclusively – captured. Conversely, looking at how closely related firms are in a technology space, or estimating directly the effect of another firm's R&D on the costs of a particular firm, comes closer to identifying knowledge spillovers and thus disembodied diffusion. Patent flows are more problematic. When used as carriers of R&D and in order to define "technological closeness", they come closer to identifying links due to knowledge spillovers. However, when they are used as proxies for input purchases, they also capture some of the embodied diffusion flows.

8. It is often confused in the literature and in policy discussions with the knowledge spillover discussed previously.
   The distinction between the two types of spillover is made in Griliches (1979).

9. There is also a technical problem here which relates to the actual measurement of total factor productivity. In the calculation of price indices by statistical authorities, new products tend to be "linked" in at their introductory price with the index unchanged. Quality adjustments in

the price index therefore reflect only the original private returns of the inventor and the consumer surplus arising from the erosion of the original monopolistic market position, rather than the full social return associated with the invention. This mismeasurement is not confined to inputs. The output of many industries in the service sector, such as health care, defence or education, is not measured directly and is instead evaluated on the basis of the cost of producing the services. Since these sectors rely heavily on research and development results obtained elsewhere, productivity growth is not measured correctly despite substantial welfare gains from indirect R&D. These types of measurement problems were discussed extensively at the June 1989 TEP Conference in the context of explanations for the "productivity paradox".

10. In modern economic theory it dates back to Schumpeter (1947), Chapter 8.

11. The weakening technological progressiveness of firms has often been attributed to conservative management approaches developed within protected oligopolistic supply structures. The case is made for US firms in Klein (1979).

12. The impact of strategic behaviour and timing on diffusion (as well as on innovation) patterns from the perspective of game-theory models in economics is discussed by Reinganum (1984 and 1989).

13. For an elaboration of this argument see Lundvall (1991).

14. This argument has often be used in order to justify unprofitable investments in areas such as semiconductors where there are repeated rounds of competition, e.g. 16K and 256K memory chips.

15. Examination of the link between technological innovation and productivity has often provided the impetus for attempts to create a measure of equipment-embodied technology. Recognising the importance for productivity of the technology embodied in purchases of intermediate inputs and in investment goods, many authors have used embodied R&D in order to assess the importance of technology for long-term movements in productivity. More particularly, this has involved examining the possibility that a slowdown of the generation or of the diffusion of new technology has contributed to the slowdown in the growth of total factor productivity (TFP). Other studies have focused on disentangling the effect of direct R&D embodied in products on the productivity of an industry or on obtaining measures of the marginal productivity of R&D expenditures or of the rate of return to R&D investments. The methodology followed in such studies has generally been to introduce (in a production function framework) a measure of embodied technology as an input into the production process, alongside a firm's (or industry's) accumulated stock of technology or expenditures on innovative activities. This measure of embodied technology accounts for the outside or "borrowed" stock of knowledge that firms take advantage of through their purchases of technologically intensive machinery and equipment or through their imitative behaviour. This allows, for example, the estimation of the contribution of the R&D of supplying industries (as embodied in intermediate and investment goods) on the productivity growth of a user industry. Both technology flow matrices based on input-output relationships and patent concordance matrices have been used in order to map embodied R&D flows and construct measures of embodied technology.

16. Hanel *et al.* (1986) evaluate the rate of return of borrowed R&D for 12 Canadian manufacturing sectors over the period 1971-82 as twice the rate of own R&D. Griliches and Lichtenberg (1984) have analysed a sample of 193 US manufacturing industries and found that the R&D embodied in purchased inputs has more of an impact on total factor productivity growth than a firm's own (process and product) R&D.

17. The methodology is based on the work of Jaffe (1986) following an idea originally suggested by Griliches. Jaffe uses patent data and develops the concept of *technological distance*. He assumes that firms which patent in the same patent class are technologically similar and that a measure of their similarity is given by the correlation of the position vectors of industries in a "technology space", where each element in the technological position vector of each sector is the fraction of the sector's R&D expenditures in a particular technological area.

18. It should be emphasised that it is not always easy to identify separately embodied R&D spillovers and knowledge spillovers in empirical work. Since both types of spillovers can occur both between industries and within industries [in the Standard Industrial Classification (SIC) sense of the word "industry"], a distinction between intra-industry and inter-industry spillovers does not help. The key lies in the particular links adopted for the analysis. Thus, using intermediate input transactions as weights for other industries' R&D when investigating inter-industry spillovers or flows of innovations suggests that one is looking at equipment-embodied diffusion; conversely, the position in a technology space or the direct (econometric) estimation of how one firm's or industry's costs are affected by the R&D done in other industries come closer to identifying knowledge spillovers. Patent flows are more problematic. When used as carriers of R&D in order to define "technological closeness", they come closer to identifying links due to knowledge spillovers. When they are used as proxies for input purchases, however, they also capture some of the spillover due to market structure and incorrect price measurement.

19. This empirical literature tests a number of hypotheses developed in stylised models of innovation in environments characterised by spillovers. Of these, Spence (1984) demonstrates analytically how spillovers can act as a disincentive to innovation. A framework that demonstrates that spillovers can spur or discourage innovation depending on the degree of complementarity between the research of rivals and on their strategic behaviour is developed in Papaconstantinou (1991).

20. The term was coined by Ergas (1987a). He distinguishes between "mission-oriented" (United States, France) and "diffusion-oriented" (Germany, Sweden) countries according to the focus that they place on diffusion policies in their overall technology policy.

Chapter 3

INNOVATION-RELATED NETWORKS AND TECHNOLOGY POLICYMAKING

ABSTRACT

Formal and informal co-operation relating to the production and the diffusion of scientific and technological knowledge and the creation of technology has become a widespread phenomenon within OECD economies. The need for co-operation stems from particular features of technical knowledge (as distinct from information) in particular its "tacit" and specific content. The marked increase in co-operation stems from the rising cost of R&D as well as the ongoing trend towards the fusion of disciplines in previously separate fields.

Informal networks, between individual researchers and between laboratories situated in different institutional settings (universities, government laboratories, industrial associative laboratories and firms) and/or in different countries are as old as organised science and technology. During the last ten to fifteen years, university-industry linkages have increased and become more formalised. They permit a division of work among scientists in very different institutional contexts. Networks among engineers in industrial technology are more selective and closed than those established in basic science, because of the tacit nature of much of the knowledge exchanged. All networks, however, require reciprocity and trust.

The growth of international and domestic inter-firm agreements, which generally build on these older networks, represents a significant and novel development of the 1980s. Firms pursue co-operative agreements in order to gain rapid access to new technologies or new markets, to benefit from economies of scale in joint R&D and production, to tap into external resources of know-how and to share risks. The goal of agreements is to organise collaboration without destroying the individuality of co-operating partners. It is assumed that parties are mutually dependent upon resources controlled by one another and that both can reap gains by pooling resources. User-producer co-operation represents a growing form of inter-firm co-operation.

Economic analysis generally considers that these agreements represent a type of organisation "lying between" market transactions and "hierarchies". This chapter argues that they represent a type of arrangement with specific features, which can be analysed in terms of transaction cost theory but must be treated as a distinctive way of organising certain types of economic relationships, notably those bearing on technology.

Networks are an important component of national systems of innovation. An important function of science and technology policy is to strengthen existing innovation-related networks and to help build networks in areas where they are lacking. While, many networks do not necessarily coincide with the boundaries of the national economies, the use of innovation-related linkages as a policy instrument can only occur on the basis of a proper identification of such linkages and their categorisation. Consequently, a possible approach to the morphology of networks is sketched out in this chapter.

INTRODUCTION

This chapter takes the analysis developed in Chapter 2 a step further. It moves from the analytical field of unforeseen R&D spillovers and related externalities, in environments shaped by competition and the overall level of technological advance, to the area of consciously formed linkages aimed at reaping returns to R&D collectively through co-operation among firms. Once inter-firm agreements and network relationships are introduced into the analysis of innovation, the traditional economic distinction between innovation and imitation/diffusion becomes even more arbitrary.

Extensive use of the terms "co-operation" and "networking" may obviously owe something to fashion, as does "globalisation" (discussed in Chapter 10). The position taken here, however, is that common use of these terms reflects an initial and still imperfect recognition of two major developments.

The first concerns the rapid and almost generalised diffusion of new information and communication technologies and the key role now played by telecommunication networks in providing essential economic inputs. The second is the fact that new technologies are less and less the result of isolated efforts by the lone inventor or the individual firm. They are increasingly created, developed, brought to market and subsequently diffused through complex mechanisms built on inter-organisational relationships and linkages. These linkages possess a number of features which make the "network mode" (Imai and Itami, 1984), a distinct form of inter-firm organisation that operates alongside and in combination with the two forms previously recognised by economic theory, namely markets and "hierarchies" (i.e. large centralised forms of corporate organisation built on extensive vertical and horizontal integration through investments and mergers)[1].

The exact content of a "network" will necessarily differ according to circumstances and its structure will be shaped by the objectives for which linkages are formed. In the case of telecommunications, for example, the need for co-operation and co-ordination stems from the particular requirements of complex sophisticated systems characterised by technical interrelatedness, systemic scale economics and the quasi-irreversibility of large investments[2]. In the case of R&D and innovation, which is the focus of this chapter, the need for co-operation stems from the particular features of scientific and technical "information" and the strong ongoing trend towards the fusion of disciplines and previously separate technical fields.

In a broad sense, actions aimed at favouring the building or consolidation of innovation-related linkages and networks have of course always been an implicit component of national science and technology policies. Today, it seems possible to approach this dimension of policy making more explicitly. As a fairly central element of government innovation policy, it could, for instance, help improve decision-making in regard to the direction and level of public financial support for science and technology. The use of innovation-related linkages as a policy instrument can only occur, however, on the basis of a systematic approach, founded on a proper identification of these linkages and a categorisation of the type discussed in Section 3.

In a highly internationalised global economy (see Chapter 10) built on open national economies and systems of innovation, a part of the "networking" arising from inter-firm co-operation will necessarily straddle oceans and national frontiers. However, it may still be important to view geographically narrower forms of innovation-related networks as one of the factors contributing to dynamic learning processes and cumulative "virtuous circles" of knowledge accumulation at local, regional and national levels. This chapter should therefore be related to Chapter 11 on technology and competitiveness; measures aimed at strengthening innovation-related networks should be viewed as one component of policies for competitiveness.

1. SECTORAL EVIDENCE OF INNOVATION-RELATED NETWORK LINKAGES

The surge of inter-firm agreements and other wider forms of innovation-related networks must be set in the context of contemporary trends in science and technology. The overall picture is one where synergies and cross-fertilisation, both between scientific disciplines and between scientific and technological advances, have played since the late 1970s an increasingly important role, notably through continual advances in computing technologies. The massive entry of computing into instrumentation has further strengthened the role of the latter. The extension of the systemic features of technologies to many areas is a necessary and inevitable outcome of these developments. Alongside these processes and as an indication of their pressure on firms, there has been a general tendency (noted in Chapter 1) towards a significant increase in R&D costs and outlays.

In combination, these developments in science and technology have created a "capability squeeze" on firms, marked by "the increase in the number of technical fields relevant to corporate growth" and "totally new requirements for significant technical advances" (Fusfeld, 1986). These pressures can be partly met by an increase of R&D within corporate structures, both nationally and internationally, or by the establishment of joint-venture corporations solely dedicated to R&D in co-operation with other firms. Often, they will require the external acquisition of knowledge, know-how and skills located in other organisations, such as universities (when the knowledge is still close to basic research) or other firms.

However, the acquisition of externally developed knowledge is not a simple process. This is due in particular to the role of tacit and specific knowledge (see Box 16).

Dosi (1988, p. 1130) has suggested that "the distinction between *technology* and *information* – with the latter being only a subset of the former – entails important analytical consequences for the theory of production". This is the analytical context, as Foray and

---

Box 16. The transmission of codified and uncodified knowledge

Recognition of the importance of linkages and relationships in everyday innovation processes has been coupled with the development by scholars and experts of a vital analytical distinction between *knowledge* and *information*. Recent understanding of the nature of the knowledge associated with engineering and production processes is at the heart of this conceptual advance (see Nelson and Rosenberg, 1990): technology invariably combines codified information drawn from previous experience and formal scientific activity with uncodified knowledge, which is industry-specific, or even firm-specific, and possesses some degree of *tacitness*.

In science-based industries and, to an even higher degree, in all other industrial sectors, codified scientific and technical knowledge is complementary to more tacit forms of knowledge generated within firms. Part of this knowledge will in time become codified within engineering disciplines. Other parts will always remain firm-specific. In each technology there are elements of *tacit* and *specific* knowledge that are not and cannot be put in "blueprint" form, and cannot, therefore, be entirely diffused either in the form of public or proprietary information. On the basis of earlier insights by Polanyi (1967), Dosi (1988, p. 1126) has suggested that "*tacitness* refers to those elements of knowledge, insight and so on, that individuals have which are ill-defined, uncodified and unpublished, which they themselves cannot fully express and which differ from person to person, but which may to some significant degree *be shared by collaborators and colleagues* who have a common experience" (emphasis added).

This kind of knowledge must be carefully distinguished from "information" in the usual sense. A recent Japanese study on systemic innovation and cross-border networks in VCRs by Imai and Baba (1991, p. 3) considers that a clear distinction should be made between the "information on the basis of which decisions are taken by firms (e.g. demand, markets, exchange rates, interest rates, etc.) and the specific 'knowledge' and 'capabilities' which allow the solution of technological and production related problems". "Information" can often fairly easily take the form of algorithms. "Knowledge" will often require more complex mechanisms of communication and transfer. It can more easily be "appropriated" privately. It requires special learning processes and can only be shared, communicated or transferred through network types of relationships.

Mowery (1990) also argue, within which inter-firm agreements and co-operation must be set.

The role played by relational and network-like forms of organisational practices, in particular with respect to the production and exchange of technology, must not, however, be restricted to studies of corporations and industry (Powell, 1990). Because contemporary networking by firms generally builds on older types of network relations among scientists and engineers, analysis must take into account earlier findings on the formal and informal patterns of communication and exchange of knowledge in the production of scientific and technical knowledge. It must also take into account research of the 1980s on university-industry relations.

Networking among scientists and engineers

Informal networks, between individual researchers and between laboratories situated in different institutional settings (universities, government laboratories, industrial associative laboratories and firms) and/or in different countries, are as old as organised science and technology. They are inherent to the existence of "communities" of scientists and of engineers belonging to the same discipline or industry and working in the same or related fields. Over the past years, there has been a significant increase in the explicit recognition of their importance for technology creation.

In a *locus classicus* on the structure of networking within scientific communities, Crane (1972) established, on the basis of empirical research, that "science grows as a result of the diffusion of ideas that are transmitted in part by means of personal influence"[3]. Scientists themselves readily acknowledged the importance of personal links for their work. The study found that while there were various sorts of interactions (informal discussion of research, published collaborations, relationships with teachers, influence of colleagues on problem choice), investigation revealed interesting patterns of strengths and directions of structural links. On the basis of a questionnaire in which scientists were asked to name contacts, small numbers of tightly-knit groups of scientists engaged in close interaction were identified, along with weaker links between the strong clusters. The most productive scientists in each group were typically in touch with each other.

The study brought convincing evidence of a position held implicitly by a large part of the scientific community, namely that these close-knit structures, completed by a wider network of the most active and productive laboratories, were essential for the development of a scientific field. Recent work on similar issues co-ordinated by Callon (1989) has suggested that initially, at least, a new scientific "fact" will often only be valued in terms of the "status" of the originating institution, including its reputation, its recognition by peer judgement and the strength and reputation of the domestic and international networks to which it belongs. These networks include what the study calls the "extended laboratory"[4], as well as the wider international network of each scientific or sub-discipline. This latter network intervenes in at least two ways: first, by the assistance it can offer the individual laboratory by ensuring that "spokesperson" or "intermediaries" obtain the best possible reception within the wider scientific community; second, by the assumption outsiders can legitimately make that the laboratory has used at least part of the scientific expertise available within the network to develop and verify the new fact or theory.

Networks also exist among scientists and engineers engaged in applied R&D in government laboratories and in industrial firms. Available case studies have shown that the networks built in these areas are much more selective and closed than those established in basic science. This is because much of the knowledge exchanged and transferred is know-how (the accumulated practical skill which allows one to execute a task smoothly and efficiently) or again "tacit", uncodified (and often uncodifiable) knowledge, e.g. practical craft details never formalised in scientific publications. Such knowledge is implicitly or explicitly proprietary; its possession represents a form of "competitive advantage" for firms and even for public sector laboratories which may be "competing" for government and industrial contracts. Informal trading of know-how, "trade secrets" and tacit knowledge involving network relationships occurs extensively and has received increasing attention. In a case study of the network set up for the high pressure gas laser, the author found that data was revealed to other laboratories on a highly selective basis. He noted that individuals and laboratories made conscious and careful choices as to what know-how would be revealed to what recipient and that "nearly every laboratory expressed a preference for giving information only to those who had something to return" (Collins, 1974, p. 59). Similar behaviour has subsequently been observed and studied among development and production engineers working in rival firms in metallurgy (von Hippel, 1987; and Schrader, 1991). The sharing of tacit knowledge was seen to follow fairly identifiable patterns based on the assurance that reciprocity and fair albeit "non-traded" exchange of knowledge will occur among participants. These sociological observations are important. They represent a "hidden" but often vital dimension of technology-sharing agreements between firms. Selection of potential partners in co-operative ventures is based on similar considerations.

Such aspects of reciprocity, obligation and trust founded on reputation shaped by peer judgement correspond to the wider findings of the theory of co-operation developed by the sociology of organisations over the 1970s and 1980s. Axelrod (1984) has summarised the most significant consequences of repeated interaction among individuals. When there is a high probability of future association, individuals are not only more likely to co-operate with others, they are also increasingly willing to punish "opportunism" (see note 1) or the refusal to co-operate. When repeated exchanges occur, quality becomes more important than quantity. The reputation of participants is the most visible signal of their reliability and bulks large in many network-like work settings.

From informal to formal networks: university-industry relations in the 1980s

The informal networks among scientists and/or engineers just described represent a major building-block and decisive ingredient for the establishment and successful operation of R&D related co-operation and communication between universities and industry. The 1984 OECD study on industry and university relationships observed that "collaboration and communication between industry and the academic community usually begin very informally and are characterised by person-to-person contacts. From this base, they may gradually become more and more formal, leading eventually to contracts and/or other forms of linkages." (OECD, 1984, p. 19). The study cites Switzerland as the most outstanding example among OECD countries of the strength and efficiency of informal relations within a wide community of scientists and engineers working in different environments.

In some OECD countries, informal university/industry relations have operated more or less extensively and effectively for a long time; over the last ten to fifteen years, they have been formalised and given greater institutional visibility. In others, where they were weak, policies have had to be devised by governments in order to help to build them up (see Section 4). Studies of university-industry linkages show that, independently of their origin, i) both the extent and the nature of university/industry relations depend upon sector and firm size; and ii) the objectives of such relationships – the function which academic institutions can perform (more or less effectively) for industry – similarly vary.

Large firms operating in high technologies almost invariably possess the widest and most efficiently managed network of linkages with the university systems, both at home and abroad. While firms of this kind generally have significant internal R&D, university research groups are considered valuable both as an extension of this research capacity and as a kind of antenna – commonly referred to by industrialists as a "window on research". When a common research interest exists, it becomes a central cog in an informal network for sharing scientific information. Firms recognise that because senior university researchers belong to well-developed scientific networks, universities have world-wide contacts and may have easier access to emerging ideas, especially from abroad.

Firms enjoying good links with academic institutions perceive that they must also keep their partners in touch with their own research activities and try to develop a true complementarity of work. Successful industrial links with universities are based upon a sense of "give and take among equals", a relationship founded on person-to-person familiarity and confidence. Consequently, research scientists employed by large firms are encouraged to play an active role in the scientific community, international as well as national, through presenting papers, publication and other contacts. Firms recognise that although in some fields they may be more advanced than university laboratories, university scientists may have the opportunity of investigating questions of interest in much greater depth than is possible in their own laboratories. A real division of labour often takes place as a result of university/industry collaboration, leading to joint publications.

During the 1980s, the creation of more effective relationships between universities and small and medium-sized firms (SMEs) became a central focus of policy in many countries. Interest has been attached in some countries to the fostering of small high technology firms in advanced sectors as "spin-offs" from university laboratories, following the highly-publicised United States' West Coast examples of Cites and Genentech (Peters, 1989). Others have not considered this approach advisable. In all countries, however, hopes have been pinned on establishing industry-university linkages for smaller firms, as a means to produce high value-added goods (and services) in new research-intensive sectors. In almost every OECD country – including the United States, where the National Science Foundation has actively promoted university-industry programmes – a range of institutions has been set up to promote university-enterprise relations, generally in close conjunction with national science and technology policies. Increasing importance is also being attached to specialist institutions at the regional or local level (for example, the Centre Québecois de Valorisation de la Biomasse in Canada), to private or public foundations (such as the Research and Development Corporation in Japan and the Technology Foundation in the Netherlands), and, more directly, to institutions such as the university-industry foundations in Spain.

The topic has ranked high on the science and technology policy agenda of the OECD. Recent work shows that after an initial phase marked by some difficulties, firms are quickly learning to make effective use of the strengths of universities, especially in fundamental research and advanced training, thereby reducing the scope for possible strategy conflicts, and that universities, under growing budget constraints, have grasped the need for relations with industry and have more clearly defined their objectives in this respect. The creation of specific university-industry units in many universities as well as in a number of large firms has no doubt made a positive contribution here. Apart from difficulties over intellectual property rights on research results, tensions which may still mark university-enterprise relations today include the competition for human resources (universities and firms are competitors in the market for scientific and technical personnel, see Chapter 6) and problems of communication between university scientists and business managers. These are most marked for SMEs lacking R&D departments, those who are in fact in greatest need of technological support from universities.

Inter-firm technology-related agreements as networks

The expressions "technical networks" and "technological networking" (see Fusfeld, 1990 and 1986) have increasingly been employed in discussions focused on technology-related agreements set up between enterprises. Use of the word "network" in this context is not devoid of ambiguity. Inter-firm agreements and alliances must also be analysed as instruments of corporate competition strategies, and some tend towards collusion in the form of an implicit protection of key technology through co-operation among a few players. A number of authors argue that this can distort competition more significantly than is often recognised (e.g. Porter and Fuller, 1986; Contractor and Lorange, 1985; and Dunning, 1988b). This is particularly true

---

Box 18. Factors explaining the preference for agreements by firms

The opportunities created by agreements are different from, and in given circumstances superior to, those associated with the "internalisation" of activities through domestic vertical or horizontal integration or foreign direct investment. In the face of financial and economic uncertainty and turbulence in the world economy and of parallel rapid and radical technological change, the new forms of agreements offer firms a way of ensuring, in a wide variety of situations, a high degree of flexibility in their operations. When technology is moving rapidly, the flexible and risk sharing (or indeed risk displacing) features of inter-firm agreements offer firms in particular a wide range of possibilities for acquiring key scientific and technical assets from outside.

In other, somewhat more strictly defined circumstances, inter-firm agreements can also provide firms with a possibility of pooling limited R&D resources in the face of rising costs. Likewise, inter-firm agreements can help create the linkages from which "systems" (e.g. market and non-market relationships of some stability among firms and between these and other forms of organisations) can emerge. As suggested in particular by Japanese authors, "systems" of this kind appear to be the basis of many key contemporary technological developments (Imai and Itami, 1984). If this is the case, then the significance of inter-firm agreements in relation to contemporary developments in science and technology may well go far beyond the already important advantages of pooling sub-critical R&D resources, or combining skills and knowledge on an interdisciplinary basis (Chesnais, 1988b).

where high levels of industrial concentration have led to the emergence of oligopolistic supply structures (see Chapter 10)[5]. Here, however, inter-firm agreements are viewed exclusively in the light of their positive features. Use of the word "network" with respect to inter-firm agreements has been said to "highlight the mobility of alliances, the flexibility of arrangements, the volatility of configurations and the multiplicity of modes of co-ordination, e.g. the fact that in some parts of the network, the market is the instrument for co-ordination, in others this involves non-market organisational mechanisms, confidence and recognition" (Callon *et al.*, 1990).

Inter-firm agreements bearing to some degree or another on technology grew quickly from the early 1980s onwards (see Box 18). As shown by Table 7 and Figure 5, figures published by data banks built up by universities and consulting firms[6] suggest a strong increase in the number of recorded agreements up to 1986. A levelling of growth followed, with the notable exception of agreements in the area of information technologies.

For the reasons given in Box 18, agreements show considerable range and variety (Chesnais, 1988*b*). Co-operation and/or technology exchange between firms (or between firms and other categories of research organisations) may take place at a single point of the R&D-to-commercialisation process or may cover the entire process. They may concern either the creation of new technology or the acquisition and use of an already existing one or both. Co-operation may concern R&D exclusively, or it may also involve arrangements for market access and joint commercialisation. An attempt to categorise the types of arrangements firms may set up with a view to producing, exchanging and/or commercially exploiting technology in common is set out in Figure 6. The two left-hand columns reflect the university-industry co-operation networks discussed previously. The right-hand part of the figure presents the most classic forms of joint production and technology transfer. The new forms are concentrated in the centre. They include:

i) The new research corporations (column C). These are private sector joint ventures financed by a

Figure 5. **Growth of newly established technology co-operation agreements**
(biotechnology, information technologies and new materials)

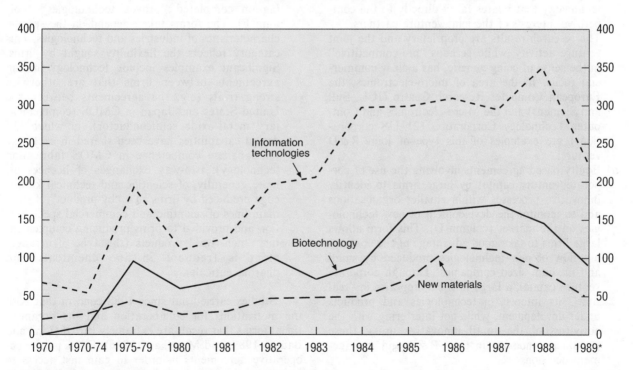

* 1989 first seven months only.

Source: MERIT-CATI data bank.

Table 7. **Trends in the growth of inter-firm agreements**

Author and type of agreement	1974	1975	1976	1977	1978	1979	1980	1981	1982	1983	1984	1985
Hladick (1985), Joint ventures by US firms in high income countries	37	14	16	15	14	27	34	40	35	–	–	–
Réseau, Milan (1985), Electronics industries alone				131					69	104	118	–
Hacklisch (1986), 41 largest world merchant semiconductor firms	–	–	–	–	2	1	4	22	19	16	42	–
LAREA-CEREM (1986), Agreements involving European firms in R&D intensive industries	–	–	–	–	–	–	15	31	58	97	131	149
Venture Economics (1986), Corporate venture capital inv. agreements	–	–	–	–	30	30	30	60	100	150	195	245
Schiller (1986), International agreements with US small biotech. firms	–	–	–	–	–	–	–	22	58	49	69	90

Source: Author's compilation of figures from studies discussed in the text or cited in the bibliography (Chesnais, 1988*b*).

number of firms on a shareholder basis. Such corporations conduct their research in separate laboratories or in other research facilities set up specifically for that purpose. They are generally staffed both by personnel assigned by the joint venture partners and by personnel hired specially for the project. Programmes generally focus on generic technology that relates fairly directly to the competitive interests of the joint venture partners.

The research results are proprietary and the joint venture activity, while perhaps "pre-competitive" in the sense of being generic, has a clear commercial focus. In the area of micro-electronics, the European Computer Research Centre (ICL, Bull and Siemens) and the Micro-electronics and Computer Technology Corporation (21 US corporations) are examples of this type of joint R&D venture;

ii) Equity based agreements involving the use of corporate venture capital by large firms to identify innovative processes within smaller organisations and to monitor the development of new technologies on the market (column D). This form allows large firms to take good advantage of the continuous flow of new technologies produced by small and medium sized companies. Through corporate venture capital, a large firm can appraise in "real time" its interest in technologies and products under development, while not interfering with the activities of the small innovative units; these remain autonomous in their R&D and management decisions.

iii) Non-equity co-operative research agreements to deal with specific research problems (column E). These are highly flexible forms of co-operation without shareholder participation, generally involving a very small number of firms possessing complementary basic in-house R&D. Agreements are of limited duration and are aimed at attaining strictly defined results. Philips and Siemens are known to have often had agreements of this type.

iv) A wide variety of technological agreements bearing on completed "proven technologies" (column F). The forms taken depend on the specific characteristics of industries and technologies. This category reflects the flexibility sought by firms. Significant examples include technology-sharing agreements between firms that are otherwise strong rivals (e.g. the agreements between the United States and Japan in CMOS (complementary metal-oxide semiconductor), in which US design capabilities have been shared in exchange for Japanese competence in CMOS fabrication technology); two-way exchanges of licences or, more generally, of scientific and technical knowledge produced by firms as a "by-product" of their main lines of scientific and commercial specialisation and provided to firms producing complementary "non-rival" products (this type of arrangement is frequent in fine chemicals and pharmaceuticals).

Studies carried out since 1988 confirm that both the motivations for collaboration and the organisational forms that result are extremely varied (Link and Bauer, 1989; and Mytelka, 1991*b*). Firms pursue co-operative agreements in order to gain fast access to new technologies or new markets, to benefit from economies of scale in joint research and/or production, to tap into sources of know-how located outside the

Figure 6. **Inter-firm research, technology and manufacturing co-operation agreements and the R&D, production and marketing spectrum**

PRE-COMPETITIVE STAGE				COMPETITIVE STAGE				
Research and development co-operation			Technological co-operation			Manufacturing and/or marketing co-operation		
A	B	C	D	E	F	G	H	I
University based co-operation research financed by associated firms (with or without public support)	Government-industry co-operative R&D projects with universities and public research institute involvement	Research and development corporations on a private joint-venture basis	Corporate venture capital in small high-tech. firms (by one or by serveral firms otherwise competitors)	Non-equity co-operative research and development agreements between two firms in selected areas	Technical agreements between firms concerning completed technology, *inter alia:* technology-sharing agreements; second-sourcing agreements; complex two-way licensing; cross-licensing in separate product markets	Industrial joint venture firms and comprehensive R&D, manufacturing and marketing consortia	Customer-supplier agreements, notably partnerships	One-way licensing and/or marketing agreements (including OEM sales agreements)
Many partners		Several partners	Few or very few partners			Few or very few partners	Few or very few partners	

Source: Chesnais (1988*b*).

Table 8. **Patterns of inter-firm agreements: general objectives**

	Number	Percentage
Simple agreements	1 500	79.6
1. Technology transfer	251	13.3
2. R&D integration	363	19.3
3. Production integration	370	19.7
4. Supply agreements	121	6.4
5. Marketing	311	16.5
6. Others	84	4.4
Complex agreements	383	20.4
of which:		
2+3	100	5.4
2+5	59	3.1
3+5	72	3.8
2+3+5	38	2.0
1+3	22	1.2
Other combinations	92	4.9
Total	1 883	100

Source: Ricotta (see Chesnais, 1988*b*).

boundaries of the firm, and to share the risks for activities that are beyond the scope or capability of a single organisation. In a summing up of findings from US literature, Powell has suggested that the exact form of the agreement that firms will choose "appears to be individually tailored to the needs of the respective parties, and to tax and regulatory considerations. The basic thrust, however, is quite obvious: to pursue new strategies of innovation through collaboration *without abrogating the separate identity and personality of the co-operating partners*" (1990, p. 315).

Access to tacit and/or proprietary technological knowledge and the preservation of the separate identity of firms explain why statistics from one of the best data banks [Futuro-Organizzazione-Ricorse (FOR), Rome] highlight the fact that:

i) R&D co-operation either in isolation or in combination with other objectives (item 2 in Table 8) represents the single most important objective of inter-firm agreements; and

ii) firms show a clear preference for non-equity agreements, since these effectively represent the least formalised, most flexible form and the one which best reflects the separate identity of participants (at least over a transitory phase).

User-producer relationships between firms

One form of inter-firm agreement on which research has yielded significant results since the termination of the OECD study on inter-firm co-operation are the customer-supplier or user-producer relationships included in Figure 6, column H.

In-depth Swedish research on inter-firm relationships connected with technical innovation has illustrated the relevance of user-producer interaction in innovation. A detailed study, carried out by Hakansson at Uppsala University, focuses on a representative sample of over 100 Swedish small and medium-sized enterprises. It is particularly useful because it sets user-producer interaction within the context of a wider variety of network relationships. The findings confirm that a significant fraction of the resources allocated to technical development involves co-operation with at least some external party and that a large part of the co-operation is largely informal and not specified in any written contract. This was the case for close to 80 per cent of all co-operation activities and suggests a strong degree of trust and reciprocity (Hakansson, 1989, pp. 112-113).

The most important finding is that in this universe of Swedish SMEs, user-producer agreements represent the most frequent form of inter-firm co-operation, with activities with customers and suppliers constituting 75 per cent of all such activities. Others are agreements with independent institutions producing new technology and science, which may in fact be regarded as parties in user-producer interaction (Hakansson, 1989, p. 98). These user-producer relationships are durable: the average age of such relationships involving co-operation in development projects was close to 10 years.

Parallel analytical work, carried out at Aalborg University in Denmark, has pointed to the factors underlying the density of user-producer interactions in the innovation process[7]. Producer firms have strong incentives to establish close relationships with user firms and even monitor some aspects of their activity. Some of these incentives, such as the attempt to ensure that innovations are not stolen by competitors, were studied in earlier technology transfer literature. Others are newer. In the case of equipment and industrial processes, knowledge produced as a result of learning-by-using can only be transformed into new products if the producers have direct contact with users. Bottlenecks and technological interdependencies, observed within user units, represent potential markets for the innovating producer. Producer firms may be interested in monitoring the competence and learning potential of users in order to understand the capability of firms to adopt new products. In turn, user firms will generally need "information" about new products. This not only means awareness, but also quite specific inside information about how new, user-value characteristics relate to their specific needs. When new user needs develop – for example, when bottleneck problems occur – user firms may be compelled to seek help from a producer in analysing and solving the problem. This requires detailed knowledge about the competence and reliability of different producers.

In situations where complex and specialised equipment and processes are developed and sold to users, direct co-operation is required throughout the process of innovation. In earlier stages, the user may present the producer with specific needs to be fulfilled. Subsequently, the producer may need to install a product and start it up in co-operation with the user. This is the phase in which the producer will often offer specific training to the user. After the equipment has been brought on-line, the producer will normally recognise obligations to the user firm regarding the de-bugging of technical hitches; the producer also has considerable interest in learning through the user's own learning process.

Lundvall (1991) argues that the pattern of user-producer relationships and the precise substance of inter-firm relationships, in terms of reciprocity and response to the needs of the other partner, could have *productivity-enhancing or inhibiting effects*. He proposes a typology of relationships that could form the basis for future analysis and case study research. The categories include: producer-dominated relationships and producer-organised markets in consumer goods; producers dominating professional users; big professional users dominating producers; stubborn user-producer relationships; and situations characterised by lack of user-producer relationships.

The overall picture: a maze of overlapping networks

The picture which emerges from the evidence discussed above is that of a maze of overlapping networks. They range from tightly knit but highly specialised and rather narrow networks among scientists in given sub-disciplines of the natural sciences or among production engineers within a narrow group of firms (in the same industry, country or industrial district) to looser and much wider networks involving firms as corporate entities.

Authors tend to agree that for firms, a useful distinction may be made between the individual network which each large corporation builds and for which "it serves as a nucleus" and the "wider set of technical alliances and consortia which provide the ties that bring into contact these individual networks". Networks of a given corporation will consist mainly of "relations among its own technical groups and laboratories; linkages between its technical groups and universities; working relationships between its technical groups and government technical organisations; trade associations; R&D consortia' (Fusfeld, 1990, p. 20). To these must be added a number of the user-supplier linkages just discussed. The wider networks include the global, transnational agreements which have recently been studied at Maastricht University (they are discussed in Chapter 10).

2. NETWORKS: SOME THEORETICAL AND PRACTICAL CONSIDERATIONS

It is now necessary to formulate some generalisations concerning the "specifiable circumstances" under which "relational", "linkage" or "network" forms of inter-firm relationships represent a "clearly identifiable form of economic exchange" in matters pertaining to the production and use of new technology (Kaneko and Imai, 1987). Although the time is ripe for the development of a full-fledged "economic theory of networks" (Australia, Bureau of Industry Economics, 1991), most studies are still couched in the framework of the Coase-Williamson theory of markets and hierarchies which serves as the starting point for the following analysis.

Network transactions in relation to markets and hierarchies

Despite early arguments by Richardson (1972) to the contrary, authors have often taken the position that agreements lie somewhere "in between" markets and hierarchies[8]. Arguing that economic exchanges can be arrayed in a continuum-like fashion, with discrete market transactions located at one end and the highly centralised firm at the other, they suggest that the intermediate or "hybrid" forms of organisation represented by agreements lie between the two traditional poles of organisation theory. Williamson, who earlier neglected linkages, is "now persuaded that transactions in the middle range are much more common" than he had previously recognised[9]. He now suggests that markets, trilateral governance (involving action by mediating third parties), bilateral governance (involving "relational contracting" between firms) and hierarchies represent four alternative forms of industrial organisation or governance. The frequency of transactions and the importance of asset specificity are now emphasised as variables affecting the predominating mode of governance (together with small numbers, uncertainty and opportunism). In a continuum perspective, one would move from the market pole, where prices capture all the relevant information necessary for exchange, towards putting-out systems, various kinds of repeated trading, quasi-firms, and sub-contracting arrangements. Contracted arrangements in the form of franchising, joint ventures and decentralised profit centres inside network corporations would be located close to the hierarchy pole.

However, the notion of the continuum fails to capture the complex realities of know-how trading and knowledge exchange in innovation. The pervasive role played by reciprocity and collaboration, as a distinctive form of co-ordination in an activity with strong economic impact, requires that networks be envisaged as a separate form. "Relational contracting" may be the right way to characterise the new forms of sub-contracting which have emerged in connection with "just-in-time" production management (see Chapter 4), but it is not appropriate for conceptualising many of the arrangements discussed above. A "markets" and "hierarchies" framework may hinder economists and technology policy makers from properly identifying the theoretical implications of the present diversity of organisational designs. Table 9 attempts to summarise some differences among markets, hierarchies and networks. Networks can no longer be considered as a hybrid and perhaps transitory form. They represent a type of arrangement with its own specific distinctive features which must henceforth be considered *in its own right*. The assertion that networks represent a distinctive form of economic organisation now extensively used for *exchanging production-related and value-creating assets possessing given characteristics* implies that this form must also be defined.

In terms of transaction cost theory, the network mode offers firms a new way of handling market imperfections, in particular those related to innovation. In certain circumstances, this mode is superior to both markets and hierarchies, but it is so because it is *different* from both and not somewhere "in between". As argued by Chesnais at the close of the OECD study on inter-firm agreements, "the complexity of scientific and technological inputs, the uncertainty of economic conditions and the risks associated with uncertain technological trajectories have reduced the advantages of vertical and horizontal integration and made 'hierarchies' a less efficient way of responding to market imperfections. But the need to respond to and exploit market imperfections in technology has also increased, and has thus pushed inter-firm agreements to the forefront of corporate strategy. In terms of transaction cost theory, earlier arguments about market failure in technology markets hold as strongly as ever, but inter-firm agreements may often be the most advantageous way of handling them" (1988*b*, p. 84).

This approach has subsequently received substantial support. Dunning argues that "to the extent that, like administrative fiat within a single firm, inter-firm co-operation is a market-replacing activity, at least some of the reasons for it should be found in the internalisation literature. But to the extent that it involves more than one firm and, abstracting from distributional questions, there is a broad coincidence of objectives among the participants, the form of governance is group oriented. To distinguish between the two forms of governance we shall refer to inter-company co-operation as *group internalisation;* such internalisation might take the form of an equity joint venture or a non-equity agreement" (1988*b*, p. 343).

Some conditions for exercising "network preferences"

On the basis of the available evidence, one can go a step further and give the many forms of exchange bearing on knowledge a more clearly defined conceptual status, along the lines suggested by the last column on the right of Table 9. Summarising a wide

Table 9. **Stylised comparison of forms of economic organisation**

Parameters	Forms		
	Markets	Hierarchies	Networks
Normative basis	Contract Property rights	Employment relationship	Complementary strengths
Means of communications	Prices	Routines	Relations
Methods of conflict resolution	Haggling; resort to courts for enforcement	Administrative fiat Supervision	Norm of reciprocity Reputational concerns
Degree of flexibility	High	Low	Medium to high
Amount of commitment among the parties	Low	Medium to high	Medium to high
Tone of climate	Precison and/or suspicion	Formal, bureaucratic	Open-end Mutual benefits
Relations between economic agents	Independence	Hierarchical	Interdependence

Source: Adapted with minor changes from Powell (1990). The initial conception of this table should, however, probably be attributed to Kaneko and Imai (1987).

range of case study material, including technological agreements between firms, produced both by social scientists and students of industrial and corporate organisation, Powell (1990, p. 13) offers a fairly comprehensive definition. He suggests that "in network modes of resource allocation, transactions occur neither through discrete exchanges nor by administrative fiat, but through networks of individuals or institutions engaged in reciprocal, preferential, mutually supportive actions. Networks can be complex: they involve neither the explicit criteria of the market, nor the well-organised routines of the hierarchy. A basic assumption of network relationships is that parties are mutually dependent upon resources controlled by another, and that there are gains to be had by the pooling of resources. In network forms of resource allocation, individual units exist not by themselves, but in relation to other units. These relationships take considerable effort to establish and sustain, thus they constrain both partners' ability to adapt to changing circumstances. As networks evolve, it may become more economically sensible to exercise voice rather than exit. Benefits and burdens come to be shared. Expectations are not frozen, but change as circumstances dictate (...). Complementarity and accommodation are the corner-stones of successful production networks."

Networks are *not* a panacea. In the case of inter-firm alliances, studies by business management specialists have shown that inter-firm agreements have *high costs* of their own, which include high bureaucratic costs related to co-ordination and numerous risks related to opportunism, non-reciprocity, defection and "Trojan Horse" effects[10]. This is why networks are for firms a response to quite specific circumstances (see Box 19).

Where complementarity is a requisite for successful innovation, agreements are formed in response to the "appropriability" of key tacit knowledge. The rather particular nature of the "information" which forms the substance of innovation-related network transactions, e.g. its more or less marked *tacit* and/or proprietary character, emerges clearly from the data presented above. As argued by Ciborra (1991), because of this exchange of tacit knowledge, technological alliances can be seen as "learning experiments".

Many authors stress *speed* among the advantages that agreements, linkages and inter-firm alliances have over internal development or acquisition through arm's length relationships. This is the factor given most emphasis by Porter and Fuller: "the time required to build expertise or gain market share internally is likely to exceed the time required with a coalition. As product life cycles have shortened and competitive rivalry has intensified, the timing advantages of coalitions are increasingly important" (1986, p. 328).

Flexibility and reversibility represent a related, but nonetheless distinct reason for preferring a network

Box 19. Conditions suitable for establishing networks

Situations in which firms may choose to form networks are those where:

i) "Complementary assets" – possessing a high degree of "*tacitness-intensity*" and thus characterised by "*appropriability*" [a] in the form of firm-specific, proprietary knowledge – represent very important inputs to the production of new knowledge and/or technology or to production processes *per se;*

ii) The exchange of such assets and the associated *learning* processes can only take place through fairly close contacts and personalised relationships;

iii) Economic instability, technological uncertainty and rapid demand change place a premium on *speed;*

iv) High R&D costs force management to seek ways of pooling resources with other firms, in some cases even with rivals;

v) *Flexibility* and the possibility of preserving the *reversibility* of decisions are of importance; and

vi) There exist at least minimum expectations on the part of would-be partners that *reciprocity* will prevail, implying the existence both of some basis for trust and some guarantee that *opportunism* will receive appropriate *punishment.*

a) Regarding "appropriability", see Dosi (1988, p. 1139ff.). Teece (1982 and 1986, p. 6) provides a recognised definition of "complementary assets".

mode. In areas such as biotechnology or new materials, they carry high premiums. Inter-firm agreements are easier to dissolve than internal developments or mergers. Their sunk costs are smaller, commitments are less definitive and inertia lower. As a result, much higher degrees of flexibility can be ensured.

The condition of reciprocity and trust raises the most serious problems in all areas and is the hardest to guarantee. It is difficult to ensure that reciprocity (or near reciprocity) will really exist or that power relationships – whether economic or institutional – will not dictate the real content of transactions bearing on tacit knowledge (e.g. between large and small firms; between full professors or heads of laboratories and associate professors or lecturers).

In the case of well-organised scientific networks, in particular in natural sciences, the profession's code of professional ethics will generally suffice to ensure reciprocity and trust, although recent cases in biology and medicine (e.g. the scientific network working on AIDS) have shown that the code may break down when financial and political pressures are strong. With respect to firms, Japanese and non-Japanese studies of research partnerships in Japan all strongly emphasise the role of trust in sustaining collaboration among competitors. While the competitive drive of each individual company is central to the success of the Japanese electronics industry, associations create a network of relationships among firms for the common good. Competitors enter into joint research without the protection of clear patent agreements, "yet secure in the knowledge that no party will cheat the others, because cheating will surely be detected and punished in the future"[11].

Further tensions are generated in this area due to the fact that perceptions related to trust and reciprocity, and more generally views about the ranking and weighting of the factors moulding the "preference for network relationships" and the best way to achieve co-operation will often differ considerably within corporations. A survey of firms recently carried out at Reading University found that inter-firm collaboration, particularly the ideal of strategic alliances with technologically sophisticated partners, is more likely to be promoted at the board level than by laboratory managers. The latter have their own informal networks for evaluating prospective partners based on peer group status. At this level, a good deal of informal collaboration, in terms of collective lobbying and avoidance of duplicated effort, appears to go entirely unrecorded, but may potentially be much more effective than formal agreements (Casson, 1991, p. 278). This corroborates earlier work by Hamel *et al.* who point out that while management may set the legal parameters for exchange what actually gets traded is determined by day-to-day interactions of engineers, marketers and product developers (1989).

Strategic alliances will be shaped by the complex game of co-operation and competition between rivals and by market access objectives as well as by considerations relating to speed, costs, and flexibility (see Chapter 1). Laboratory managers and engineers will continue to be guided by a set of considerations largely based on scientific networks. There will thus be two sets of partly overlapping network relationships, at times in harmony, at times partially in conflict.

3. THE MORPHOLOGY OF NETWORKS AND NATIONAL SYSTEMS OF INNOVATION

Actions aimed at favouring the building or consolidation of innovation-related linkages and networks have of course always implicitly been a fairly important component of national science and technology policies. Among other things, these policies foster the development of basic information, strengthen existing networks by remedying their weak points, help build networks in areas where they are lacking and provide the externalities that will encourage their development. A more explicit set of conceptual tools and indicators is now being explored in work supported by the OECD and can help meet these objectives more effectively.

The need to identify national systems of innovation (see Box 20) and to understand the way they work, has strengthened this requirement[12]. Even the narrower R&D components of national systems, present in different combinations in all OECD countries (see Box 21), can be seen to possess systemic features, to a greater or lesser extent, on account of two main factors: first, the financial circuits originating in corporate investment and public funding mechanisms, which ensure the "system's" stability and capacity to grow; second, the linkages and inter-institutional relationships which bind the components together. An assessment of the qualitative features of these relationships and their effectiveness calls for a set of concepts and

---

Box 20.   A definition of the national system of innovation

"The network of institutions in the public and private sectors whose activities and interactions initiate, import, modify and diffuse new technologies may be described as 'the national system of innovation'" (Freeman, 1987a, p. 1).

"The rate of technical change in any country and the effectiveness of companies in world competition in international trade in goods and services, does not depend simply on the scale of their Research and Development and other technical activities. It depends upon the way in which the available resources are managed and organised, both at the enterprise and at the national level. The national system of innovation may enable a country with rather limited resources, nevertheless, to make very rapid progress through appropriate combinations of imported technology and local adaptation and development. On the other hand, weaknesses in the national system of innovation may lead to more abundant resources being squandered by the pursuit of inappropriate objectives or the use of ineffective methods" (p. 3).

indicators drawing on the sociology of institutions as much as on economic theory.

Concepts and indicators for the analysis of innovation-related networks

The configuration of innovation-related networks obviously varies considerably. A more systematic analysis of their morphological features may help identify the poles around which they are organised and the nature and intensity of the links that connect the components. It may also allow identification of their strengths and weaknesses. One aim of such analysis is to answer questions which may be of considerable interest to policy makers. Can links be substituted or are they complementary? Are they co-ordinated? If so, are they formalised?

Component elements of networks

The links between actors and economic agents in innovation-related networks may be arranged in a wide variety of configurations. Networks for given technologies or industrial sectors encompassing the whole innovation process (see Figure 1 in Chapter 1) will involve the existence of three main "poles":

– A scientific pole which generates knowledge and produces documents in the form of scientific papers and trains personnel. This pole is thus divided into a research component and a training component whose proximity will vary. This pole will include external (public and private) research centres, universities, and some firms' laboratories (depending on

the degree to which their activities resemble those of university or public research centres).

– A technical or techno-industrial pole which generates artefacts, pilot projects, prototypes and test stations, and which produces drawings, patents, standards, rules of the art, etc. It includes pilot plants built by firms, vocational technology centres, etc.

– A market pole, which corresponds to the universe of users and the professionals' or practitioners' market. The structure of this pole, and therefore its relationship with other poles, may vary. It is clear, for example, that the relationship between a major industrial group and the final consumer will differ according to whether the consumer buys the product from a network of distributors working in collaboration with the producer or whether the producer is forced to retail his goods through a concentrated distribution network (hypermarkets and their procurement agencies) that may impose its own standards on the producer (see also the discussion of user-supplier linkages in Section 1).

"Intermediaries" and "translation" processes

The creation of linkages and the forms of co-operation established by the various participants in the innovation process will involve a series of "intermediaries". These are not limited solely to material or capital goods. Four other categories of intermediaries enter the linking process: disembodied information or documents of any kind (books, articles, patents, notes, diskettes, letters, etc.); technical artefacts (scientific instruments, machines, robots, materials, etc.); trained personnel and their skills (knowledge, know-how, etc.); funds supporting innovation (subsidies, credits, orders).

The structure and efficiency of innovation-related systems may depend on the production, transformation, circulation and use of these "intermediaries". In addition, the various actors who communicate, co-operate, confront and mutually define each other through interposed "intermediaries" should be included. From a sociological viewpoint, networks possess two complementary dimensions. Signposts are provided by the flow of "intermediaries" and their paths but also by the identity, location and behaviour of the "actors" that use and transform these intermediaries.

Because of the variety and heterogeneity of the numerous types of network relationships and linkages, gaps in "compatibility" may occur that require *bridging*. In this case, a "translation" or transposition process must take place. In working towards a given strategic objective, an actor will "translate" scientific facts, instruments and intermediaries; that is, he will re-interpret them and divert them from their intended purpose in order to produce new texts, new instruments and new machines. In other words, the transposition of information or experience originating with a researcher or engineer in one field must be mirrored by a similar action by a researcher in another domain at the interface between their respective fields. The length and complexity of the operation may vary. It may involve a series of mutually interrelated intermediaries arranged in subtle feedback loops and retro-actions. This operation will introduce continuity between the various actors, thereby creating a common language. As a result, the network will become a space in which information intelligible to all its members will circulate. The network of agreements set up as part of Japan's Very Large Scale Integration (VLSI) project illustrates the results of these successive "translations" (see Box 22).

The morphology of networks

From the point of view of policy, it may be important to distinguish among the different forms taken by networks. They may be converging, dispersed or integrated; they may be long, short, or incomplete.

The characteristic feature of strongly converging networks is their connectivity, or close linkage between different actors. Co-ordination may be highly formalised and accompanied by mechanisms or rules linking actors and activities with varying degrees of constraint. In a strongly converging network, "any actor belonging to the network, whatever his position within it (researcher, engineer, sales staff, user, etc.), can at any time mobilise all the skills within the network without having to embark upon adaptations, 'translations' or costly decoding" (Callon *et al.*, 1990). In a strongly converging techno-economic network, a researcher working in the basic sciences knows that the problems on which he is working have direct implications for a network of expectations and demands ready to take up his findings as soon as they are released by the laboratory.

Dispersed networks lie at the other end of the spectrum. Here, relations between actors exist (between poles or within the same pole), but they have low density. The "translations" that allow change-overs from one register to another, from science to technology, for example, are not yet firmly in place. In contrast with a "converging network", it is difficult for an actor in a "dispersed" network to mobilise the rest of the network.

In addition, networks may be long, short or incomplete. Long networks are relatively rare; they run uninterruptedly from basic research to users. A short network does not, but is not "incomplete". Missing links in a short network do not reflect a weakness to be corrected but are a structural characteristic. The taxonomy established by Pavitt (1984, p. 353 ff.) has shown for instance that innovations in some industries must be based on basic research, while others procure technology from suppliers without embarking on intense research programmes. In incomplete networks, instead, one or more poles are either missing or under-developed. While relations among actors are sufficiently well defined to speak of a network, they are poorly developed. When they are somewhat more developed, there is what might be called a "chained network".

At the national or international level, comparison of the morphology of networks should help to explain why either the same or different sectors or branches exhibit differing technological performances or innovative capacities. Analysis of the networks in manufacturing industries should also make it possible to determine whether or not the location of poles or links of particular importance has shifted, an occurrence which has significant implications for national technology policy.

The dynamics of networks

A morphological analysis (composition of networks, nature and density/intensity of linkages) may be supported by a parallel dynamic analysis aimed in particular at establishing whether networks are static (technology is stable or evolving along a still profitable trajectory), flexible, or in the process of branching out (Imai and Baba, 1991).

In practice, analysis of networks reveals the presence of actors with evolving strategies. They fluctuate between "attachment" (remaining in contact with a network of firms, suppliers, university departments, research centres, etc.) and "detachment" (disengaging from a network in order to join another, construct a different one, etc.). Attachment includes notions of competence, procedural rationality, limited yet perfect information, predictability, and incremental innovation, and, in another analytical framework, the concept of technological trajectory. Detachment is linked to the concept of flexibility and to notions of uncertainty, risk, variable factor combination, radical innovation and market creation.

The behaviour of actors can bring about changes in a network, by lengthening or shortening it or by eliciting a change in the composition and/or configuration of its poles. As a result, network connectivity may increase or decrease. Thus, the transformation of these links spurs, and after a fashion shapes, the technological evolution of the productive sub-systems to which these networks relate.

Innovation-related networks and national systems of innovation

When analysed in network terms, the notion of "national systems of innovation" gains somewhat in substance and intelligibility. The constituent elements of such "systems" and the linkages between them can be identified and the degree to which they are organised into a coherent whole can begin to be determined. Network analysis may permit identification of the actors, of the type of information they exchange and of their position (within the sector, trading sphere, national economy, etc.).

As has been made clear, every identifiable network has arisen from the need to meet a specific technological objective, and it is the common goal that brings unity of purpose to the various sub-objectives pursued by each of the actors within the network. While a coherent national system of innovation will necessarily include a series of more or less co-ordinated innovation-related networks, this does not mean that analysis should be carried out at the level of a "national innovation-oriented macro-network". The loss of information would make such an analysis meaningless.

Network analysis must take place at lower or more intermediate levels of the productive system, since the composite network system is not simply a mosaic of sub-networks. Networks both juxtapose and interpenetrate each other. Depending on the size of the productive sub-system studied, there may be one or more techno-economic networks involved in a system of innovation and, depending on the relevant field of technology, such networks may either be interlinked or relatively separate.

The results of this type of analysis supplement traditional data regarding the origin and volume of resources devoted to different types of R&D. For example, when analysed in terms of innovation-related networks, the conundrum of why the policy pursued by the Japanese authorities should be so effective – in view of the parsimonious funding by the Japanese government – becomes less of a mystery. Japanese superiority in certain areas comes from a capacity to mobilise networks of firms and laboratories to meet specific objectives agreed upon by common consent. Furthermore, analysis of major Japanese industrial groups reveals levels of versatility, strategic and technological complementarities and synergies which few of their counterparts in other countries have been able to match. The results achieved by Japanese industry, notably its ability to catch up and then overtake the technological performance of some of its competitors, may be explained by its organisation into large, diversified groups. This has led to the emergence of dense, interrelated networks in which individuals and information circulate; in addition, these networks, without harming the vital competitive basis, encourage firms to behave in a manner more beneficial to the economy as a whole than the highly individualistic strategies adopted by firms in other countries.

Globalisation has made it harder to identify the contours of national systems; it has also made their monitoring more difficult (see Chapters 10 and 11). The boundaries of many key innovation-related networks for a given national system of innovation no longer necessarily coincide with those of the nation-state. Analysis of alliances between firms shows that in many instances these networks are largely transnational. A national system of innovation will be composed of local networks that may or may not be interlinked and that may or may not interrelate with global networks. From an economic standpoint, it is clear that technology poles and market poles are crucially important to the final results stemming from innovation. Thus, university research centres, notably in small countries, may be wholly outwardly oriented, regardless of their geographical location, due the fact that the locations of the technology poles to which they are related have been displaced. In a world in which "distance" has been rendered unimportant by advances in telecommunications, the creation of national centres of scientific excellence no longer constitutes *per se* a sufficient force of attraction for productive investment, except perhaps where the labour force must have specific skills. In this context, determination of the morphology of networks and the location of their various poles take on heightened importance.

4. SOME POLICY CONCLUSIONS

This chapter shows the extent of innovation-related network formation among researchers, institutions and enterprises and examines the conditions which must be satisfied in order for network relationships to emerge and flourish. Some overall conclusions can now be drawn.

Co-operation and its implications

The analysis sheds light on issues related to the appropriability of R&D and technological know-how by individual firms. An increasing number of studies confirm the overall pattern of the rapid transfer or "leakage" of proprietary industrial information from firms which generate it to others (see Chapter 2). This may indicate the presence of certain basic forms of network formation and know-how trading among firms, which can viewed in a rather different light than an uncompensated "leakage" of such proprietary knowledge to the detriment of the originating firms. If the information transfer is simple leakage without compensation, then innovators face serious appropriability problems (Mansfield, 1985, p. 213). However, if, as case studies show, it is the result of formal and informal co-operation and two-way know-how trading, then it is a "phenomenon which actually increases firms' ability to appropriate rent from technical know-how" (von Hippel, 1987, p. 301).

This advantage only accrues to institutions and firms that demonstrate a capacity for and readiness to co-operate. It is a collective rather than an individual form of rent appropriation and requires social and cultural preconditions which some nations possess more widely than others (see Ferguson, 1991; and Hollingsworth, 1990),[13]. But the other side of the coin cannot be ignored: firms which form technological alliances and coalitions are able to act anti-competitively, e.g. by placing restrictions on the access to technology by firms which are not parties to such coalitions. It is not certain, however, whether this latter argument is sufficient to condemn inter-firm co-operation as some authors have done (Porter, 1990a).

The advantages stemming from co-operation and networks today make the creation of networks both an objective and an instrument of government policy.

Network creation and management in science and technology policy

Government research programmes have as one of their objectives the creation of links among laboratories with a view to achieving significant advances in certain fields and generating new information. Seen in this light, the concept of a research budget takes on a new meaning. Over and above the intrinsic mass of the budget, which must be commensurate with the goal pursued, are the size and quality of the networks that it will help take root.

Government policy objectives in fields relating to science and civil and/or military technology have in fact often only been met through the creation of networks. "Concerted actions" in certain fields of technological research in France and the Apollo space research programme in the United States were completed by mobilising the various skills needed in specific areas; some were drawn from industry, others from universities and public laboratories. Most cooperative research projects, also known as pre-competitive research programmes, initiated by the European Community (ESPRIT, BRITE, RACE) are further examples of a strategy designed to encourage the formation of networks aimed at producing both new scientific information and new tacit knowledge that could subsequently be applied in an industrial or commercial context.

At the other end of the spectrum are to be found government measures to stimulate industrial research that may be seen as attempts to initiate the formation of techno-economic networks. Tax incentives offered to employers to hire researchers are a means to bring network "nodes" into firms (notably SMEs), since every scientist or engineer hired by a firm brings with him some kind of network of university relations which can, if necessary, be mobilised on behalf of the firm. Research tax credits also help promote the expansion of existing networks.

As further elements of science and technology policy, research evaluation programmes may act as instruments for determining the quality of the networks that emerge from research programmes and for identifying and remedying weak links. To take the case of university-industry relations, for example, if the actors involved engage in behaviour that stifles initiatives and restricts collaboration, a stable and efficient relationship between public research or teaching departments and industry may prove impossible. Such failure may explain weaknesses perceived in the technological performance of a given techno-economic network. By identifying the weak link in the network (in this instance university-industry relations), the cause of the failure can be established and a remedy sought. This may take the form of redrafting of certain regulatory, legislative or statutory provisions. A French technology-related agency, the Agence Française pour la Maîtrise de l'Energie (AFME) has used a network-based approach to assess the relevance of its programmes.

Supplying network externalities

Creating a network involves a process of trial and error, because at the outset, no one can identify "good" actors or formulate issues with the requisite degree of accuracy. The costs involved may therefore slow progress when expected benefits are not high or strong enough to encourage actors to continue building the network. Government bodies can play a role by providing some kinds of support. "Flexible technology programmes" (Callon *et al.*, 1990) provide the basis for an infrastructure that will accommodate subsequent initiatives from the private sector. The European Community's EUREKA programme, for example, seems to offer a supply of externalities sufficient to encourage initiatives from the firms concerned.

"Science parks" are sites created by local, regional or national authorities in order to promote and "informally institutionalise" exchanges of information. The authorities provide externalities in the form of land, equipment and a particularly sophisticated telecommunications infrastructure, so that links between public research centres and private firms, or synergies between local firms and the subsidiaries of national or international groups, may be fostered. Examples of such "science parks", "technology poles" or "technopoles" and "technology cities" abound: Tsukuba in Japan, Sophia-Antipolis in France, the Cambridge Science Park in the United Kingdom. The rationale behind the provision of such infrastructure is to encourage the development of innovation-related networks[14]. Siting major items of equipment for basic scientific research at the same location will also be beneficial for industry.

Special emphasis should be laid on infrastructure in the form of telecommunications networks, which now represents an important supporting factor for many innovation-related networks. The decisions that public authorities take in favour of providing this externality may influence the architecture of innovation-related networks, strengthen existing networks and encourage the development of new ones.

NOTES

1. See Williamson (1975 and 1985). Williamson strongly contributed to establishing the theory of economic organisation as an area of economics in its own right, albeit not a "main stream" one, if only because it is difficult to formalise and to treat with econometric techniques. However, the origins of this branch of economics are older.

   In an early article, Coase (1937) presented an account of the firm as a *governance structure,* thus breaking with prevailing neo-classical presentations of the firm as a "black box" production function. Coase's key insight was that firms and markets were alternative means for organising similar kinds of transactions. This provocative paper went unnoticed until it was rediscovered by Williamson and other proponents of transaction costs economics in the 1970s. This current of thought has seriously investigated the notion that organisational form matters a great deal, and in so doing has moved the theory of economic organisation much closer to the fields of law, organisation theory, business history and technological innovation. In the 1980s new interfaces have been created with work done in the social sciences, game theory and, of course, the economics of technological change.

   The core of Williamson's argument is that exchanges which are straightforward, non-repetitive and require no transaction-specific investments will occur in markets, but that transactions between firms that involve uncertainty about their outcome, recur frequently or require substantial "transaction-specific investments" are more likely to take place within hierarchically organised firms. In particular, transactions are moved out of markets into hierarchies as knowledge specific to the transaction (asset specificity) builds up. When this occurs, the bureaucratic costs of large scale corporate organisation will be preferred to market transactions. Williamson offers three reasons for this: *i)* the imperfect character of markets in which asset specific transactions take place and the fact that discretionary behaviour by traders becomes possible, even frequent; *ii)* "bounded rationality", defined as the inability of economic actors to write contracts that cover all possible contingencies; when transactions are internalised, there is little need to anticipate such contingencies since they can be handled within the firm's internal "governance structure"; and *iii)* "opportunism", defined as the rational pursuit by economic actors of their own advantage, with every means at their disposal, including guile and deceit. Within the same corporation opportunism is mitigated by hierarchical relations and by the stronger identification that parties can be presumed to have.

2. See Antonelli (1991*b*). Work carried out by the OECD on telecommunications and the convergence of computer, communication and control technology has of course entailed attention to the word "network" as understood in this context. It also draws attention to the role played by the enabling infrastructural conditions for some of the relations and linkages which form part of innovation's related networks. See, for example, OECD (1989*b*) and Kimbel (1988). Within the OECD, use of the term "network" in relation to innovation-related activities dates back to a study on industry-university co-operation in research (OECD, 1984). Subsequent work by the OECD on inter-firm co-operation in technology (Chesnais, 1988*b*) confirmed the extent to which by 1986-87 the pooling and exchange of knowledge by rival or competing firms had become a central characteristic of corporate behaviour in the creation and use of technology. Papers prepared for TEP-related conferences or symposia have confirmed this trend and also shown the importance of user-producer relations and linkages in the innovation process (e.g. Ferguson, 1991; and Lundvall, 1991).

3. For a recent appraisal, see Blume (1990).

4. The notion of "extended laboratory" refers to the fact that no category of "research related actor" will make an isolated contribution to the work of the laboratory. Individual research workers are a microcosm of the network of universities which trained them; they can call on contacts and partners who participate either directly or indirectly in the development of research and the assessment of findings. Documents and instruments, without which certain experiments could not even be attempted, exhibit the same powers of extension. They include the team which designed them and with whom contact must be maintained to solve eventual problems. The same principle of extension also applies to those who commission work and who embody the expectations of the public or private institutions. The concerns of this category of actor will also be expressed at various stages in the work of the laboratory.

5. The OECD report on competition policy and joint ventures (OECD 1986b, pp. 93-94) concludes that "on the one hand there are numerous advantages which attract firms to joint ventures. These advantages include the sharing of costs and risks of projects beyond the reach of a single enterprise. This feature is particularly important in the areas of R&D, natural resource exploration and exploitation, and large scale engineering and construction projects. A joint project may also create efficiencies through economies of scale, rationalisation or specialisation (...). On the other hand, joint ventures may have

harmful effects if actual or potential competition is eliminated, *in particular when the parents to the joint venture command relatively high market shares*. Market foreclosure can pose a problem when there is a vertical relationship between the venture and the parents, and competing firms are at a disadvantage concerning access to an essential input. Competition may also be harmed if the parties bind themselves by ancillary restraints which are not necessary to reach the legitimate purpose of the joint venture or share information which can lead to collusion in areas outside the scope of the project."

6. See in particular the following data bases:

   a) Wall Street Journal data base on inter-firm agreements (raw data, no processing of information);

   b) FOR (Montedison), Rome, (Ricotta and colleagues);

   c) ARPA (Advanced Research Programme on Agreements), Politecnico di Milano (Mariotti and colleagues);

   d) LAREA/CEREM, Paris-X University, Nanterre (Delapierre and colleagues);

   e) MERIT, University of Limburg, Maastricht, (Hagedoorn and colleagues).

   For full references and a discussion of some weaknesses in data bases, see Chesnais (1988b).

7. A range of papers reflecting the work of the Aalborg group are published in Freeman and Lundvall (1988).

8. This position was taken by Mariti and Smiley (1983) in the first extensive industry survey to be carried out on interfirm cooperation. It continues to be adopted mechanically by others (see, for instance, Malmberg, 1990). It overlooks Richardson's conclusion in 1972, namely that "firms are not islands of planned co-ordination in a sea of market relations but are linked together in patterns of co-operation and affiliation. Planned co-ordination does not stop at the frontiers of the individual firm but can be effected through co-operation between firms. The dichotomy between firm and market, between directed and spontaneous co-ordination, is misleading; it ignores the institutional fact of inter-firm co-operation and assumes away the distinct method of co-ordination that this can provide".

9. Williamson (1985, p. 83). See also Favereau (1989). Some limits of a discussion couched in terms of transaction costs are discussed in Foray (1990a).

10. On the side of innovation theory, studies often take a rather simplistic "all co-operation is good" approach.

Political scientists and business management specialists are more hardheaded. Studies by good business management specialists (see in particular Hamel, *et al.*, 1989). Porter and Fuller (1986) list the following main costs and drawbacks: bureaucratic costs of co-ordination between partners, contingent upon their respective strategy and configuration; lack of trust between partners that makes co-ordination more difficult; erosion of competitive position – alliances and agreements can strengthen the position of the allied competitor; adverse bargaining position – alliances can expose one of the partners to extraction of profits by the other because of a weaker bargaining position.

11. See Ouchi (1984, p. 122). For two recent comprehensive reviews of the literature and findings on Japanese co-operative research, see Sigurdson (1986) and Levy and Samuels (1991).

12. The term "national system of innovation" was coined by Freeman in his study on Japan (1987a) and later used in Part V of Dosi *et al.* (1988). It is now the object of two major international projects. The first is led by Nelson and Rosenberg under the auspices of Columbia University. The second is based at the Institute for Production at Aalborg University (Lundvall, Andersen and Dalum).

13. Hollingsworth states that: "Social systems of production operating in volatile markets with rapidly changing and highly complex technologies require a different type of work organisation, work skills, system of control, relations with suppliers, and competitors from a system when markets are stable and technologies are less complex and relatively unchanging (...). To compete effectively in international markets, a diversified quality system of production requires co-operative relations with one's competitors, one with a highly developed association system, well developed networks, communitarian obligations and collective resources that individual actors cannot normally generate alone, even if they recognise their importance" (1990, p. 15).

14. To quote a US study (David *et al.*, 1988, p. 11): "Just as the development of innovations based on the research findings of the laboratories requires sustained interaction between basic and applied research projects and personnel, the development of state-of-the-art equipment for accelerators also relies heavily on an interactive, two-way flow: of ideas and data. The interactive nature of this relationship underlines the importance of geographic proximity for the exploitation of the economic benefits of High Energy Particle Physics."

Chapter 4

TECHNOLOGY AND CORPORATE ORGANISATION

ABSTRACT

New forms of work organisation and corporate management emerged during the 1980s, in response to the pressures and opportunities of new technologies and the instability of the international macro-economic environment. The fordist model, which dominated industrial production for many generations, encountered growing difficulties from the late 1960s in the United States and from the early 1970s in other major OECD countries. The priority given to economies of scale obtained through undifferentiated mass production entailed a high degree of rigidity in the face of rapidly changing demand and markets.

A new model of corporate organisation developed in Japan often known as "toyotism" affords increased flexibility and quality of production while retaining the main advantages of standardisation. Systems of "just-in-time" production and "total quality" manufacturing, which require close co-operation between firms and their qualified long-term sub-contractors, have been adopted. R&D, design, engineering and industrial manufacture have been closely integrated, and greater attention is paid to demand characteristics and users' needs. A new paradigm of human resource management has also emerged. On-line management and worker responsibility has led to some decentralisation of decision-making regarding production; there is a growing recognition that workers' skills require constant improvement.

A second and complementary change in corporate organisations is the emergence of the "network firm". The origin of this notion is traced back to the Japanese corporation and its particular structure. Genuine network firms have also emerged elsewhere, in particular in Italy. The network firm is characterised by increased centralisation of allocative and management functions along side new forms of linkages ("electronic quasi-integration") between the core (or "hub") corporation and decentralised semi-autonomous and even independent production units. Organised on the basis of opportunities created by the new technologies, the network firm makes considerable use of various forms of inter-firm agreements. Japanese experts anticipate a further evolution for MNEs, which would involve the establishment of "sub-core" enterprises at regional levels within the wider global network. Assignment within this global structure of quasi-autonomous management functions to these sub-core firms and the particular networks they may build is viewed as a key dimension of this new form of MNE organisation.

The need for the organisational and managerial changes also concerns small and medium enterprises (SMEs). Today these firms are in a paradoxical situation. Technological developments have rarely seemed to be so favourable to small-size production, yet small firms remain highly vulnerable and are often in obvious need of assistance. A closer look at the advantages and disadvantages of small size in relation to the new technologies may help to explain why the nature and attributes of the linkages experienced by SMEs with other firms and institutions will often determine their long-term performance and their capacity to stay in operation. The approach may also offer a sound foundation for public policies addressing the viability problems of industrial SMEs.

INTRODUCTION

The central role played by firms in the innovation process and their function as the main, and often the sole, institution bringing new technology to the market has been stressed throughout the previous discussion. In the first three chapters, the analysis of the interactive nature of innovation, the nodal character of industrial R&D, the blurring of the distinction between the production of technology and diffusion, and the adopting of network relationships by and among firms have all placed the enterprise firmly at the centre of contemporary changes. This chapter now discusses the implications of these changes for corporate organisation *per se*.

These implications concern both large firms and SMEs. With respect to large firms, two changes have been singled out as warranting particular consideration. The first concerns the adoption in corporate organisation of practices, organisational structures and managerial approaches that differ significantly from previous management models. These new forms have often been inspired by the success of Japanese corporations in reaping the benefits of new technology. They are frequently referred to as the Toyota Production System or more briefly "toyotism" [1]. The second

relates to the emergence of new forms of "network corporations" alongside the classic management patterns of multinational enterprise[2] (examined in Section 2). The performance of small firms and their capacity to use technology and adapt their modes of management cannot be disassociated from the setting and linkages within which they operate (see Section 3).

This chapter continues the discussion of co-operation begun in Chapter 3. The analysis moves from innovation-related inter-firm networks to issues concerning the relationships between assembler firms and their suppliers and, more decisively still, co-operation within firms. A key strand here, entirely consistent with the conclusions of the Sundqvist Report (OECD, 1988*b*, Chapters 3 and 4), is the necessity for firms to view co-operation as a central managerial and organisational concept which is not limited to their external environment but also holds for the internal governance of the firm's activity. Chapter 5 on intangible investment and Chapter 7 on education and training acquire their full significance only if the notion of human capital and the role of "humanware" in production and innovation is accepted as a central component of the current industrial context (Shimada, 1989).

1. TOWARDS A NEW MANAGEMENT PARADIGM FOR LARGE FIRMS

This section focuses primarily on large corporations. The data on corporate world market shares in several key industries (discussed in Chapter 10) serve to underscore the place occupied by such firms. They are also the home of large R&D industrial laboratories, and they account for a very high percentage of business enterprise R&D. Some indications of the scale of corporate R&D within the largest firms, in comparison with other institutions, are given in Table 10. As can be seen, the United States Department of Defense R&D budget is in a category of its own. But in other cases, corporate R&D spending by the largest MNEs can be as large as, and sometimes larger than, the R&D budgets of major government

departments. Figures published by the National Science Foundation at the end of the 1970s (*Science Indicators,* 1976, pp. 103-104 and 1978, p. 82) show that well over 50 per cent of United States industrial R&D was carried out by the first 20 spenders. Data published in the 1980s on the level of concentration of industrial R&D in Japan confirmed this trend and showed that it was not specific to the United States (OECD, 1989*e*).

Since the mid-1980s, significant effort has been directed towards explaining the productivity paradox, namely the relatively poor productivity performance in many countries and industrial sectors, despite what seem to be potentially major productivity-enhancing

Table 10. **R&D budgets of selected major agencies and large corporations in 1989**

A. United States ($ million)

Government agencies		Private firms		Universities and colleges	
1. Department of Defense	38 076.2	1. General Motors	5 247.5	1. Johns Hopkins University	648.4
2. Department of Health and Human Services	7 981.4	2. IBM	5 201.0	2. Massachussets Institute of Technology	287.2
3. Department of Energy	6 065.6	3. Ford	3 167.0	3. Cornell University	286.7
4. NASA	5 913.0	4. AT&T	2 652.0	4. Stanford University	286.0
5. NSF	1 724.0	5. Digital Equipment	1 525.1	5. Wisconsin (Madison)	286.0
6. Department of Agriculture	1 127.5	6. Du Pont	1 387.0	6. University of Michigan	281.0
7. Department of the Interior	479.8	7. General Electric	1 334.0	7. University of Minnesota	258.6
8. Department of Commerce	407.0	8. Hewlett-Packard	1 269.0	8. Texas A&M University	250.7
9. Environment Protection Agency	380.3	9. Eastman Kodak	1 253.0	9. California, Los Angeles	227.8
10. Department of Transportation	318.1	10. United Technologies	956.6	10. University of Washington	221.7

B. Other OECD countries ($ million ppp)

Government agencies		Private firms	
Japan			
Education	4 192.6	Hitachi, Ltd.	1 693.1
Science and Technology Agency	2 290.0	Matsushita Electrical Industrial Co., Ltd.	1 653.8
MITI	1 146.6	Toyota Motor Corp.	1 472.3
Defence Agency	456.7	NEC Corp.	1 374.1
Agriculture, Forestry and Fisheries	333.9	Fujitsu Ltd.	1 349.6
		NTT	1 177.8
		Toshiba Corp.	1 128.7
		Nissan Motor Co., Ltd.	1 055.1
Germany			
Fed. Ministry of Research and Technology	3 296.0		
Fed. Ministry of Defence	1 418.2		
Fed. Ministry of Education and Science	497.3		
Fed. Ministry for Economic Affairs	417.7		
France			
Defence	4 666.0	Aerospatiale	1 655.6
Research and Technology	3 488.7	CGE	1 625.6
Posts and Telecommunications	1 174.3	Thomson	1 264.3
Industry	735.5	Renault	842.9
		SNECMA	767.6
		Peugeot	767.6
		Rhône-Poulenc	636.7
		Bull	587.0
		Elf Aquitaine	481.6
		Matra	391.3
United Kingdom			
Ministry of Defence	3 529.4	ICI	894.5
Department of Trade and Industry	488.6	Shell Transport and Trading	758.2
Department of Energy	271.2	Unilever	654.0
Ministry of Agr., Fish and Food	182.7	Glaxo	639.6
Departments of Education and Science	128.2	Smithkline Beecham	630.0
		General Electric	625.2
		British Petroleum	527.4
		STC	434.4
		Rolls-Royce	379.9
		British Telecom	365.5

Sources:
Germany: *Statistische Informationen. Wissenschaft, Forschung und Entwiklung, Ausgarben des Bundes 1987 bis 1990,* December 1990.
US: Firms: *Inside R&D's* annual report on the 100 biggest R&D spenders in US industry, June 1990. Government Agencies: *Federal Funds for Research and Development;* Fiscal Years 1989, 1990 and 1991, NSF 90-327, Table 3.
France: Government agencies: *Finance Bill for 1990. Supplementary Report on the State of Research and Technology Development,* Imprimerie national Paris, 1989.
Japan: Government agencies: *Indicators of Science and Technology,* Science and Technology Agency, 1990. Firms: *Trends of Principal Indicators on Research and Development activities in Japan,* April 1990, Agency of Industrial Science and Technology, Japan.
UK: Gouvernement agencies: Cabinet Office 1990, *Annual Review of Government Funded R&D,* Table 1.1. Firms: *The Independent,* 10/6/91.

investments. One explanation proposed by a number of experts and scholars points to the need for organisational and institutional changes corresponding to the requirements of the new technologies. This argument was advanced most strongly within OECD by the Sundqvist report (OECD, 1988b); it is to be found notably in the work of scholars interested in the socioeconomic dimensions of growth in a Kondratiev-type long wave approach, such as Perez and Freeman (1988).

A still-evolving hypothesis, for which there is a growing consensus (see the beginning of Chapter 7), argues that the widespread diffusion of new computerised technologies and the efficient use of these technologies can only come about in the wake of an effective shift to new systems of work organisation, skill formation, product development and management strategies which allow firms to survive and prosper in the new competitive conditions.

For the automobile industry, for example, a study of corporate organisation and technological potential (Womack *et al.* 1990) has shown the impact of different levels and kinds of automation on productivity (worker hours per automobile produced) and on quality (defects per vehicle) in plants in the United States. Plants that have adopted high levels of automated assembly but have not changed their human resource management strategies, have not improved productivity and quality over those which have not automated. Gains have been experienced particularly in plants which have adopted reformed human resource and labour management strategies (fewer job classifications, flexible work organisation, extensive communication), whether technology upgrading strategy has been moderate or extensive. Overall, productivity and quality are most improved by adoption of a strategy that combines advanced technology with higher worker participation and higher skills. The next most productive arrangement combines lower technology with high worker participation and skill. Low skill and participation give poor results at all levels of technology.

Seen in this light, the interest generated by the success of Japanese management practices over the last few years appears well justified. "Toyotism" – or the variant "hondaism" or "sonyism" which close observers of the Japanese scene have begun to identify – may represent a major component of the "adequate socio-institutional framework" required to "unleash the full potential" of the new technologies (Boyer, 1991). The expression "toyotism" is, of course, simply a convenient abbreviated term similar to universally accepted terms like "taylorism" and "fordism"[3].

Fordism and its limits

The content and meaning of toyotism and its considerable novelty can be appreciated more fully by recalling the content of fordism and the nature of the difficulties that the fordist management model ran into from the late 1960s onwards. The fordist model of industrial management and work organisation (called "taylorist" by some authors) is based on a few simple building blocks (see Box 23). It was highly successful through the first part of the century, but encountered growing difficulties from the mid-1960s in the United States and from the early 1970s in other major OECD countries.

Increased mechanisation and corresponding rises in capital/output ratios ceased being converted into higher productivity. Declines in the rate of growth of labour productivity, as well as in total factor productivity (TFP), set in from the mid-1960s onwards in the United States. Because of the "catching up" process which was still at work, it was only in the early to mid-1970s that this productivity slowdown reached Japan (where firms reacted rapidly) and the main industrialised countries of Western Europe (where firms were very slow to react, behaving much like their United States counterparts). Three broad sets of factors can account for this slowdown.

The first was the progressive exhaustion of the main clusters of innovations which furnished the technological foundation of the fordist industrial management paradigm in "scale intensive" mechanical and electrical engineering and automotive industries. More generally, fordism (in the wider sense of the "*école de la régulation*") waned. The phenomenon was first observed in heavy chemical plant engineering and wrongly characterised by some authors (Giarini and Loubergé, 1978) as expressing declining returns to R&D across the whole system; in fact, it represented first and foremost declining returns to the particular technological paradigms underlying the post-war boom. The industrial pattern of "technological exhaustion" has now been documented by Patel and Soete (1987); it has principally occurred in the industries most closely associated with the fordist paradigm and its period of success.

The second factor concerns the breakdown of worker acceptance of the work relationships at the heart of the fordist approach to the organisation of production. This point is now well documented for Europe and the United States. From the mid-1960s onwards, low-skilled blue collar workers in the automobile industry started to rebel against the monotonous character of assembly-line tasks, as they increasingly recognised the discrepancy between the de-skilling tendency of taylorist manufacturing techniques and rising social expectations regarding quality and initiative as aspects of work.

A third major vulnerability of the fordist paradigm became apparent following the recession triggered by the change in oil prices in 1974-75, the 1979-82 recession and the onset of turbulent and uncertain macro-economic conditions in the 1980s. It stems from the priority given to *scale economies* obtained through undifferentiated *mass production,* leading to *very high rigidity* in the face of uncertain and rapidly changing demand and markets.

This rigidity is not associated simply with the scale of investment or the organisation of the large assembly line. It characterises the whole approach which was considered in the heyday of fordism to be an irreversible phenomenon (see Galbraith's study of corporate planning in a context of complete supplier domination over consumers in *The New Industrial State,* 1967). The inertia of mass production lies not simply in the quantities of standardised products produced but also their "quality" (i.e. their numerous defects) and their lack of versatility and flexibility in the face of changes in consumer demand. Thus, the

traditional very long lag between the perception of and/or decision to shape a new demand, the conception of products aimed at meeting this demand, their design, testing and eventual appearance on the market were part of the overall rigidity of fordist industrial management which Japanese management practice has revolutionalised.

The decline and crisis of fordism was apparent from the late 1960s onwards in the sharp drop in the rate of growth of labour productivity and of TFP in the United States and later in Europe (see Chapter 8), as well as in the parallel and related drop in the rate of return to capital. These factors might not have sufficed to challenge the fordist paradigm of industrial management and work organisation deeply or rapidly, had Japanese competition not entered Western markets, in particular the United States domestic market[4].

In the United States, fordist forms of work organisation, more oriented towards management control than towards maximum participation, have continued to be adopted, particularly in mass production assembly operations, but less so in batch production. There is evidence of general upgrading of employment quality, but the directive management tradition, combined with uncertainty over job loss due to technological change, has not led to clear-cut, widely accepted new directions for re-organising work (Dertouzos *et al.*, 1989).

The emergence of toyotism

Toyotism began as the Japanese response to the weaknesses of fordism. At the start, it *exclusively* took the form of a set of organisational changes or innovations using essentially the same basic technology as did the fordist assembly line (Jones, 1988 and Coriat, 1990). Computer-assisted distribution and computer-assisted manufacturing (CAD/CAM) and

flexible manufacturing systems (FMS) came later. Its major departures from fordism consisted in:

i) Adoption of networking, sub-contracting and "just-in-time delivery", a basic reversal of the trend to ever greater vertical and horizontal integration which had become an important dimension of United States and European fordism;

ii) Reorganisation of work at the factory and production levels (see Coriat, 1990; Shimada, 1991; and Osterman, 1990); and

iii) Significant reductions in the compartmentalisation and hierarchical organisation of R&D, design, production engineering and marketing within firms (Clark *et al.*, 1987).

In combination, these three changes lead to totally new levels of *flexibility,* while retaining the main advantages of standardisation and setting new standards for quality ("zero defects") (see Figure 7). A Japanese study for the Helsinki TEP Conference argued again convincingly that "organisation counts

Figure 7. **The fordist and toyotist models of industrial governance**

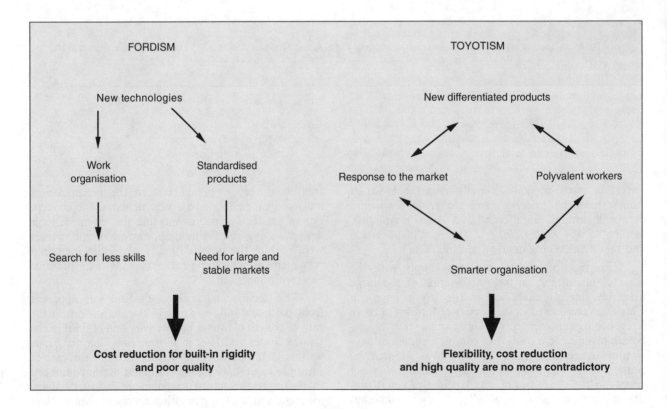

Source : R. Boyer (forthcoming).

94

more for productivity than micro-electronics *per se*" (Watanabe, forthcoming).

The historical forms of the Japanese enterprise, namely the *zaibatsu* or *keiretsu* type of large group structures, have been used to build up stable long-term inter-firm contractual relationships and have led to "network" types of corporate organisation (see Section 2). These collective or "social" forms of Japanese organisation bring substantial benefits to each individual profit-making centre and represent one of the main sources of Japan's present overwhelming industrial competitiveness.

Box 24. Hierarchies based on vertical integration versus horizontal informal co-ordination

The differences between the long dominant and still largely prevalent United States hierarchical approach (See Chapter 3, note 1) and the Japanese approach can usefully be discussed with reference to "transaction costs" theory, which provides a theoretical justification for concentration, integration and consolidation. The stress laid on transaction costs reflects considerable mistrust of "arm's length" market relationships upstream of final consumer markets, for instance with suppliers of intermediate inputs to production. It is coupled with underestimation of the collective benefits which can accrue to individual firms through efficient inter-firm co-operation (see Chapter 3) and through constant re-evaluation of *intra-industry* and *inter-industry* divisions of labour. In the United States and Europe, the accent has long been placed on concentration and integration. External sourcing through inter-firm co-operation and sub-contracting has been treated as an ancillary and very subordinate complement to integration and has been organised by the larger firms on an extremely unfavourable basis for the smaller ones (for a sharp critique, see Dertouzos *et al.*, 1989).

In contrast, Japanese firms have recognised that:

i) United States type corporate structures based on the "hierarchies" model are sources of extremely high "bureaucratic" costs; and

ii) An inter-firm division of labour can be organised to reap the benefits of specialisation, of cost and risk-spreading, so as to bring large "collective" sources of productivity to individual firms (notably those at the centre of industrial networks) without incurring the risks of unstable market relationships.

Japanese industrial economists have sought to explain the originality and success of this particular form of group structure with reference to the theory of "transaction costs" and the conceptual distinction between markets and hierarchies (see Box 24). They suggest that the *keiretsu* offers a range of opportunities which differ both from markets and from hierarchies. As one author has put it: "From the standpoint of the firm, by forming or joining a group, it can economise on the transaction costs that it would have incurred if the transaction had been done through the market, and at the same time, it can avoid the scale diseconomies or management control loss which would have occurred if it had expanded and performed that transaction internally" (Goto, 1982, pp. 53-54). A *keiretsu* thus forms a durable framework for joint development, production and trading activities, ranging over the whole value chain. For example, in the automobile industry a typical *keiretsu* may consist of an auto-assembler, a bank, a trading company, an insurance company, materials processing companies, and a host of manufacturing enterprises in different sectors, tied together through equity cross-links and a common commitment (see Section 2).

The organisation of manufacturing and approaches to labour management

The strength of Japan's industrial competitiveness, along with the spread of Japanese direct investment and the introduction of new technologies, has led to the progressive adoption of new organisational approaches to industrial production and marketing, whose value firms in the United States and Europe have begun to recognise. Among them may be counted the Japanese organisational principles of "kan-ban" ("just-in-time" – and initially manual – management of stocks and delivery from component manufacturers and sub-contractors). The advent of computer-aided "real-time" techniques has brought these principles within the reach of United States and European management. Similarly, computer-aided techniques on the shop floor have made it considerably easier to begin moving away from assembly lines towards decentralised work organisation.

The application of Japanese management principles and the potential of new technologies have given rise to a system which involves extensive use of quasi-integration but also makes a greater demand on market forces to provide many functions, services and products formerly produced internally. Suppliers become increasingly responsible for all aspects of component and sub-assembly development, design, production and delivery. In turn, they gain *economies of scale* by integrating many of their own functions over a larger range of similar products and *economies of scope* by covering more of the production cycle (particularly on the development and design end of production)[5]. Finally, in addition to its contributions to management of materials and intermediary product flows, for work management at the factory level, the "toyotist" model has significant implications for R&D and product development (see Box 25).

The move to more decentralised market-driven structures within firms has also led to a major change in the organisation of work and in the approach to the management of a firm's human resources. Japanese experts have proposed the term "humanware", defined as the source of productivity stemming from the quality of relationships and interactions between human resources and other basic elements of the production process, such as machines and software (Shimada, 1991). It is related to an approach to production, in the widest sense of the term including R&D and product design, which lays a high premium on labour's involvement in decision-making at the plant level, human control and co-operation. The potential of computer-controlled equipment to perform many tasks flexibly and the challenge of organising development, production and delivery to take advantage of this potential have changed much of the way that work in manufacturing and many service industries is now being organised. Responsibility for many operations is being placed on the shop-floor and this effectively provides a less hierarchical organisation of production within large firms. With flatter hierarchies and less direct control from the top come greater flexibility of work methods, delegation of responsibility in organising work, ensuring quality, performing preventive maintenance and other tasks.

At least two series of tasks traditionally performed by white collar staff working close to management are now best performed in the workshop (Coriat, 1989). The first is production planning, generally considered to be a key attribute and prerogative of management. In a classic plant, this is performed by white collar technicians in a separate office and then a shop supervisor gives the orders and distributes the work to the workshop. The new information technologies offer the possibility of doing this at the workshop level using computer terminals and electronic equipment. This is often more economic and efficient than the classic division of labour between conception (in this case planning) and execution. Yet again, the choices made on skill distribution and training are crucial.

The second, classic management task is the calculation of the cost of production, again generally performed in a separate office using analytical or other accounting techniques. Using workshop computers, it is now possible to do this on-site and obtain better results in respect of control objectives such as the diminution of production costs, the optimisation of energy and raw material consumption, the maximisation of the rate of utilisation of machines and equipment, etc.

"Functional flexibility" – based on the quality of work inputs, higher ternal mobility across tasks, reduction of job boundaries, greater flexibility in job design, and training and retraining – appears to have clearly outdistanced the fordist over-emphasis on "numerical flexibility" and the quantity of labour; controlling the number of hours worked and the organisation of working time; regulation of hiring and dismissals; and use of part-time and temporary employees. The poor results obtained from strategies based on numerical flexibility became particularly apparent with the adoption of advanced manufacturing technology (AMT), where the inherent flexibility of the equipment is crucially reliant on a system of work organisation using highly skilled and multi-skilled workers (Vickery and Campbell, 1989).

As a result of these changing practices, a new management paradigm with some degree of application in all the principal industrialised countries has emerged. OECD firms now appear to be moving, somewhat hesitantly, towards principles of industrial management which incorporate the lessons learned from the toyotist model and which will lead to a new post-fordist management paradigm (Boyer, forthcoming). Box 26 summarises these twelve "principles", and Table 11 gives a good idea of attempts to move towards the new management paradigm in five major OECD countries. Outside Japan, Germany is the only large advanced industrialised country where the management style and performance of corporations can be related fairly directly to the overall structural competitiveness of the national economy. Further evidence suggests that the situation of Swedish firms is not quite so favourable as it appears here. United Kingdom and Belgian firms would probably get marks similar to those of their French counterparts; Italian and Dutch firms would probably improve a little on that performance, but would nonetheless be far from the Swedish and certainly the German ones.

Box 26.   **Twelve "golden principles" for renewing industrial organisation**

1. The global optimisation of production flows, e.g. the generalised adoption of the Japanese "*kan-ban*", "just-in-time" and "total quality" system based on qualified sub-contractors and leading to the elimination of raw material and component stocks.
2. The close integration between R&D, design, engineering and industrial manufacture, an area where Japanese firms also have the lead but have been followed by German, Swedish and now a few United States large corporations.
3. The establishment of new, closer relationships with users, principally industrial users but perhaps also final consumers, although there is still little evidence that the latter type of interaction really exists.
4. The establishment of a new pattern of production combining lower costs with much higher levels of quality and "zero defects". This again would represent a generalisation of Japanese performance.
5. The incorporation of a correct definition of demand characteristics and their evolution into design and production strategies.
6. Higher levels of decentralisation of decisions about production with on-line management and worker responsibility: this is a particular Japanese pattern, with some few extensions into contexts of co-management by unions and of social-democratic consensus-seeking as in Germany and Sweden.
7. Decentralisation of supply through networking and joint ventures with component and materials supplier firms upstream and retailers downstream.
8. Long-term and co-operative sub-contracting with smaller firms.
9. A lower level of division of tasks within firms and the organisation of work on a team or "circle" basis. This stems from point 6. and represents the result of re-examination by Japanese automobile engineers of "taylorist" principles of "scientific management".
10. Higher priority and higher private outlays (as well as public educational investment) for vocational training. This stems from and partially overlaps 6. and 9. and is discussed below in Chapter 7.
11. The enhancement of workers' and employees' skills as a source of commitment, competence and productivity.
12. A new approach to employment, long-term contracts and wages. This might be a generalisation of Japanese practice to other countries at a time when it appears to be coming under heavy strain in Japan.

Table 11. **Five OECD countries facing the challenge of the new management policies**
A very provisional and tentative synthesis

Principles	France		Japan		Sweden		United States		Germany	
P1: Global optimisation	−	Recognised, but difficulties in implementing it	++	Strong e.g. capital and inventories output ratio decline	+	Present e.g. Uddevalla low break even point	−	Hindered by fordist inertia, succeeded by Japanese firms	+	Exists but some conservatism
P2: R&D and production integration	−	Efforts but lagging organisation	++	Leader role e.g. shorter design times	++	Very dynamic product and process R&D	0	Difficult in spite of successful examples	0	Follows the old model with minor exceptions
P3: User-producer interaction and diversification	−	Fairly low e.g. failure of the equipment goods industry	+	Important for equipment goods	+	Limited due to the role of external markets and the size of the economy	−	Fairly low e.g. quasi none for equipment goods	+	Significant via servicing of equipment goods
P4: High quality at low cost	−	Recent efforts but still quality problems	++	A key feature of Japanese styles	0	Quality of servicing but extra cost of customised goods	−	For quality and relative high costs	0	High quality but not clear cost advantage
P5: Productive versatility to demand	−	Traditionally low	++	Important e.g. short lag in the car industry	+	Existing, even if not very fast	−	Sluggish in mature industries, present in high-tech industries	+	Average, along the previous model
P6: Production decentralisation	−	Typical centralisation in large firms	+	Significant if not general e.g. impact of micro-electronics (ME)	+	Well-known experiments (from Kalmar to Uddevalla) but taylorism still exists	−	High centralisation in spite of Japanese successful experiments	+	Significant responsibility for skilled workers, but rather centralised management
P7: Horizontal co-ordination and net-working	−	Emerging, but not very strong	++	Large e.g. Kan-ban and now ME	+	Exists at the plant level	−	Used to be forbidden by antitrust laws, now reversed	+	Yes, at the regional level
P8: Long-run, comparative subcontracting	−	Emerging, for example in the car industry	+	Applied to first tier subcontractors	0	Not clear	−	Idem	+	Role of professional mobility
P9: Recomposition of production – maintenance – programming	−	Not very easy due to hierarchical barriers	++	Very significant	+	Active field in some key experiments	−	Rather difficult	+	Exists, but not very strong
P10: General education and on job training	0	Average performance of both general education and training	+	Very significant in large firms, lower in other firms	++	Active role of public authorities in retraining and upgrading skills	−	One of the poorest performances for OECD countries	++	Excellent system combining general education and practical learning
P11: Workers' competence and commitment	0	Implicit: now major concern, but practice is lagging	++	Usually strong e.g. support to the "firm culture"	+	In order to fight against turnover and absenteeism	−	A tradition of adversarial relations, control and financial incentive	+	Clear for high- and medium-skilled workers
P12: Long-term compromise over job tenure and/or good wages	0	Only marginally present in some firms; no longer existent at the national level	+	Implicit, covers only large firms employees	++	At the national level: compromise over a maintained quasi-full employment	0	Marginally exists short-sighted capital-labour relations	++	Active negotiations about new technologies, wages, work duration

GENERAL SUPPORT AND CLOSENESS TO THE NEW MODEL	FORDIST AND CULTURAL INERTIA	JAPAN INVENTED IT (TOYOTISM, SONYISM) "Diversified quality mass-production"	AN ORIGINAL VARIANT (VOLVOISM) "Customized quality, competitive medium sized production"	FORDISM NOSTALGIA "The hindrance of having been too successful: a very difficult transition to a new regime"	AN ORIGINAL VARIANT "Quality competitive medium- or mass-production"
SYNTHETIC INDEX*	−0.375	0.80	0.54	−0.50	0.50

* Obtained by algebraically summing up all the "pluses" (+) and "minuses" (−) and dividing by the maximum score (12 x 2) in order to get an index, between 1 (complete support) and −1 (complete opposition to the new model).

Source: Boyer (forthcoming).

2. THE EMERGENCE OF THE NETWORK FORM OF CORPORATE ORGANISATION

The term "network firm" owes its origin to studies of the Japanese corporation and the Japanese approach to the management of resource-sourcing, manufacture and marketing of products. Consequently the analysis must begin by looking at the Japanese experience. In contrast with Japan where "networking" is rooted in the historical development and structure of industrial enterprises, the rise of the network firm in the United States and Europe is recent and still somewhat novel. It is directly related to the establishment of world-wide telematic communications. Outside Japan, multinational network firms are still uncommon in manufacturing, although much more developed in some service industries (e.g. tourism).

There are two roads towards the network form of corporate organisation. The hardest involves the *transformation* of strongly centralised and hierarchical MNEs into decentralised firms based on intra-corporate networks. The second road is *creation*, from the ground up, of genuine network firms, through the adoption of this model for the growth and international expansion of a medium-sized domestic firm. This road has been successfully followed by a few Italian firms.

Both forms rely on intra-corporate networks. These have been defined by Bressand and Distler as electronics-based "distribution and data management systems within a single firm". Their aim and outcome is "to improve co-ordination of day-to-day activities by enabling end-users within the same company to share information and processing capacities. Electronic networks allow *real-time* transmission, processing and distribution of information related to customers, as well as remote interaction between plants. They play a key role in making it possible to manage global networks of plants as integrated production facilities and in moving from the traditional 'product out' vision (in which products are created and then marketed) to the 'market in' vision in which production is a direct function of demand" (Distler, 1989, p. 17).

The emergence of genuine "network firms", in countries other than Japan, along with parallel changes in the corporate organisation of large traditional United States and European multinational enterprises, point to the elaboration of similar forms of multinational corporate organisation which may become widely adopted. The previous configuration of competitive advantages held by MNEs (firm-specific ownership advantages, location advantages and the special advantages arising from the capacity to informalise part of their transactions) may change as a result of the "new dimensions created by networking". Dunning suggests that "the choice of network partners, and the nature, form and outcome of the relationships formed, may decisively influence a firm's competitive

position – particularly where it is a core enterprise whose advantages lie as much in its ability to co-ordinate external, as internal, transactional relationships" (1988*b*, p. 345).

The network firm: a traditional Japanese institution

The term "networking" was first applied in studies of the *keiretsu,* or more strictly speaking *kigyoshudan,*

Box 27. Some characteristic features of the *keiretsu* or *kigyoshudan*

A *keiretsu* is a group of enterprises tied together through reciprocal share-holdings, credit relations, trading relations, and interlocking directorships. The members of such a group will not necessarily be subsidiaries or minority-controlled associates of particular parents. Many are firms with no single dominant share-holding interest. But in these enterprises, other members of the *keiretsu* will be among their largest shareholders and will collectively hold substantial blocks of shares. This distinguishes such enterprises sharply from those which have been described as controlled through constellations of interests. In a controlling constellation the participants comprise a diverse collection of interests among which any coalitions are unstable and short-lived. In contrast, the controlling participants in many *keiretsu* form a stable and durable coalition with other *keiretsu* members. Such enterprises can be described as controlled through coalitions of aligned participations which are dominated by their 20 largest shareholders. The system of aligned participations which runs through the core of the *keiretsu* is reinforced by the provision of loan capital by the group bank and the handling of sales by group trading companies, and this allows mutual insurance companies and enterprises controlled by families and other corporate interests to become affiliated to the *keiretsu.* All members are united through these economic bonds and by their common orientation to the idea of the group. This group orientation is expressed in the power accorded to the presidents' club (*shacho kai*). This "club" is, in fact, a committee, made up of the presidents of the major member enterprises, which meets regularly, generally monthly, to evolve a group policy (Scott, 1986, pp. 167-168).

---

type of industrial group (see Box 27). Observers have compared and contrasted the long-term inter-corporate industrial relations which this form enhances with the essentially finance-motivated "conglomerate" corporations which spread in the United States from the mid-1960s onwards and later became increasingly frequent in the United Kingdom and some other countries. The network form of corporate organisation is rooted in Japanese history and can be traced back to the pre-Meji merchant houses (Scott, 1986). Today, it is often used by Japanese industrial economists as a way of describing two distinct, but evermore closely related phenomena: i) the set of relationships between a "lead" or "hub" manufacturer working on a "just-in-time" basis and the associated web of smaller supplier firms; and ii) a particular Japanese approach to the management of the multinational corporation which stresses "on the spot activities" connected to "on-the-spot information" (Imai, 1988c; and Imai and Baba, 1991).

On the pattern of the broad set of intra-group "network" relationships, stemming from the keiretsu form, Japanese firms developed inter-firm relationships with component suppliers and industrial sub-contractors and also made widespread use of inter-firm co-operation bearing on R&D and technology (see Sigurdson, 1986; Chesnais, 1988b; and Levy and Samuels, 1991). It also explains, as Imai argues (1988c), why major Japanese corporations will generally make concomitant use of three types of networks in their domestic operations (see Figure 8):

– the wider intra-group "affiliated" or keiretsu networks;
– hub firms' supplier networks;
– R&D networks, consisting of co-operative R&D relationships, which can be (either separately or simultaneously) inter-group, intra-group or external (e.g. with universities or public laboratories).

As multi-nationalisation and globalisation proceeds, large Japanese firms are also developing a further set of network structures to meet the requirements of production and marketing in a global setting. These are examined below.

Figure 8. **Internal networks within the Japanese keiretsu-type group structure**

Inter-organisational networks

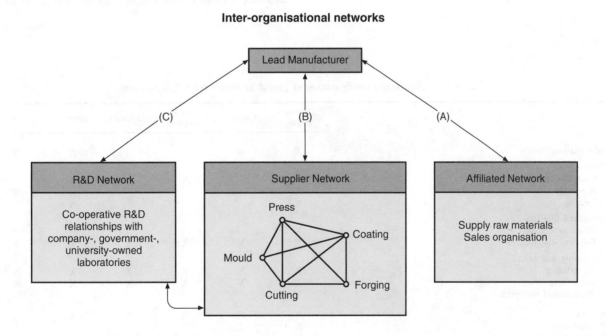

Source: Imai (1988c).

Changes made in corporate organisation by multinational enterprises (MNEs)

Since their emergence in the early part of the century, the internal organisation and management structures of MNEs have gone through several phases[6]. During the decades of growth of the classic MNE, two principal forms were available: one, vertically integrated, suited *single product firms* even after they had grown very large and gone international; the other, multi-divisional, suited *multi-product firms*. United States companies were usually organised in the first of these forms when they acquired their first foreign subsidiaries. As foreign operations were consolidated and extended, the MNE generally established an *international division* to co-ordinate certain functions such as transfer pricing between affiliates, finance, and the distribution of exports among production units.

By the mid-1970s, the limitations of this organisational structure had become clear. Domestic divisions were still usually organised by product, whereas international divisions had responsibility for overseas production of these same products. Problems of "sub-optimisation" were generated, and firms began moving to alternative organisational structures to resolve this problem:

i) One solution was to organise the whole corporation into world-wide *or global* product divisions. This ran into different problems of sub-optimisation: problems common to overseas operations were handled in separate divisions which had few incentives to co-operate.

ii) A second solution was to break down an international division into *area divisions,* each responsible for all operations in a given foreign area. This solution was often adopted when foreign subsidiaries supply one another with components or intermediate products and thus require close co-ordination.

The choice between global product divisions and foreign area divisions represented a balancing of advantages among imperfect alternatives. Studies by Caves (1982) and Michalet (1985) noted the prevalence of global product divisions in firms with heavy R&D spending, which demands close global management of product-specific intangible assets. By contrast, international area divisions were adopted by firms whose mature product lines were supplied to homogeneous common end-user markets and where the MNE's main co-ordination problem lay in its regional marketing organisation.

The advent of telematics offered the opportunity to evolve new forms of management systems and make progressive changes in the organisation of functional specialisation. A mid-1980s survey of the changes in corporate organisations which followed the adoption of telematics in 40 United States and European MNEs identified effects on performance and significant changes in management practices. The internal organisation of MNEs was found to be increasingly characterised by a *world-wide network* of *near-to-market affiliates*. They sold the products of a few highly efficient and specialised factories, managed by *global* product divisions. The use of telematics seemed to be

Table 12. **Illustrative configuration of global activities for a US company**

	United States	Canada	United Kingdom	France	Germany	Japan
Inbound logistics	X		X		X	X
Operations						
Components	X		X			
Assembly	X				X	X
Testing	X				X	X
Outbound logistics						
Order processing	X					
Physical distribution	X	X	X	X	X	X
Marketing and sales						
Advertising	X	X	X	X	X	X
Sales force	X	X	X	X	X	X
Promotional materials	X					
Service	X	X	X	X	X	X
Procurement	X					X
Technological development	X					X
Human resource management	X	X	X	X	X	X
Firm infrastructure	X					

Source: Porter (1986).

Table 13. **Configuration and co-ordination issues by category of activity**

Value activity	Configuration issues	Co-ordination issues
Operations	Location of production facilities for components and end products	Allocation of production tasks among dispersed facilities. Networking of international plants Transferring process technology and production know-how among plants
Marketing and sales	Product line selection Country (market) selection Location of preparation of advertising and promotional materials	Commonality of brand name worldwide Co-ordination of sales to multinational accounts Similarity of channels and product positioning worldwide Co-ordination of pricing in different countries
Services	Location of the service organisation	Similarity of service standards and procedures worldwide
Technology development	Number and location of R&D centres	Allocation of research tasks among dispersed R&D centers Interchange among R&D centers Developing products responsive to market needs in many countries Sequence of product introductions around the world
Procurement	Localisation of the purchasing function	Locating and managing suppliers in different countries Transferring knowledge about input markets Co-ordinating purchases of common items

Source: Porter (1986).

paving the way to greater affiliate specialisation. At the same time, as a consequence of new on-line transmission systems, new units became *centralised* functional service centres to provide trading, financial, technological and electronic data processing (EDP) services to all the affiliates. The survey concluded that "the new emerging structure seemed to be characterised by a combination of product divisions responsible for production tasks at a global level, domestic product affiliates restricted to performing commercial tasks, and central corporate functional units which manage the international flows of inputs, transport, technical and financial resources" (Antonelli, 1986, p. 5).

Although not a full shift to the network form, a structure of this type represents a significant departure from previous forms of corporate organisation. There is a shift from a phase of internationalisation where multinational enterprises operate within the context of industries still organised on a "multi-domestic" basis to one where, on the basis of the opportunities created by the new technologies, they can significantly modify the organisation of their "global reach" and give industries the configuration of "global industries". As defined by Porter, these are industries "in which a firm's competitive position in one country is significantly affected by its position in other countries or vice versa"; the international industries cease being "merely a collection of domestic industries to become 'a series of linked domestic industries in which the rivals compete against each other on a truly worldwide basis" (Porter, 1986, p. 18).

Telematics offer firms operating in global industries the opportunity to establish a new division of functions on the basis of the radically new means now at their disposal for ensuring co-ordination and control. A typical division of functions (or configuration of value-related activities) between the parent company and its affiliates and also *among* the affiliates is shown in Table 12. Table 13 lists configuration and co-ordination issues, which are dealt with through a large degree of decentralisation or dispersion of near market activities, through much higher degrees of centralisation or concentration for manufacturing operations (coupled with a range of sourcing arrangements) and through complete centralisation of vital functions (finance, R&D, etc.). These changes have given rise to new configurations, but how they will evolve is still to be seen.

Building network firms from the ground up: some Italian examples

Firms which have succeeded in forming "intra-corporate networks" to an extent which allows identification of genuine network firms are often "*ex nihilo*" creations, formed by a domestic core firm through the grouping of previously quite small firms. They result from recognition of the special opportunities offered by telematics in combination with recent changes in the characteristics of consumer demand. These new network firms are often founded on the exploitation of production and sales capacities located within already existing SMEs.

This second road to the network form of organisation has been observed primarily in Italian firms with a strong regional base which have established varieties of networks, ranging from loose consortia to high levels of "quasi-integration", in order to market and develop new products in industries such as textiles and engineering equipment[7]. They have reaped two different kinds of economies. For small firms in rapidly evolving market-dependent industries (textiles is the classic example), there are economies of scale in developing joint solutions to technology diffusion (new process equipment), technology development (new materials, dyes, etc.), and marketing (trade fairs, export arrangements) while continuing to compete vigorously. Economies of scope have been reaped by small machinery-producing firms which band together to provide installation of medium-sized integrated manufacturing systems (assembly, handling and control equipment), well beyond the means of the individual firms. In both cases, a combination of technology development activities and operational activities (logistics, marketing and services) are tackled through small-firm networking.

Networking gains economies of scale or scope by changing the balance of activities between those carried out within the firm and those which are performed externally. This is done by creating an intermediate group of activities under the partial control of the individual firm but also with reliance on intermediate market forces (external collaborators or controlling organisers). The overall result has been the growth of organisational forms which have some of the characteristics of integration combined with decentralisation, in a few cases with shared or more collaborative control.

Networking with high levels of "quasi-integration" is found in the textile and garment complex organised by firms like Benetton (Italy) which has combined *three separate elements in one single "network"*. The first is Benetton's strong autonomous in-house capacity for design, styling, fashion market forecasting and advertising. The second is strongly decentralised manufacturing: the seven fully-owned factories account for less than 20 per cent of output; the rest is produced by over 350 small and very small enterprises and artisan shops, almost all of which existed prior to

their entry into the Benetton network. The third is a highly decentralised sales network, with a two-tier structure consisting of some 75 firms operating as agents for the group, and which gather orders, supervise and promote sales of the myriad of retail outlets set up in dozens of countries.

This third aspect is the most novel, still little known to Japanese firms which use *keiretsu* trading companies (*sogo soshas*) for their marketing. About 2 500 independent firms own the 4 200 shops that distribute the Benetton products. A large number of these firms were active prior to entering the Benetton network. Although no formal direct link exists between them and the group, the shops must follow style and display directions given by the parent company through agents. They are also obliged to order products from area agents and sell only goods with the Benetton brand name. This is not a franchising relation, since no specific charge is made for using the brand name. Shops and commercial agents are connected to the Benetton Group through the same kind of relation as exists between the company and its manufacturing sub-contractors. The centre elaborates strategy and controls the critical resources of all the autonomous firms involved in a system of which the core corporation is the leader and the principal beneficiary.

The emergence of network firms in Italy bears a direct relationship to overall Italian industrial organisation: a *strong industrial dualism,* where a few large modern multinational firms flanked by medium-sized firms with a strong regional base coexist with a very large number of small traditional firms. The network firm was initially the result of the introduction by large firms of new communication-intensive techniques in the governance of "vertical" transactions with smaller industrial sub-contractors and with retailers (as in the Benetton case). But network firms also cluster around communication channels introduced to manage "horizontal" transactions between firms of the same industry in order to achieve a better division of labour.

Quite new patterns of "functional specialisation in manufacturing and marketing" occur within firms as the network form of organisation becomes established. "Each unit resembles more and more a 'quasi-firm' which internalises a limited amount of resources/opportunities, without the constraint (e.g. the organisational and financial burden) of a fully integrated structure" (Rullani and Zanfei, 1988, p. 68). "Functional specialisation" occurs both at the core and at the periphery of the network firm. At the centre, overall planning, strategic management and monitoring of manufacturing divisions, commercial units, and the electronic network – as well as other specialised functions, notably R&D, design and advertising – are provided by a central corporate staff. At the periphery of the system, functional specialisation is much more modest e.g. the manufacturing of precisely specified components, products or sales.

New style multinational enterprises and further Japanese developments

The network form of corporate organisation is marked by an above average call on joint ventures and other forms of formal or informal inter-firm agreements. Its development appears to be more rapid and radical when it occurs in new or previously small structures than when it concerns large, well-established multinational corporations with well-established corporate traditions and strong organisational structures. Its evolution is characterised by changes in functional specialisation: increased centralisation of allocative and management functions, and new forms of linkages ("electronic quasi-integration") between the "core" or "hub" corporation and decentralised semi-autonomous and even independent production and/or commercialisation units (see Box 28).

The outcome of the process is the "new style multinational", likened by Dunning to "the central nervous system of a much larger group of interdependent but less formally governed activities, whose function is primarily to advance the global competitive strategy and position of the core organisation. This it does, not only, or even mainly, by organising its internal production and transactions in the most efficient way; or by its technology, product and marketing strategies, but by the nature and form of alliances it concludes with other firms" (Dunning 1988b, p. 327). Only a few such firms exist, but their sphere of influence is increasing. Their growth represents one of the features of the globalisation process viewed as a new phase in the organisation of international production (see Chapter 10).

Japanese research discussed during the TEP (Imai and Baba, 1991) regarding the advent of a more developed and original form of network firm, has exlored the hypothesis that further developments may be on the way, as a new, more advanced form of network firm evolves, parallel to existing forms which would retain their relative efficiencies in some industries and product areas. The new "cross-border" and "multi-layer" network form (see Figure 9) would be built on functional specialisation and would be marked by the emergence of strong regional entities and a totally new form of corporate centralisation, characterised by co-ordination rather than hierarchy.

This would entail numerous, radical departures from the area division structure adopted by other MNEs. In its simplest form, the development of cross-border multi-layer network firms would combine two apparently contradictory movements. First, it would encourage widening regional sub-networks, including many sectors and small-sized firms, as well as increasing multi-sector interactions. Second, it would move

---

Box 28.   A conceptual view of the network firm

Findings about the configuration, functional division of labour and co-ordination and control mechanisms of older MNEs moving away from earlier management approaches, coupled with the experience of the new *ex nihilo* network firms in Italy, have led to an initial theoretical generalisation for non-Japanese firms (Antonelli, 1988a). In this view, a blurring of the frontiers of firms leads to a blurring of the distinction between the "internalisation" and the "externalisation" of activities. Stress is placed on the new and more efficient means at the hands of management for co-ordination and control, leading to a new mix of decentralisation and centralisation, the overall result of which is a much more highly centralised and more closely controlled "global" multinational enterprise: in short, larger, more global, but with more flexible and efficient "hierarchies".

Telematics appear to have induced adoption of forms of "electronic quasi-integration" which make possible, as firms grow, greater internalisation of important network externalities. The downward shift of average co-ordination costs through implementation of new communication intensive organisational techniques delays attainment of the maximum efficient size and thus enables larger firms to operate efficiently. The network firm can continuously assess the vertical and lateral organisation of the productive process as a whole and evaluate each business unit, or technically separable production activity, according to the opportunity cost of market transactions, as opposed to administrative co-ordination and new forms of electronic quasi-integration.

The Italian evidence in fact suggests that the adoption of telematics and network firms makes it possible: i) to rejuvenate traditional "low-tech" manufacturing activities such as textile, garment, footwear, and obsolete tertiary sectors such as small shop retailing by introducing "high-tech" innovations and thus increasing total factor productivity without dramatic disruptions; ii) to blend traditional industrial structures based on small firms with modern trends towards customer sophistication and market segmentation, both in consumer and capital goods markets; iii) to combine the advantages of large scale regarding co-ordination, with those of small size in manufacturing customised products (Antonelli, 1988a, p. 28).

Figure 9. **Cross-border network**

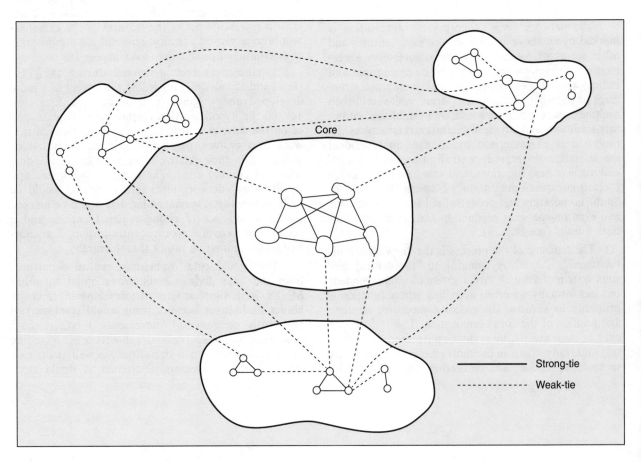

Source: Imai and Baba (1991).

towards more global co-ordination of key parameters among regions. The former moves in the direction of decentralisation and autonomy, while the latter moves towards centralised management. This network form would facilitate the development of industrial networks crossing national boundaries, through self-organising interaction among the quasi-autonomous participants.

This "true network firm" could transform the nature of hierarchies. For the intense feedback from the satellites to the core would be substituted the circulation of on-the-spot information among all the actors as a basis for control co-ordination. Crucial differences ensue: emphasis is placed on process co-ordination and team entrepreneurship rather than on planned control and a single decision maker. In addition, the information infrastructure is used to create a feedback loop within the network rather than as a means of control. In very recent work, Imai (1991) suggests that there are in fact considerable difficulties in extending multi-level learning and the chain-linked information exchange created by the *keiretsu* in their home base to global network firms.

3. THE CHALLENGES AND PROBLEMS FACING SMALL FIRMS

The organisational and managerial changes required on the part of small manufacturing firms to meet the opportunities and constraints of technological change in the context of the increased openness of domestic economies are a vital issue in many OECD countries. The positive cumulative feedback effects conducive to technological accumulation and structural competitiveness (see Chapter 11) which occur at the level of *industry clusters* and of regional or local *sites* can never be founded solely on the existence of large firms. Such processes require a much wider industrial base, marked by the presence of small firms, both in an independent capacity and in a range of inter-firm relationships with the larger corporations.

The tendency towards industrial concentration appears to have been clearly reversed since the mid-1970s, as witnessed by the increased numbers of industrial SMEs and other industrial *establishments* (which may be part of a larger firm) having fewer than 100 employees[8]. This trend is attributable, to a large extent, to economic and technological developments that have reduced some earlier barriers related to scale economies and are well suited to the flexibility of the smaller organisation (Acs and Audretsch, 1990). A long-term strengthening of the place of SMEs in domestic industrial structures is not necessarily guaranteed by these developments. Small manufacturing firms are faced with a situation where the new opportunities offered by the fall in production thresholds due to economies of scale are offset by the stringent requirements of new technologies with respect to the level and sophistication of intangible investments and the marketing of products on a customised basis.

Today small firms are in a paradoxical situation. Technological developments have rarely seemed to be so favourable to small size production, yet small firms remain highly vulnerable and are often in obvious need of assistance. This paradox deems further exploration. A closer look at the advantages and disadvantages of small size in relation to the new technologies may help to explain why the nature and attributes of the linkages experienced by SMEs with other firms and institutions will often determine their long-term performance and their capacity to stay in operation. The approach may also offer a sound foundation for public policies addressing the viability problems of industrial SMEs. Work focusing on organisational solutions for SMEs in relation to the new technologies has not received the same support as have studies addressing similar questions in large corporations. The OECD has recently begun work which should give studies in this area an international basis.

New technology and the advantages and disadvantages of small size

As a result of the diffusion of micro-electronics within manufacturing, smaller product series have become increasingly competitive and allowed a larger product variety than before. In the late 1970s and even more so in the 1980s, a rapid diffusion of micro-computer technologies in production took place in the form of computer-aided design, computer-aided manufacturing, robotics, artificial intelligence applications and flexible manufacturing systems. This lowers the overall set-up costs considerably and thus enables the producer to switch more often from one product to another. Increasingly it seems that efficiency and variety may no longer be rival objectives. In a number of industries, ranging from clothing and leather products to steel and metal products, it becomes increasingly profitable to produce small batches with a wide spectrum of combinations of properties.

In the materials industry, in its widest sense the emergence of the "new materials" (see OECD 1990g) is part of the same trend. While some adaptation in the materials used has always been necessary, it was usual practice to choose the material whose features best suited the most important technical requirement and then took the other features of that material as constraints upon design and processing. Now the materials choice itself has become increasingly an endogenous design variable, which can be subjected to computer programming. This has become possible due to both a growing understanding of the microscopic properties of matter and the use of new computer-controlled technologies. The small scale but high value-added character of new materials manufacturing is why the spectrum of firms operating in this area includes a large number of very active small firms.

Firms working with advanced technologies in areas where technology is evolving rapidly, as in most sectors affected by IT and in new materials, must be capable of keeping abreast of technological developments. The *knowledge intensity of production* does not necessarily imply the capacity of every firm to carry out in-house R&D, but it certainly requires firms to belong to one or several networks where R&D is being done and *technological information* is being processed and circulated. Equally, the range and variety of intangible investments (see Chapter 5) which the small industrial firm must be capable of financing are the same as those of larger firms. However, the financial and human resources of the small firm are often too small to do this or to establish on its own the range of external sourcing relationships which large network corporations are now setting up (see Box 28).

Small-scale production based on the new technologies also makes severe demands on firms with respect to marketing. In batch production, in mechanical engineering for instance, the loss of scale economies must be compensated by economies of scope arising from the possibility of using the same capital equipment to manufacture a broad range of different products. If this does not occur unit costs of production will rise steeply. But manufacturing a broad range of products implies *marketing* these goods. Another dimension of what Willinger and Zuscovitch (1988) have named the "informational viability problems" faced by small firms, pertains to the information gathering and other marketing costs incurred by selling small quantities of output on "micro-markets". This now applies to a growing number of industries. In a "production régime" based increasingly on variety and customised demand (Cohendet *et al.*, 1991) marketing costs cease to be fixed or fairly fixed costs and can become a significant burden on profitability. Collecting and processing appropriate information about micro-markets and then acting on this information is a costly business for SMEs. Obviously this is a major reason why small manufacturing firms in textile and clothing may see many advantages in entering a network such as that of Benetton.

Different settings for the activities of SME

A key to understanding the wide variance in the viability and performance of manufacturing SMEs lies in the nature and quality of their relationships and ties with other firms and public and private sector institutions and the extent to which these links offer these firms adequate ways of overcoming the disadvantages defined above.

The range of possibilities is considerable. Leaving aside for the moment the support SMEs may receive from government, and other public authorities (see below), they include:

– arrangements made by firms (possibly with the help of municipalities and local banks) on some sort of co-operation basis for the creation of communal support systems, in particular information networks regarding markets and technology (this was the aim of the Italian industrial districts: Becattini, 1987);
– the emergence in the vicinity of manufacturing sites either of business service firms (which may themselves be small or medium sized) or the antennae of larger technical institutions, capable of compensating for the lack of in-house expertise by industrial SMEs and supplying them with specialised business consultancy (Peyrache, 1990) as well as with necessary linkages to specialised information networks, on a basis and at a price compatible with the financial resources of these firms;

– long-term, stable contractual relationships with large firms of a type which relieves the small manufacturing firm of the need to continually search for markets while at the same time offering such a firm the means not only of surviving but of progressing technologically and improving its human capital resources and capacity building (see Chapters 5 and 7).

These possibilities are of course not in any way mutually exclusive. On the contrary, within regional environments one may see SMEs enjoying the benefits of all three developments. In many cases, however, even when SMEs can count on a satisfactory range of institutional linkages, the key factor in their environment will be their relationship with large firms.

Small firms and large firms

Often the viability of the small firm will depend on the nature of this relationship and on the strategies adopted by large partners (see Dertouzos *et al.*, 1989). Large firms have a strong position. Micro-electronic information and control technologies, and the new corporate management paradigm offer large corporations solutions to many of their organisational problems. They combine high levels of concentration/centralisation of capital with a much less vertically integrated approach to production. In the mechanical and electrical engineering industries, the end-result after nearly twenty years of experience in Japan and eight to ten years in other countries, is *flexible or differentiated mass production.* As Sciberras and Payne (1985) and Watanabe (1983) foresaw, on the basis of careful case study analysis of the effects of the diffusion of numerically-controlled machine-tools, the new management paradigm helps large firms reconcile the imperatives of variety, flexibility and economies of scope, while still benefiting from scale economies obtained through the standardisation of component and part production produced in decentralised firms and plants. In industries where "just-in-time" production prevails (as in automobiles) the initiative in partnerships lies firmly in the hands of the large firms and places the small ones in a position that has been defined as "quasi-integration" (Leborgne and Lipietz, 1989).

In the area of consumer goods, differentiated, or indeed *customised,* mass-production controlled by large firms (as in the case of Benetton discussed in Section 2) is also gaining ground. All this is rather different from the model announced by Piore and Sabel[9] or by observers of the Italian industrial districts (Bagnasco, 1977; Becattini, 1987; and Brusco, 1982). These studies presented micro-electronics based manufacturing as offering a major opening for really independent small firms and a much higher degree of industrial democracy which could bring in its wake a "new industrial divide". Some measure of flexible specialisation has occurred but not on the scale predicted nor in ways which have reduced the economic vulnerability of SMEs significantly.

In the last fifteen years, a shift from the earlier forms of sub-contracting to a new form of industrial partnership between small and large firms may have started to occur (European Commission, 1989). However, the extent of the shift and the precise content of the new partnership remain to be assessed carefully. In some industries there may really be a change in the quality of partnerships. In others, the main and perhaps sole changes for SMEs may simply be in the nature of the status offered them as subordinate entities within the wider "network" corporate structures discussed above.

Public support for SMEs

For the above reasons, government support for the management of changes in technology and demand should concentrate on encouraging the effective transfer and assimilation of technological and market information. This is also increasingly true for support from regional and local authorities. Technological information is costly and requires a long-term view. Larger companies obtain it via their research centres and other specialist services. SMEs have greater difficulty in creating, obtaining and evaluating such information because their resources are limited. Technological information may derive from a variety of sources, including the SME's own internal R&D potential but also public research centres, universities, and other public and private agencies. Various networks such as technology parks, technopoles, technology co-operation networks etc. can help to disseminate it. Supplementary resources are required to enable firms to integrate it in their short and long-term strategies.

Ongoing OECD work suggests that, in most countries, after a spate of measures and programmes to help the SMEs, governments are increasingly concerned with the development of success strategies by the SMEs, and are seeking to define policies based on market-efficiency criteria. A number of institutions and programmes have been set up to facilitate the transfer of technological information and maintain the SMEs' competitiveness. In some countries it is even possible to speak of an excessive number of programmes. These may now require harmonisation and re-organisation to increase effectiveness and economic efficiency. Since SMEs are particularly vulnerable in

the start-up and structural transformation phases, programmes in support of SME competitiveness should be directed at these phases.

For the reasons just discussed, high priority should also be given to measures aimed at improving the capacity of management and the firm's relationship with its environment. This relationship increasingly depends on *networks* based on a partnership (large firms, suppliers, distributors, etc.) expertise (consultancy services, public research establishment, universities) and the support of national, regional and local authorities. Government policies which are general and "across-the-board" in their conception may not be very effective to the heterogeneity and diversity of situations. Programmes that have well-defined objectives and were adjusted to a firm's specific environment should now preferably be geared to regional and local intermediate structures.

The work recently undertaken at the OECD underlines the value and importance of these intermediate structures – or interface bodies or "brokers" – for the transfer of technological information. The latter include contract research organisations (see Chapter 1) and units based in universities and public research centres. These "brokers" should enable the SMEs to find the "right" technological sources and the additional resources required for their adjustment. Government support programmes for these operations can be particularly effective. To quote a few examples, such is the case of the OTTO programmes in the United States, the CRITTs in France, the local technological institutes in Japan and the centres specialising in technological information in Quebec.

Finally, there is the question of finance and the particular responsibility of national and local banking institutions (see also Chapter 11, Section 3). Financing is a decisive element of SME development. They are frequently in an unfavourable position with regard to debt and equity capital and access to financial markets and this affects their investment capacity. There are several basic financial factors which contribute to their creation, start-up and development, their ability to adapt to changing markets and capacity to keep ahead of the increasingly rapid changes imposed by the economic environment and technological progress. These include the amount of financial resources available to SMEs and the nature and quality of the financial instruments and institutions in question. Governmental fiscal support policies for SMEs will be successful only to the extent that banks also accept to assume special responsibility for these firms.

4. SOME POLICY CONCLUSIONS

The organisational shifts towards the network form of corporate organisation, now taking place in a growing number of OECD countries through different routes and on the basis of different historical legacies, attempt to meet the challenges and opportunities discussed throughout this report. They can be described in terms of the relative efficiencies and constraints of both the internal and external markets of firms. Firms seek to achieve economies of scale and scope by better organisation of the external sourcing of key resources to production (notably technology). Coupled with this are the relative efficiencies to be found in highly centralised hierarchical structures versus those found in more decentralised structures with greater competition between entities (Grandstrand *et al.*, 1990). The returns to be gained from the adoption of such forms are particularly significant for the international production carried out by MNEs.

In the case of large corporations, governments can adopt the position that they can fend for themselves and that there is no particular role for governments to play in their transition to the new organisational paradigms. This is only true within limits. Change in corporate organisation represents one important dimension of the overall process of social change associated with technological change. Change within the corporation cannot succeed completely independently of other components of change at the national level, and corporations themselves may be tempted to take a narrow and short-term view of the issues and the ways to tackle them.

The conclusions of research undertaken at Wisconsin University on corporate governance within US industry point to the systemic dimension of corporate change, its implications for competitiveness, productivity and growth and some of the tasks to be confronted with other social partners and government. The argument made is that: "Social system of production operating in volatile markets with rapidly changing and highly complex technologies requires a different type of work organisation, work skills, system of control, relations with suppliers, and competitors from a system when markets are stable and technologies are less complex and relatively unchanging. Systems capable of producing diversified, high quality products require high levels of social peace and redundant investments in the skills of employees which are likely to occur only if the labour force has high levels of job security. To compete effectively in international markets, a diversified quality system of production requires co-operation relations with one's competitors, one with a highly developed association system, well developed networks, communitarian obligations and collective resources that *individual actors cannot normally generate alone,* even if they recognise their importance. In short, high levels of growth and productivity in manufacturing sectors with high quality products and volatile markets will not be enhanced by an environment which has a low level of institutional developments in unions, business associations and promotional networks." (Hollingsworth, 1990, p. 24).

The issue is not an academic one, but has consequences for the stability of international relations. Differences in capacity to monitor change in corporate organisation and related social changes many have strong effects on the structural competitiveness (see Chapter 10). Observers may be led to define as "competition between systems" (Ostry, 1990*a*), situations which really reflect differences in the success of countries in identifying and confronting the organisational and social dimensions of technical change.

With respect to SMEs, the public policy implications of the analysis have been abundantly spelt out in Section 3. The only point to be stressed in conclusion is that the performance of SMEs is highly dependent on the economic and institutional setting of their activity (see Section 3) and that the most effective policies may in fact be those which address this question.

NOTES

1. The term "Toyota Production System" was chosen by the second report of the MIT International Motor Vehicle Programme. See Womack *et al.* (1990). On the notion of toyotism at OECD, see in particular Boyer (forthcoming).

2. The choice of these two broad processes reflects the focus of the TEP Conferences held in Member countries (for instance those in Helsinki, Stockholm and Tokyo) which paid particular attention to the emergence of new forms of work organisation and management practices within larger firms and to new forms of corporate organisation involving different concepts of corporate frontiers or boundaries, in particular for multinational enterprises (MNEs). The proceedings of the Stockholm Conference are available in Deiaco, *et al.* (1990).

3. A caveat may be in order. As Porter (1990*b*, p. 4) has pointed out, an exaggerated focus on the American management paradigm in past decades and on the Japanese model since the 1980s carries the danger of overlooking a variety of other less extensively used management approaches (relating in particular to small and medium enterprises and industries with strong customer-specific traits) which have emerged within some national contexts and can be identified as representing a source of structural competitiveness (see Chapter 11). With respect to the organisation of work in manufacturing and the management of supplies and stocks, it does seem, however, as argued by Womack *et al.* (1990), that a totally new approach has emerged within the "scale-intensive" mechanical and electrical engineering industries and is likely to receive wide application.

4. Evidence for this assertion can be found in the extremely conservative reactions of many of the major United States and European firms which operated up to the mid-1970s within very strong domestic oligopolist structures. These acted as effective entry barriers, both in themselves and with additional protection from trade policy. In steel, consumer electronics and automobiles, but also in food processing and many other industries, firms reacted to declines in productivity growth and profits as if they were due to purely cyclical characteristics. They thus employed management policies founded on their previous international domination and market power: price mark-ups, minor product innovations, larger outlays for advertising and calls on governments for retaliatory trade measures of various sorts against "unfair competitive practices" were the main responses of many large United States and European firms to the failure of their management paradigm and to Japanese competition.

5. Economies of scope exist where activities are interrelated horizontally by technological proximity, not necessarily by industry. For example, the production of motorcycles, agricultural equipment and marine equipment is linked by the application of powerful light-weight engines, and there are economies of scope to be derived from developing products and processes for all three applications. Other important technological developments of the 1980s such as the fusion of information technologies and composite materials technology have simultaneously opened up new vistas for "economies of scope" and new opportunities for "variety" (Willinger and Zuscovitch, 1988). Technological developments have also strengthened the need for firms to establish closer relationships with users and customers and incorporate highly differentiated and flexible potential demand into corporate planning.

6. For overviews, see the classic studies by Stopford and Wells (1972); Michalet (1985); and Caves (1982).

7. The information in this section is drawn from Rullani and Zanfei (1988).

8. The available data is set out and discussed in a systematic manner in Segenberger *et al.* (1990).

9. Piore and Sabel (1984). For a critique, see Leborgne (1987); Coriat (1990); and Cainarca *et al.* (1990).

Chapter 5

THE GROWTH AND MANAGEMENT
OF INTANGIBLE INVESTMENT

ABSTRACT

Intangible investments are growing rapidly and, by some measures, industrial intangible investment now exceeds physical investment. Intangible investments and their complementarities with physical investment are increasingly recognised as key determinants of competitiveness, growth and productivity.

Intangible investments are divided into four groups. "Intangible investments in technology" develop the knowledge and competence base to introduce new products and processes (R&D including software R&D, technology acquisition, design, engineering, and scanning and search activities). Investments in human resources, organisation and the information structure can be grouped together as "enabling intangible investments". Market exploration and market organisation are treated separately, as is software (particularly when it is integrated in equipment).

Intangible investment has been driven by management recognition that technology, skills and organisation determine competitiveness, and corporate strategies and organisation are being built on greater inputs of intangible investments. Structural changes in industry are also contributing. High rates of investment are being undertaken in advanced computer-controlled equipment and high technology industries which require high levels of innovation, skill and information intensive. Rationalisation and replacement investment in manufacturing is growing, accompanied by declining investment in plant and structures.

The share of R&D in firm-level activities and GDP has increased. Moreover, the share of R&D financed by industry has risen substantially. Firms also make more use of technology from other sources.

The growth of enabling intangible investments is difficult to analyse because of lack of data. With technology and equipment requiring new skills, firm-based training and skill formation are more important. Training is concentrated in large firms, high technology industries and at higher skill levels. Shortages of information technology personnel and engineers persist in many countries, as do widespread shortages of general training, particularly in small firms.

Work on the production floor and relations between firms are being transformed, placing a competitive premium on appropriate organisation. Market exploration gains importance in a period of shortening product cycles and greater competition. Finally, software is an important factor of competitiveness. Software embedded in products is increasing, as are software intensive computer-controlled manufacturing processes, quality control, testing, storage, handling, sales and delivery.

Although intangible investments have increased, they are still not measured adequately or comparably across countries. Harmonisation of definitions, accounting conventions and data collection are important for policy development. The balance between intangible and physical investment, incentives and disincentives to different investments and the necessary complementarities between them are crucial areas for policy.

INTRODUCTION

The relative weight accorded to intangible and to physical investment has now become an important issue in corporate and public decision-making; this factor also affects economic performance. Intangible investment covers all long-term outlays by firms aimed at increasing future performance other than by the purchase of fixed assets. In addition to investments in technology (R&D expenditures or the purchase of its results), it includes investments in training, work organisation, labour relations, management structures, the formation of technological and commercial links with other firms and with suppliers and customers, market exploration and the acquisition and exploitation of software.

Between the more traditional investment in physical assets and intangible investment there is a complementary relationship with at least two dimensions. One concerns the social and organisational conditions for growth. Intangible investments play a crucial role in the realisation of the economic potential of technological change by creating a conducive "socio-economic" context[1]. Investments in marketing or in the management of scientific and technological information, for example, are an important element in the interaction of demand and supply for goods and services. They develop the necessary information structures that help identify market signals and transform them into instructions for the deployment of productive resources. These same structures are also used to stimulate consumer response to innovative efforts by firms.

A second dimension of this complementary relationship between physical and intangible investment relates to the creation of productive resources and lies in fact at the source of the growth process itself. The "new" theories of economic growth developed in recent years focus on the importance for long-term growth rates of generating both physical and intangible investments[2]. They suggest that the process of undertaking investment itself creates and reveals further opportunities. Thus, growth depends on the accumulation of a stock of knowledge and competences, itself the outcome of physical and intangible investments whose social returns (as discussed in Chapters 2 and 3) tend to be higher than the private returns captured by individual investors.

1. TOWARDS A BROADER DEFINITION OF INVESTMENT

Complementarities and interrelatedness among different kinds of investment raise a range of strategic, conceptual and measurement issues. For firms, expenditures for tangible and intangible assets and their organisation and management to meet strategic goals are of key importance for survival and growth. But these expenditures may not be reflected adequately, or even included separately, in project funding, company decision structures or company accounts (see Box 29). For analysts and policy-makers, industrial statistics and national accounts have not kept pace with technological change, and better data are needed on the growth and structure of intangible investments and the shifting complementarities between intangible and physical assets.

The key role of investment in driving the results of technical progress into use thus demands a wider definition of investment to encompass both:

- intangible investment (R&D, design and engineering, patents and licences, training and human capital formation, organisation of production and labour relations, market exploration, software); and
- tangible investment in physical assets (machinery, plant, buildings).

Although there has been an upsurge of interest in intangible investment, no consensus has been reached on a full list or a system of categorisation, and few studies cover all of R&D, training, marketing and software.

Figure 10 attempts fuller coverage (also see the list in Box 29). A first set of intangible investment activities can be described as "intangible investments in technology". They develop the knowledge and competence base to introduce new products and processes (R&D, including software R&D, technology acquisition via licences and patents, design, engineering, and scanning and search activities to keep up with competitors).

Other investments, in human resources, organisation and the information structure, are the major part of "enabling intangible investments". They are essential for the success of investment in fixed assets. For example, the effective integration of isolated pieces of manufacturing equipment into a computer-aided manufacturing system relies on a range of complementary investments in this area (Rush and Bessant, 1990).

Market exploration and market organisation are shown separately, as they are usually of a different kind and involve different actors and different techniques than do other intangible investments. However, this differentiation does not always hold, as market testing and market research are closely related to some intangible investments in technology (design and engineering) and to some enabling investments (building information structures).

Software is a separately identifiable fixed asset. It may be integrated in equipment or be an essential adjunct to it and is thus shown alongside physical investment in Figure 10[3]. Information systems in firms may also be set up independently of hardware investments, and they may be classified as own-account intangible investments for in-house use. They result in intangible assets identified separately from software production and included with investments in organisation.

Figure 11 gives a dynamic picture of the structure of intangible and physical investments during preparation and commercialisation of a new product. It does not attempt to show the interaction between market opportunities, technological capabilities and management strategies (see Figure 1 in Chapter 1) and understates the importance of market opportunities and market signals in driving innovation. However, it shows clearly the shifting importance of different kinds of investment as products and processes are developed, tested, introduced, and superseded.

– During the early part of the product life cycle, intangible investments in technology (R&D, design and engineering, and patent and licence work) predominate, driven by market opportunities and market exploration.

Box 29. Corporate accounting and the changing structure of investment

The changing balance between physical and intangible investment has stimulated a great deal of interest in the way in which firms should account for investment in intangible assets. In accounting terms, physical capital is essentially long-lived (it can include directly incorporated software). Intangible capital consists of long-lived rights or assets without physical substance, with future economic benefits which the entity can control (for example a licence arrangement). Despite conceptual and practical difficulties involved in defining intangibles and adapting accounting rules to take them into account more adequately, the economic importance of intangible investment for the performance and competitiveness of enterprises means that they will be progressively absorbed into accounting systems. The OECD suggests that an intangible asset should be recognised i) when it is separable, that is, it can be separated from all other assets without compromising the activities of the enterprise; and ii) if its value can be determined by either its purchase cost, or by allocation of part of a global cost, or by its production cost to the enterprise.

An indicative list of intangible assets includes:

- Research and development expenditure;
- Know-how;
- Industrial patterns and design;
- Patents, licences;
- Artistic creations, copyright;
- Right to receive royalty payment;
- Training and other investment in human resources;
- Market share;
- Product certification;
- Customer lists, subscriber lists and lists of potential customers;
- Product brands and service brands;
- Software and similar products.

Figure 10. Classification of intangible and tangible investments

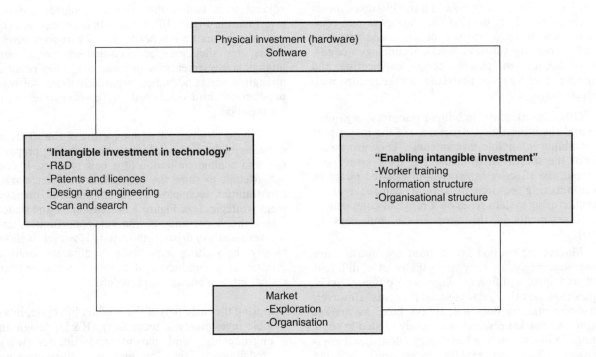

Physical investment (hardware)
Software

"Intangible investment in technology"
-R&D
-Patents and licences
-Design and engineering
-Scan and search

"Enabling intangible investment"
-Worker training
-Information structure
-Organisational structure

Market
-Exploration
-Organisation

Source: OECD, see text.

Figure 11. Distribution of intangible and tangible investments over the product life cycle

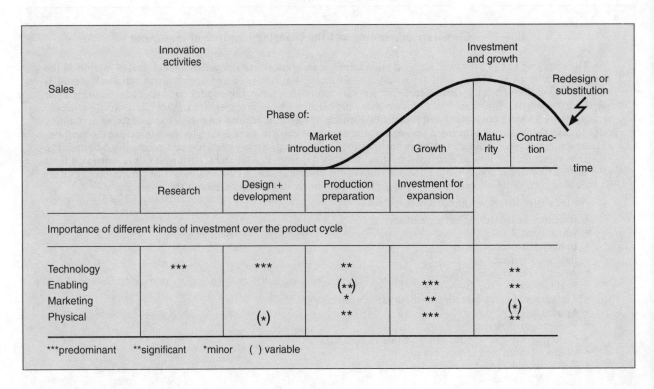

Source: OECD, adapted from Scholz (1990).

- As new products near the launch phase, intangible investments in technology continue, but market testing and market development increase. Worker training and the beginning of organisation of production (enabling intangible investment) accompany process innovation and physical investment in new equipment.
- In the market growth phase, capacity investment and expansion are major investments. Extensive worker training and organisation of production and distribution will be accompanied by marketing and advertising associated with large-scale production.
- Finally, as the product matures and markets are saturated, efforts to adapt and renew the product may be made, accompanied by further training, work re-organisation and investments in plant and equipment for rationalisation and modernisation.

2. INVESTMENT IN INTANGIBLES: STRUCTURE AND LINKS WITH INNOVATION

Sources of statistical data for investment in intangibles by industry are still very spotty. Few surveys have been undertaken and it is rare that more than one or two of the categories mentioned above are covered. Further information can be obtained, directly or indirectly, from innovation surveys, which are being more widely undertaken and are becoming more comparable, despite differences in definitions and coverage (OECD, 1989c). They include information on: the cost of the various activities leading up to success/first marketing; factors which influence the decision to innovate; and the effects of the innovation on production activities (see Chapter 1). Coverage varies from that for total intangibles in the following ways:

- Innovation expenditures only include intangible investment made in connection with the introduction of new products and processes. In the case of Finland, which has undertaken both innovation and intangible surveys, only around half of intangible investment is undertaken in connection with innovation (Figure 12).
- The non-innovative intangible investment is mainly in training and marketing.
- Innovation expenditures include all investment in fixed capital required to introduce a new product or process. In Finland this is about half the total; this is also true in Germany and Italy.

Finland is one country that has undertaken a relatively comprehensive survey of investment in intangibles for 1987 (Finland, Central Statistical Office, 1989). Overall, the survey showed that physical investment accounted for almost three quarters (72 per cent) of total physical and intangibles investment. The high rate of tangible to intangible investment is probably atypical and reflects the high share of industry operating in capital-intensive industries (pulp and paper, metals). Within intangible investment, Mk 3.3 billion were devoted to R&D, Mk 1.8 billion to marketing, Mk 1.2 billion to education and training and Mk 1.9 billion to other intangible investment.

Strong investment in "other intangibles" has complemented Finland's low but rising R&D expenditures to underpin the country's good growth and productivity performance during the 1980s. The survey further showed that intangible investment was concentrated (72 per cent) in large enterprises. R&D was the most important category of expenditure overall and in the first two size groups (firms of over 100 employees). Small firms spent more on marketing and "other intangible investments" (organisation, goodwill, patents, subscription to data bases, remuneration for innovative ideas, cost of exploring for natural resources, other human resource development, etc.).

Innovation surveys have been more widely undertaken. They have now been made in about half of the OECD Member countries, and they have been carried out regularly since the late 1970s by the IFO Institute in Germany (see Table 14)[4]. As might be expected, R&D and design and engineering are the two most important categories of intangible investment in innovation, with R&D generally the larger. (This finding is confirmed by results of other surveys, e.g. in the Nordic countries.) The Italian concentration on design is no surprise.

Table 14. **Distribution of innovation expenditures Selected survey results**

Percentage distribution

	R&D	Marketing	Design/ engineering	Patents Licences	Physical investment
Germany (1988)	26	4	22	2	46[1]
Finland (1988)	39	4.5	6.5[2]	4	46
Italy (1981-85)	18	5.5	25[3]	..	51.5

Definitions and expenditure groups are broadly comparable but not identical.
1. Process development, 28 per cent; production preparation, 18 per cent; including labour training, reorganisation and pilot production, 3 per cent.
2. Including training and organisational development.
3. Including patents and licences.
Source: See note 6.

Figure 12. **Innovation and investment expenditures in Finland, 1987/88**

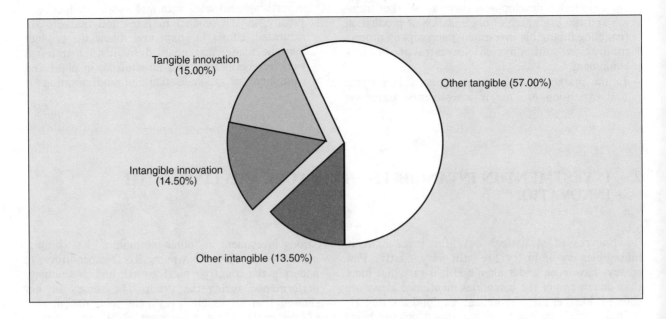

Tangible innovation
(15.00%)

Other tangible (57.00%)

Intangible innovation
(14.50%)

Other intangible (13.50%)

Source: Adapted from Finland Central Statistical Office.

Further evidence must be gleaned from measures of intangible investment compiled by summing expenditures data for the component categories taken from different surveys. The relative importance of training is shown in provisional data for the United Kingdom in Table 15. These data show that manufacturing training expenditures are almost two-thirds of R&D expenditures and one-third of gross domestic fixed capital formation (GDFCF). For all of the economy, training expenditures from all sources are equivalent to almost one-half of GDFCF, further evidence of the importance of intangible investments relative to physical investment. In 1987, Swedish manufacturing investments in intangibles (R&D, advertising, software) were greater than manufacturing GDFCF. In the engineering industries, intangible investments were almost twice as large as physical investment. Firm-based training and organisation (enabling investments) and innovation-related engineering and design have also grown strongly, but recent data on these intangibles are unavailable[5].

Table 15. **Expenditures on intangible assets: United Kingdom, 1987**

Percentage of value added

	Manufacturing	GDP
R&D	6.3	2.3
Training	3.9	8.0
Advertising[1]	2.4	1.4
Total	12.6	11.7
For comparison:		
Gross domestic fixed capital formation	11.5	17.6

1. All advertising, including advertising of imported goods.
Source: OECD, drawn from data supplied by United Kingdom Department of Trade and Industry. All expenditures included. For example, total training expenditures include funding by public and private sectors.

3. COMPLEMENTARITIES BETWEEN INTANGIBLE AND PHYSICAL INVESTMENT

Linkages and interactions between physical and intangible investment drive total investment and productivity growth. Distinctions are often made between training and physical investment and between marketing and R&D, because of the division of functional responsibilities in many firms and government departments, not because they are not interlinked. This section discusses evolving complementarities among different investments and makes some comparisons based on the product cycle relations presented in Figure 11.

If the full value of investments in new equipment is to be gained, then physical and intangible investments should be closely linked. In-firm training and investments in the re-organisation of work and in software should accompany physical investment at firm level, to ensure that equipment is used effectively and that the productivity potential of the equipment is reaped (see Chapters 4 and 7). In the absence of detailed firm-level data to investigate the sequence of activities which lead to productivity growth, sectoral data can be used to examine the relations between some kinds of investment.

For three countries with published detailed sectoral data on training by firms (France, Japan and the Netherlands), the following statistical results are obtained (See Box 30):

- sectoral training activity is closely related to physical investment (industries which have high expenditures or participation in training have relatively high expenditures on physical investment);
- sectoral training activity is more closely related to sectoral R&D expenditures than it is to physical investment (industries which spend relatively more on R&D also train their employees more);
- the relations between physical investment and R&D are less consistent;
- the relation between prior growth and sectoral training activity is strong in the Netherlands but shows no relation between previous sectoral growth in output and the pattern of in-firm training in either France or Japan.

The results show a high level of complementarity between training activities and physical investment in manufacturing, and they are consistent with the product life cycle distribution of investment presented in Figure 11. Industrial sectors – such as electrical machinery and electronics, non-ferrous metals, chemicals and petroleum-refining and automobiles – which invest most in physical assets (principally machinery and equipment) are also the industries which make the greatest efforts to train their staff in order to reap the full benefits of physical investment. This group of industries includes flow process industries and assembly industries which produce a range of investment, intermediate and consumer goods. High levels of training coupled with physical investment thus do not appear as the exclusive domain of industries with a unique set of characteristics. On the other hand, complementarities are not uniformly strong across all manufacturing industries. Some, such as pharmaceuticals and precision instruments (professional, scientific, measuring and controlling equipment), have relatively high training expenditures but relatively low investments in physical assets.

Complementarity also exists between firm-based training activities and industrial R&D expenditures. This suggests that firms attach a balanced importance to both areas. In France, where government finance of industrial R&D is relatively high in some strategic sectors, the importance of a firm-based decision-making and financing process is underscored by the fact that the correlation between company-financed training and company-financed R&D is stronger than the correlation between company-financed training and R&D expenditures financed from all sources, including government. Chemicals and pharmaceuticals, precision instruments, electrical machinery, electronics, transport and non-ferrous metals all show relatively high levels of both training and R&D activities. These industries are not characterised by a particular kind of manufacturing; they cover investment-intensive industries, skilled labour-intensive industries and scale-intensive process industries. Thus, the driving force for competitiveness appears to reside in a highly and broadly skilled work force (to develop and produce manufactured products), combined with substantial innovation and product development (to extend the product range and production technology).

The weaker relations observed between R&D and physical investment in the same year are to be expected, as R&D does not necessarily lead to physical investment. Even when R&D activities are directly aimed at developing new industrial products and processes, the chain of events between R&D and investment is long. Many other intangible investments, such as design and production preparation and marketing, intervene before the launching of new products or the installation of new processes. In addition, R&D is often an exploratory or scanning activity designed to provide information to management. A relatively high degree of improvement activities in R&D similarly does not lead to new products. Furthermore, countries with high levels of government defence and large-scale research projects predictably have fewer direct links between research activities and industrial investment. This helps explain weaker links between R&D and investment in France than in Japan and the

Box 30.   Training, R&D and physical investment in manufacturing: statistical results

Training, R&D and physical investment in the same year

France (1986):   Spearman rank correlation

Training: gross wages / Investment: value added[a]	0.386 (N = 15)
Training: gross wages / Total R&D: value added	0.671
Training: gross wages / Company-financed R&D: value added	0.786
Investment: value added / Total R&D: value added	0.104
Investment: value added / Company-financed R&D: value added	0.218

Japan (1985)[b]:

Training: labour cost / Investment: value added	0.529 (N = 18)
Training: labour cost / R&D: value added	0.628
Investment: value added / R&D: value added	0.393

Netherlands (1986):

Trainees: employees / Investment: value added	0.729 (N = 16)
Trainees: employees / Machinery & equipment: value added	0.688
Trainees: employees / R&D: value added (1985)	0.821 (N = 7)

Training compared with 1979-85 manufacturing growth rates

France:

Training: gross wages (1986) / Growth value added (1980 prices)	0.020 (N = 14)
Training: gross wages / Growth production (1980 base)	0.129 (N = 13)

Japan:

Training: labour cost (1985) / Growth production	–0.055 (N = 13)

Netherlands:

Trainees: employees (1986) / Growth production	0.800 (N = 11)

Note: N = number of manufacturing sectors.

a) Different aggregations

Training: gross wages / Investment: value added	0.450 (N = 15)

b) Different aggregations

Training : labour cost / Investment: value added	0.430 (N = 20)
Training: labour cost / R&D: production	0.462 (N = 16)
Investment: value added / R&D: value added	0.499 (N = 23)

Data for statistical testing was drawn from the following sources:

France: Training: Expenditures by firms on professional training as a percentage of gross wages. On-the-job training expenditures not included. CEREQ, *Statistique de la Formation Professionnelle Continue Financée par les Entreprises,* various issues, particularly December 1988.

R&D: La Documentation Française, *Recherche et Développement dans les Entreprises,* various issues, particularly Résultats 1986.

Japan: Training: Cost of education and vocational training calculated by the OECD as a percentage of average monthly labour cost per regular employee. Employee wages while training and on-the-job training expenditures not included. Ministry of Labour, *Yearbook of Labour Statistics,* various issues, particularly Section IV, "Welfare Provisions", 1986.

R&D: Statistics Bureau, Management and Co-ordination Agency, *Report on the Survey of Research and Development,* No. 61, 1986.

Netherlands: Training: Participants in company training and external training as a percentage of the number of employees by branch of activity, Centraal bureau voor de statistiek, "Bedrijfsopleidingen in Nederland: 1986", Voorburg/Heerlen, October 1988.

For all other data the following sources were used:

Investment, machinery and equipment investment, production, value added and employment are all from the OECD/ISIS Data Base, published by OECD, *Industrial Structure Statistics,* various issues. The growth in production is taken from *OECD/Indicators of Industrial Activity.*

R&D from the OECD/STIID Data Base, except some data for France and Japan from original sources cited above.

Original data sets for training, R&D and investment for the same country were not directly comparable, so *ad hoc* conversion keys were drawn up to aggregate data into comparable sets across the three variables (training, R&D, physical investment) for as many manufacturing sectors as possible.

Netherlands. And for many industries, weak relations are due to the fact that technological spillovers spur investment (i.e. new textile machinery is developed by the machinery industry, not the textile industry, and new composite materials used in furniture manufacture do not usually originate in that industry).

Finally, the relatively poor relation between preceding sectoral growth and firm-based training can be explained in two ways. First, the growth period (1979-85) was marked by slackening activity in energy-intensive (metals-refining) and petroleum-related industries. Both of these industries continued to devote large efforts to training to ensure the smooth operations of capital-intensive process plants. On the other hand, growth was moderately good in some industries which did not subsequently train extensively. So, for many industries, there was a divergence between the relative growth path and training efforts. Second, the period in the mid-1980s for which detailed training data is available was marked by high unemployment. The shift of resources to provide a better match between growth and internal training may have been delayed due to the external availability of skilled labour requiring little training.

In France and Japan, other long-term factors influence firm-based training and help insulate it from short-term economic performance. In France, a tax incentive has focused business attention on the importance of training. Small firms and lagging industries are encouraged to increase their training efforts and as a result, differences between large and small firms and between leading and lagging industries in this area are reduced. In Japan, a long-term focus on human resource management and improving skills at all levels and in all industries has made training expenditures less sensitive to economic fluctuations and to intersectoral differences in effort.

Overall, the results suggest that, at sector level, there are important complementarities between intangible investments in firm-based training and expenditures on research and development and on physical investment. They also suggest the increasing importance that firms attach to the interrelations between different kinds of investment. Further investigations must be undertaken at firm level to disentangle these relations and to examine how they have changed over time. At sector level, the aggregation problem (hazy boundaries between industries, industrial activities with different characteristics classified together, firm-specific and size-specific differences) leaves little to be gained by more detailed examination of the small amount of available sectorally disaggregated firm-based training data.

4. INVESTMENT IN INTANGIBLES: GENERAL TRENDS

Intangible investments in R&D, technology purchase, software, market exploration, training and organisational adaptation are growing rapidly, but their magnitudes and the relations between them, and between them and physical investment, have been relatively little explored. Overall trends in intangible investments are considered below, with some discussion of the factors influencing these trends. Considerable conceptual and analytical work will be needed before any measures and comparisons of "stocks" of the corresponding intangible assets are available.

The analysis is based on trends in spending at current prices. All data except for software are expenditures by performers or purchasers of the services and goods involved. For example, employee training is expenditure by firms on training their employees in-house or by purchasing external training services. The data for software are the revenues of the industry selling software services and exclude all own-account software production for in-house use. Other comparable data are not available.

In five major OECD countries the share of intangible investment in total physical and intangible invest-

ment (R&D, advertising and software – not including training and expenditures on organisational change) rose by over 40 per cent from the mid-1970s to the mid-1980s (see Table 16 and note 5). The share of GDP going to intangible investment rose significantly,

Table 16. **Trends in investment**[1]

Percentage of total

	1974	1984
Expansion investment (industrial plant)	40.4	34.1
Modernisation investment (machinery and equipment)	45.0	45.0
Intangible investment (R&D, advertising, software)	14.6	20.9
Total	100.0	100.0

1. Average for five major OECD countries: United States, Japan, Germany, France and the United Kingdom.
Source: Adapted from OECD (1988*d*).

while total physical investment in non-residential construction, machinery and equipment declined (see Table 17). Physical investment in the business sector subsequently revived strongly, but net physical investment as a proportion of output is still lower than in the 1970s, and physical investment is still being outstripped by high rates of investment in intangibles.

Table 17. **Investment in tangible/intangible assets**
Percentage of GDP

| | Tangible[1] | | Intangible[2] | |
	1974	1984	1974	1984
United States	14.2	13.2	4.4	6.2
Japan	26.9	22.9	2.4	3.5
France	16.8	13.4	2.3	3.1
Germany	15.0	13.8	2.4	3.6
Italy	18.1	14.7	1.0	1.9
Netherlands	16.1	13.5	2.6	3.7
United Kingdom	16.3	13.5	3.1	3.8
Average	17.6	15.0	2.6	3.7

1. Tangible investment = machinery and equipment + non-residential construction.
2. Intangible investment = expenditures for R&D, advertising and software. R&D capital expenditures included. Training, expenditures for organisational change not included.
Source: Adapted from OECD (1988*d*).

Overall physical investment nonetheless remains higher than individual intangible investments. For all major OECD countries investment in machinery and equipment in the manufacturing sector still outweighs (mostly manufacturing) R&D expenditures financed by business enterprises. Even in Japanese manufacturing, which in 1985 had the highest ratio of business R&D to manufacturing investment, R&D was less than one-half of physical investment in machinery and equipment.

Important technology-related shifts have helped to drive the share of physical investment downwards and intangible investments upwards (see Box 31). Changes in Japanese industrial structure illustrate some of these trends. Shifts from declining sectors to more rapidly growing ones have required increasing expenditures on intangible investments. The export-oriented machinery industries, for example, which have grown at the expense of primary metals industries, have higher relative expenditures on R&D, training, and software. Other countries are also making these shifts, so that some European countries and the United States now have an investment structure closer to that of Japan. The relative levels of intangible and physical investment, the sectoral structure of these investments and the speed at which the investment structure can change to accommodate shifts in patterns of demand and consumption all determine long-term competitive performance.

Box 31. **Factors in the changing balance of physical and intangible investments**

A variety of factors affect the changing balance of physical and intangible investments. Among them may be counted:

- Increasing realisation by managements that technology, skills and organisation determine competitiveness. As these production factors increasingly determine comparative advantage, there has been rapid growth in the intangible components of investment.
- High rates of investment in advanced computer-controlled equipment. Prices of computer-related equipment have fallen rapidly while performance has increased even more dramatically, boosting the importance of computer-related investment. This has been mirrored by a general shift towards flexible manufacturing methods which emphasise skill, organisation and development over capital, labour and natural resources.
- The structural shift to high technology industries which are innovation, skill and information intensive rather than capital intensive. Machinery industries (electronics, electrical, non-electrical and transport equipment) increased their share of total manufacturing investment from 25-35 per cent in major OECD countries in the mid-1970s to around 50 per cent by the late 1980s. These industries increased intangible investments (R&D, design, training, marketing) as their share of physical investment increased. Sectors with declines in investment shares were primary metals, chemicals, textiles, wood products and non-metallic minerals. With the exception of chemicals, they are not major investors in intangibles. There has also been a shift towards service industries which are often major investors in training and other intangibles.
- An extended period of investment in rationalisation and replacement in manufacturing (machinery and equipment) and a relative decline in expansion (new plant and structures). Rationalisation investment often emphasises training, R&D and technology scanning to maintain competitiveness during adjustment. Greenfield (new site) investment, with considerable investment in structures, will also include relatively high training (enabling) and engineering (technology) intangible investments.

5. THE EVOLUTION OF INVESTMENTS IN VARIOUS INTANGIBLES

This section traces investment experience in several of the main categories of intangible investment defined in Section 1, within the limits imposed by the available data. Among intangible investments in technology it emphasises R&D and technology licensing, and among enabling intangible investment it treats worker training and organisational structure, before turning to market exploration and software. R&D and worker training are the most important of the categories discussed, but others, such as software, are growing rapidly.

Research and development

For the major OECD countries, growth rates of R&D investment have generally outstripped the growth in physical investment measured by investment in machinery and equipment and by imports of capital equipment (see Figure 13). The United Kingdom stands out by its poor R&D performance in the 1980s. In the United Kingdom and the United States, very rapid growth in imports of capital equipment has surpassed the growth of intangible R&D investment, but this shows the increasing reliance of UK and US industry on imported capital equipment, rather than a fundamental structural difference in the importance of intangibles.

Research and development are important components of technological innovation. They link market opportunities and market needs via business strategies to final sales and consumption. In terms of performance, R&D is carried out by business enterprises (the principal R&D performer), higher education, the government sector and other minor performers (mainly private non-profit bodies). There are two principal sources of finance: business enterprises (the major financer) and governments (almost as important). Other national or foreign sources of finance are considerably less important. A large share of government finance is transferred to higher education performers and, in some countries, to business enterprises. Different dimensions of R&D and differences between countries in the institutional patterns of research are discussed in Chapters 1 and 3. Only the main trends are discussed here.

In almost all countries, R&D has increased rapidly, and its share of firm-level activities and GDP has increased. Business R&D growth is shown in Figure 14. In smaller economies and in some of the less R&D-intensive countries, it has increased at a faster rate than in major countries, but in most cases from a lower base. Expenditure data include capital expenditures on R&D equipment, buildings and land. R&D

capital expenditures in OECD countries ranged from 9 to 25 per cent of gross R&D in 1987 or the most recent year available, and from 9 to 23 per cent of business enterprise R&D costs. R&D capital costs are highest in less advanced OECD countries. Growth in the share of researchers in the labour force has matched increasing expenditures. Japan and the United States have around 65 research workers per 10 000 labour force. Most other major and mid-sized economies have around 35-45 researchers per 10 000 labour force.

The share of R&D financed by industry has also increased substantially in most countries (Figure 15). The shift towards a larger share of business funding has been most marked in the smaller economies which formerly relied heavily on government financing of R&D. Their firms have been forced to sharpen competitiveness and expand their international operations, in part by increasing R&D financed and performed by business. France and the United States have been the exceptions to this trend among the major countries. They have continued large government-financed projects on defence and space research, among others. The same industries are the most research-intensive in most countries. They are, in descending order of intensity: aerospace, computers and office machinery, electronics and components, drugs, scientific instruments, and electrical machinery. These industries not only spend the highest share of value added on research, they are also the core industries from which new products and processes spread to other industries (see Chapter 2).

Technology is driving the long-term growth in business financing of R&D. Awareness of its importance as a key strategic and competitive variable for new and improved products, processes and ways of organising production continues to grow. All industries are becoming increasingly research-intensive. High-technology industries such as electronics, communications, aerospace and pharmaceuticals are devoting a large and in many cases increasing share of their resources to R&D. This is occurring in tandem with a structural shift in manufacturing away from less research-intensive traditional industries towards industries which are more research-intensive.

Because of the importance of commercial R&D to growth and competitiveness, all governments give some kind of support to business R&D, either through direct subsidies or through the taxation system. The magnitude and pervasiveness of government expenditures for R&D has led to considerable analytical work seeking to assess the results of these expenditures and their efficiency. The method of financing used and the relationships between private and public performers of

Figure 13. R&D and tangible gross fixed capital formation (GFCF)
Manufacturing industry
Volumes 1980 = 100

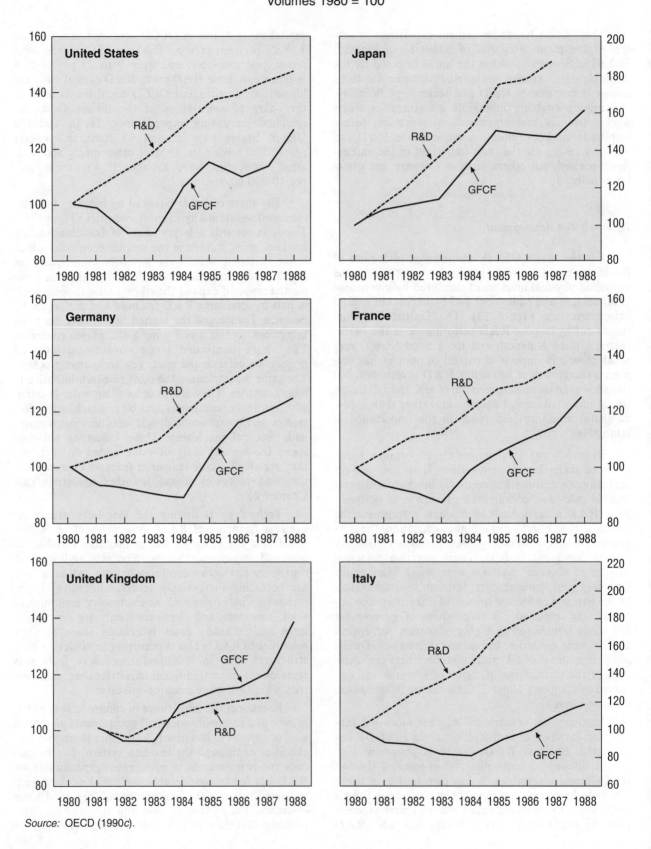

Source: OECD (1990*c*).

124

Figure 14. Percentage of gross domestic expenditure on R&D financed by the business enterprise sector

Major economies

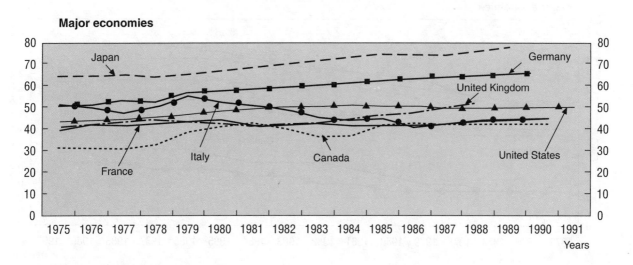

Medium economies

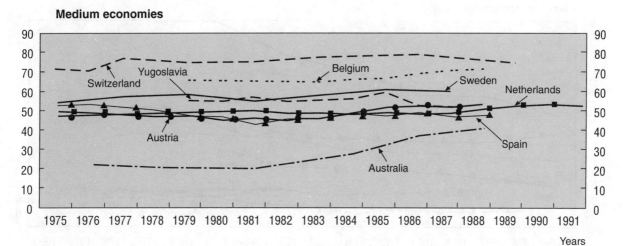

Small economies

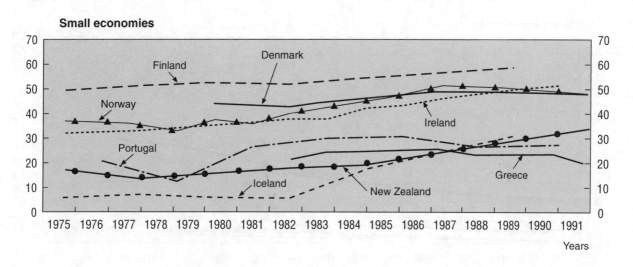

Source: OECD, STIID Data Bank, June 1991.

Figure 15. Business enterprise R&D as a percentage of domestic product of industry

Major economies

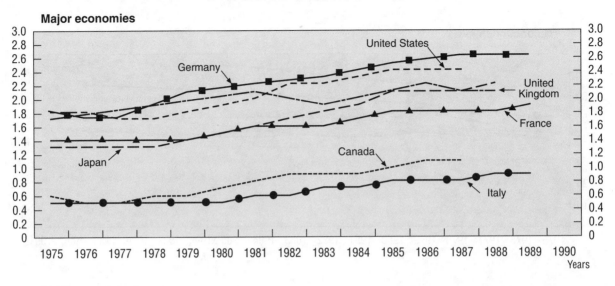

Medium economies

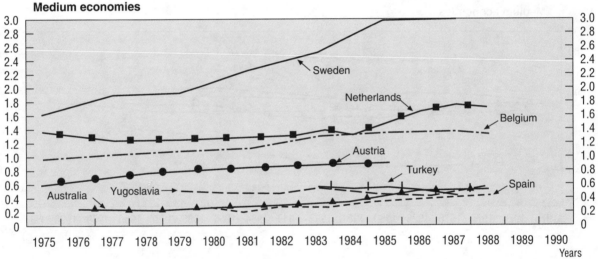

Small economies

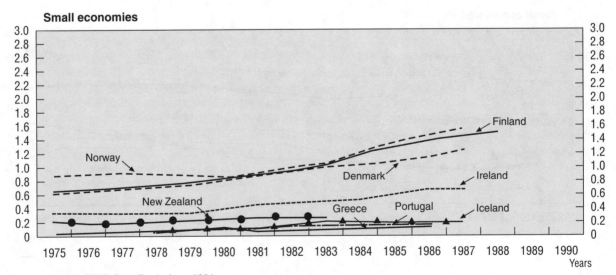

Source: OECD, STIID Data Bank, June 1991.

R&D influence both the impact of government support and the resulting intangible investment.

Studies carried out in the United States and other large countries suggest that publicly-financed R&D is not as effective at boosting growth and raising productivity as business-financed R&D. These results are not surprising in large countries with major government-financed defence and prestige research projects, which do not primarily aim to produce commercial products and whose civilian spin-offs are fortuitous.

The effectiveness of government finance may be greater in smaller countries or those with relatively smaller R&D defence expenditures. In some, significant government R&D funding supports both the research infrastructure and commercially oriented research with large social returns and/or potentially large spillovers to other enterprises or sectors, such as research in SMEs or joint research between groups of firms or between firms, universities and public research institutions.

In Denmark, Germany and Sweden, for example, a well-organised network of publicly backed applied technology institutions provides effective contract research services and consulting to SMEs. Analysis of government subsidisation of micro-electronics applications in Germany and the United Kingdom suggests that government supported projects generally resulted in technical success, but commercial success was more elusive. In Austria, relatively few publicly supported research projects in small firms were commercially successful but projects which were successful gave significant returns. Project additionality is one of the aims of government support – that is, the project would not be undertaken without government support, or is significantly enhanced by support. This is difficult to prove. However, careful analysis of the German scheme subsidising salaries of research personnel in SMEs showed that in the early part of the programme additional R&D and innovation expenditures amounted to around 60 per cent of government funds. This compares favourably with estimated price elasticities of R&D of around 30 per cent (a $1 drop in the user cost of R&D increases R&D expenditures by 30 cents) (OECD, 1990c).

There may also be strong multipliers between publicly-funded institutional research and commercial industrial development, when appropriate links have been formed between institutions and industry. In Norway, for example, publicly financed strategic R&D in universities and technological institutes significantly stimulates subsequent applied industrial R&D, at least in some industries. Also, public R&D is used by industry to explore interesting areas where commercial risk is too high to commit funds until the initial ground has been broken. However, in some industries the nature of publicly funded research and the mechanism of execution can substitute private R&D, not augment it (K. Smith, 1990; and OECD, 1990c).

International technology licensing and purchase

To complement their own development activities, firms use technology developed by others. They may license or purchase the technology, as well as join research associations, consortia or joint ventures or participate in other kinds of technical co-operation. Licensing allows technology to be used in return for royalties; stable intellectual property protection often enters the arrangement. Although licensing and technology purchase have not seen the rapid growth of R&D or of other technology agreements, they have grown steadily and are an important form of technology transfer in some industries, for small firms and for technologically less advanced countries[6]. Only international technology licensing and purchase data are regularly collected for many OECD countries, although domestic licensing is probably significant.

From the mid-1970s to the mid-1980s, annual growth in international payments for patents, licences and technical know-how was around 2 per cent in real terms, with international licence receipts worth around US $11.5 billion for reporting OECD countries in 1985 and $23.5 billion in 1989. Five countries, the United States, the United Kingdom, France, Germany and Japan, are the major suppliers (85 per cent of transactions). For major performers, international receipts and payments for technology are of the order of 10-20 per cent of business enterprise R&D. Countries such as Australia and Portugal rely extensively on foreign technologies and make larger payments for technology, often from foreign subsidiaries to parent firms. But for large OECD countries, production under independent licensing agreements is less important than production from foreign affiliates, and sales under *independent* licence (between firms with no equity holdings in each other) are estimated at around 5-10 per cent of sales from foreign subsidiaries.

Firms often adapt imported technology to local conditions and in many cases improve on it to build up their own technological base. At the same time, much licensed technology is actively used by the licenser, which means that it is likely to be undergoing incremental change and improvement, and licensers usually want to exert tight control over the use of the technology. In all cases, firms licensing in have a large incentive to boost their own development efforts rather than passively accepting licensing; major technology importers such as Japan and Korea have built extensive technological capabilities in this manner.

Investment in firm-based training and skill formation[7]

The use of computer-based advanced manufacturing technology (AMT) and office equipment has emphasised increasing and retaining skills. "Functional" flexibility of the labour force (mobility, better job design, training and retraining) is replacing numerical" flexibility (temporary/part-time employees, working time adjustments, hiring/dismissal regulations) (see Chapter 4). But the development of functional flexibility has varied greatly. Where there are flexible labour markets external to firms (e.g. the United Kingdom and the United States), less emphasis has been given to developing long-term functional flexibility. Where inflexible external labour markets prevail (e.g. Germany, Sweden and Japan), long-term development of internal functional flexibility within firms to ensure the supply of skills has been greater.

In addition to these systemic differences, education and training efforts in all countries vary greatly among firms, industries and groups of workers (see Chapter 7). The skills necessary for the efficient application of new technologies may be developed in the workplace through special vocational training in-house and on-the-job training or in conjunction with outside institutions and training by technology suppliers. Training is generally concentrated in large firms, in high technology industries and at higher skill levels. Concentrated patterns of R&D and use of AMT accentuates the degree of advanced training. Higher intensities of training are associated, for example, with introduction of new technologies in high-technology industries or with special training programmes to ameliorate the effects of structural adjustment in declining sectors (e.g. basic steel and non-ferrous metals industries).

Small firms generally train less than large firms, although between 1972 and 1988 training at least doubled for all firm sizes. Employees in French enterprises with less than 50 employees are only one-fifth as likely to receive enterprise-financed training as employees in enterprises with 2 000 or more employees, and small firms spend less than one-third of the share of gross wages that goes to training in big firms. Traditional industries spend only one-third as much as electronics industries on training, so that small firms in traditional industries spend only one-tenth of the high-technology sectoral average. Finally, intensive firm-based training is often directed at higher level employees. In France, managers, engineers, supervisors and technicians are at least twice as likely to receive training as less skilled workers (France, Centre d'Etudes et de Recherches sur les Qualifications, 1989).

Large Japanese manufacturing firms also make much greater training efforts than small ones. When compared with total labour costs, formal training expenditures are four times greater in large firms (5 000+ employees) than in small ones (30-99 employees). The difference between large and small firms is much more marked in manufacturing; overall, large firms have only twice the expenditures of small ones. In Japan, small firms also trail large ones in

the use of best practice techniques and associated staff development. A very low share give technical and engineering training, and less than 10 per cent train employees to use advanced technologies and equipment (computer controlled production systems, computer numerical control machine tools, industrial robots) compared with 30 per cent of large enterprises. These observations suggest that large manufacturing firms are far more likely to have formal training programmes, but it also highlights the leading role that large manufacturing firms have overall in the Japanese economy.

The Japanese data also show that the manufacturing industries most advanced in process applications (electronics, machinery, metal refining and chemicals) spend considerably more per employee on education and vocational training than established industries such as textiles, clothing, wood products, leather, paper, ceramic and stone products (Table 18)[8]. However, Japanese utilities and finance sectors also have large formal expenditures on training and education. The same group of industries have consistently been the major investors in training, while industries which invested relatively little in the past have, in general, not improved their relative position.

Patterns are similar in Australia, Canada, the Netherlands and the United Kingdom. In 1989, Australian employers with more than 100 employees spent on average over three times more of gross wages and salaries on training than small ones (1 to 19 employees). Most large firm training takes place in-house in formal programmes. But small employers who do train (many do not) spend a higher share of gross wages (including wage costs) on training over longer periods than medium and large ones. As in Japan, utilities and communications were intensive providers of formal training.

In Canada, in 1984-85, over four-fifths of firms with more than 500 employees had formal training programmes, whereas only one-fifth of firms with less than 20 employees did. Manufacturing firms were more likely to train, with emphasis on managerial, professional and technical occupations. Large firms were much more likely to retrain employees in response to technological change.

In the Netherlands the same pattern is evident across all industries, with employees in firms with more than 500 employees being almost five times more likely (seven times more likely in manufacturing) to participate in training than in firms with less than 100 employees. Small firms are less likely to have formal training activities, but even when they do their employees are less likely to participate. Employer training expenditures (including worker time spent on training) were estimated to be close to 1.5 per cent of gross aggregate wages in 1986, approximately two to three times greater than government higher vocational education expenditures, and growing rapidly.

Table 18. **Employers' expenditure on training by industry: Japan**
Percentage of average labour cost

	1981 %	Rank[1]	1988 %	Rank
Food	0.31	6 (6)	0.33	5
Textiles	0.10	18 (13)	0.14	19
Clothing	0.22	10 (15)	0.30	7
Leather & leather products	0.18	16 (19)	0.26	12
Wood & wood products	0.10	19 (17)	0.14	18
Furniture and fixtures	0.14	17 (12)	0.32	6
Pulp, paper, paperboard	0.18	15 (11)	0.23	16
Printing and publishing	0.18	14 (18)	0.22	17
Chemicals	0.30	7 (5)	0.38	3
Petroleum and coal	0.33	4 (4)	0.36	4
Rubber and plastics	0.18	12 (10)	0.24	14
Ceramic, stone, clay	0.18	13 (16)	0.26	11
Iron and steel	0.39	2 (2)	0.29	8
Non-ferrous metal products	0.22	9 (9)	0.24	15
Fabricated metal products	0.22	11 (14)	0.26	10
Machinery	0.43	1 (7)	0.53	2
Electrical machinery	0.33	3 (3)	0.63	1
Transportation equipment	0.32	5 (1)	0.24	13
Precision machinery	0.29	8 (8)	0.27	9
Total Manufacturing	0.29		0.38	

Note: Employee wages while training not included in training expenditures, on-the-job training costs not included.
1. Figures in brackets are the 1975 ranking.
Source: Calculated by OECD from *Year Book of Labour Statistics.* Statistics and Information Department, Ministry of Labour, Japan, 1975, 1981, 1988. Average monthly cost of education and vocational training divided by total average monthly labour cost per regular employee, by industry. Data drawn from tri-annual surveys of "Welfare provisions".

In 1986-87, UK employers spent about £18 billion on training (both on and off the job, including labour costs), and total training was worth close to one-half of fixed capital formation. Patterns of training are roughly the same as in other countries, except that very small private sector establishments (less than 25 employees) have only slightly less than average training time per trainee receiving training. In Australia also, where small employers on average are less likely to train their employees and train them less, those who do train do so more intensively than the average. On the other hand, the UK public sector trains more than average, but very small public establishments train considerably less. Training is weighted towards the young and towards those with existing qualifications and managerial and professional workers who are likely to benefit from training. Firms with skill supply problems and those facing international competition are much more likely to train, but surprisingly few UK firms had any sort of plan for training or had undertaken formal training needs analysis.

Despite increased training efforts, there are still shortages of specific training courses and skills in particular technologies. A comprehensive survey of the use of computerised automation in metal-working industries in the United States showed that in 84 per cent of plants with programmable machines, no classes were given by employers for the development of the new skills. In-firm training was more likely where there were established apprenticeship programmes and where workers already had higher qualifications, or in high-technology firms for white-collar employees. Broad-based training was generally not provided (Kelley and Brooks, 1988).

However, many firms are placing a greater premium on training, particularly in high technology sectors where there are internal and external skill shortfalls. In the early 1980s, studies showed very high returns on training investment at Motorola in the United States (Hooker, 1990). Returns were greatest when a full range of training was undertaken, and training is now integral to firm strategy. Up to 85 per cent of productivity gains come from training, improved skills and changes in the organisation of work associated with new product introduction. Factory redesign and reorganisation of work structures and links between development, production and suppliers has raised productivity dramatically. It is estimated that increased skills and organisation changes gave five times the improvements expected from technological changes alone. To achieve these objectives, a minimum of 1.5 per cent of payroll and 2 per cent of employee time is devoted to training. Training has shifted from being an expense to being an integral part of investment.

Overall, the evidence suggests that formal firm-based training is concentrated in large high-technology firms and in a few manufacturing and service sectors and that it is further concentrated on higher-level employees. But in many countries there are still considerable shortages in information technology (IT) personnel and engineers as well as widespread shortfalls in general training, particularly in small firms (Pearson *et al.*, 1988). To the extent that small firms are less structured than large ones, more informal on the job training may occur, and formal training may be less essential. But small firms have been responsible for a large share of employment growth and application of new technologies, and their relatively poor training record suggests that when they neglect their human capital, they are limiting their own ability to develop new products and fully utilise new equipment.

This suggests that government action should focus on the training infrastructure and on incentives to firms to develop an appropriate spread and mix of training. Government training initiatives are often complementary to firm-based vocational training and can give an important boost to firms' own efforts by reducing the relative costs to firms.

Organisational change

Effective organisation combined with the efficient use of skills determine firm-level competitiveness and long-term macro-economic performance. Despite the qualitative attention paid to new forms of corporate organisation (Chapter 4), there is little information on the costs of the constant reorganisation necessary to sustain the new flexible structures towards which most firms are evolving. Some estimates suggest that around 25 per cent of labour cost expenditures in large Swedish industrial firms are devoted to co-ordination and organisation activities. On the other hand, according to survey data, narrowly defined organisational development to introduce new products is of the order of 3 per cent of total innovation costs in Finland and Germany.

No matter how they are measured, substantial and rising expenditures and management time are devoted to developing intangible assets in flexible organisations. A major service industry now supplies business consultancy services and expertise to assist organisational change. Some governments subsidise the costs associated with this change, particularly in SMEs.

Marketing

In a period of shortening product cycles and greater competition, market development takes on increasing importance. In addition, the response to market signals and recognition of market opportunities is a key factor driving innovation, while lack of adequate marketing hampers new product introduction and the rapid use of new technologies. The organisational structures of firms are important for the successful marketing of new products and also affect how well new products and production methods are linked to markets. The new flexible structures seek to obtain more rapid response to customers and markets and to gain economies of scope and scale by spreading marketing efforts over a larger range of related products, so as to improve the benefit/risk relation. Much of the perceived success of Japanese *keiretsu* structures (discussed in Chapter 4) is based on the ability to link production successfully to markets.

The lack of such vertically and horizontally integrated marketing structures poses a problem for small firms. The problem is particularly acute when, as in Ireland and Australia, small high technology firms without the benefit of large integrated enterprise structures must make the leap to foreign markets very early in their product and service development cycle because of the relatively small size of domestic markets. Even Germany experienced shortages of expertise to translate technically successful micro-electronics products into marketable ones and then to market them. Gov-

ernments have begun to tackle these problems: in Denmark, for example, the government is promoting marketing networks among small firms to answer the challenges of market development and distribution.

Measures of expenditures on market exploration and development vary widely. In Germany, sales preparation, market testing, advertising and sales organisation to launch new products are around 4 per cent of innovation expenditures or around one-sixth of industrial R&D. In Finland and Italy, market development is also around 5 per cent of total innovation expenditures; it is equivalent in Italy to one-third of R&D while in Finland it is less than one-tenth (see Table 14). In Finland, long-term marketing expenditures – the market costs of innovations *plus* the costs of gaining new market areas and improving the enterprise's competitive position – were estimated in 1987 to be over a fifth of industrial intangible investment, equivalent to around 0.7 per cent of industry turnover.

Software[9]

With investment shifting towards human resources, software creation and computerised production, the role of software is becoming increasingly important. The value of software embedded directly in

Table 19. **World software market**

$ millions

	1985	1987 (estimates)
Australia	500	737
Austria	188	275
Belgium	300	440
Canada	598	783
Denmark	209	316
Finland	186	282
France	2 159	3 157
Germany	1 864	2 730
Japan	2 861	3 999
Italy	1 071	1 677
Netherlands	575	833
Norway	188	300
Spain	305	518
Sweden	344	534
Switzerland	341	501
United Kingdom	1 831	2 771
United States	16 546	23 610
Total OECD	30 065	43 464
Other countries[1]	860	1 652
Total	30 925	45 116

Note: Includes consultancy and software bundled together by hardware manufacturers.
1. Brazil, South Africa, Israel, Mexico, South Korea, Singapore, Taiwan, India.
Source: Schware (1989) and OECD. Original sources mostly unofficial.

products is continually increasing, and indirectly used software is increasing through software-intensive production processes, quality control, testing, storage, handling, sales and delivery. At the same time, data on software investment remain inadequate for a number of reasons (see Box 32).

In 1987, the software engineering services industry was estimated at some $45 billion, with the United States taking up about one-half of this value. The industry has expanded at 20-30 per cent per year, so that in 1990 the industry was around $100 billion (see Table 19). These data cover the dedicated supply industry and are not total expenditures on software development and purchase.

Structural shifts in the industry are illustrated by the United Kingdom computing services industry, which more than doubled in current values over the 1985-89 period. Software development activities expanded (software products, software support and maintenance, development of custom software) and there was a sharp decline in hardware-related activities (leasing, maintenance). This decline matches the shift by users away from expensive centralised mainframe operations towards more localised systems based on microcomputers, personal and minicomputers, which

are more dependent on once only inputs of software. Bureau services (computer service firms performing particular tasks such as preparing accounts for customers) have increased somewhat, as have consultancy and training (see *British Business,* 29 September 1989).

Four main forces are currently shaping the software supply industry:

– rapid declines in the price of hardware and even more spectacular declines in price/performance ratios, driven in part by low-cost supply of computing equipment and peripherals from Japan and East Asian assemblers, with low productivity in programming and supply of software solutions standing in contrast to this, so that an increasing share of the costs of computing and computer-related systems are in software, and hardware suppliers are increasingly selling software and associated training and consultancy services;

– the issue of incompatibility between different computer systems, the promotion of "open systems" designed to operate according to internationally agreed standards to ensure compatibility between different products, and the battle over different basic standards for the operating systems which drive all computers and computer-based equipment;

Box 32. Difficulties in measuring software investments

Software research and development expenditures take a significant share of total R&D, but software activities are still inadequately measured, in part because of definition problems[a]:

- Expenditures on software are of two different kinds. The first is expenditure to purchase and apply developed software, a form of capital expenditure often associated with equipment purchase. The second is the cost of developing entirely new software (software R&D). If the second kind is included both under software and R&D, they may be important sources of double counting. For example, in 1988, 23 per cent of total Canadian business R&D was software R&D and in firms performing software R&D the share was 38 per cent.

- A large part of software development is performed outside the dedicated software supply industry. For example, the Canadian business machinery industry was responsible in 1988 for only 16 per cent of all software and the computer services industry only 15 per cent. The biggest single spender was the telecommunications and electronic machinery industry with 44 per cent of the total: research on telecommunications switching systems, new semiconductor components and integrated manufacturing systems is largely software design and development.

- The software supply industry provides software engineering services (systems software and applications software), but definitions and survey coverage vary between countries, boundaries of the industry are poorly defined and statistics are not comparable across many countries[b]. All activities of computer service firms are often described as software activities, but they usually also include other activities.

---

a) Software is the instructions that direct the operations of computer equipment and the information content (data) that computers manipulate. Software engineering is the systematic approach to the specification, development, operation, maintenance and retirement of software. Systems software manages the components of a computer system, and applications software applies computer power to the performance of required tasks. Definitions adapted from *The Institute of Electrical and Electronics Engineers Standard Glossary of Software Engineering Terminology* (1983), quoted in Schware (1989).

b) The current system of National Accounts will have to be enlarged to include own-account production of software for internal (in-house) use and for commercial use. Software investment (whether acquired or created internally) will most likely be classified as an input to production in the forthcoming revision of the UN System of National Accounts.

- the use of a wide variety of different software and hardware products from multiple suppliers to meet particular task requirements and the shift away from using proprietary software on a single vendor's hardware;
- the move by hardware suppliers and systems users to supply integrated systems (for example AMT or management systems) on the basis of their experience in assembling systems for their own use.

Non-specialist firms developing software face challenges in human resources, quality control, software inventory and cost accounting. Control and transparency are particularly important issues when firms have a rising share of product development costs devoted to software; they can be tackled, for example, by setting up software sub-firms. Skills shortages have also led to a wide range of internal and external education and training strategies. Overall, better systems of software development and improved management of software resources and human resources are among the most important challenges facing firms over the medium term.

6. SOME POLICY CONCLUSIONS

Corporate strategies and new corporate structures are being built on greater inputs of intangible investments. The level and composition of intangible investments and the complementary way in which they are employed and used in conjunction with physical investment are the key elements determining firm competitiveness, growth and productivity performance.

- Increased R&D, effective innovation, and the introduction of new products, processes and ways of organising production and the delivery of services are adding to the stock of usable knowledge and are one of the cornerstones of competitive performance.
- Skills and competence are needed to drive this whole process. The rapidly evolving world competitive system is driven by the availability and use of highly trained management and workers.
- The organisation of work on the production floor and the relations between firms are being radically transformed, thus placing a competitive premium on the appropriate forms of firm organisation.
- Market exploration and advertising are necessary to ensure the coupling of market opportunities, technological effort, production and final consumption.
- Finally, software is essential to drive increasingly computer-based production and services, and to organise the internal and external information structure to enable flexible response to changing market conditions in the competitive battle to retain and expand markets.

All of these changes explain rapid increases in intangible investments and the increasing attention paid to them. The data presented in this chapter give a clear picture of the rapid growth of intangibles, their increasing importance relative to physical investment and some of the complementarities among them. However, clarification and harmonisation of definitions, accounting conventions and data collection will be important to further policy development. Some areas for policy are discussed below.

Governments supply a large share of the services which contribute to the creation of intangible assets. Considerable R&D is financed by governments, and in many countries the technology infrastructure surrounding firms (technological institutes, testing, consultancy and engineering services, standard-setting bodies) is also supported and shaped by governments. The biggest intangible inputs in most countries are the education and training services supplied through government-financed education and vocational training institutions. These institutions build the intangible assets in human resources on which firm competitiveness is based. Government-backed service-supplying institutions and mechanisms thus provide both specific inputs to firms and build the general technological and commercial infrastructure for the whole economy. Particular attention is already given in many countries to areas – trailing regions and the small firm sector – where these basic services and intangible investments are weak or where there are market failures in their delivery.

Increasing inputs of R&D and computer-based automation accompanied by declining productivity gives rise to what has been called the productivity paradox see Chapter8). Explanations are usually of two kinds: long lag times in institutional adjustment or inefficient investment. Analysis of both has focused on shortages in skills and poor organisation, shortcomings which must be overcome if total productivity performance is to improve. The problems are, in essence, due to insufficient intangible investment at the public level (poor education, ill-adapted institutions, slow standard-setting mechanisms) as well as in private firms. These key complementary assets to physical assets require boosting.

A range of intangible investments (particularly R&D and training, but other investments in information and organisation) are becoming increasingly important relative to physical investment. Most countries have had investment incentives for physical investment (tax incentives such as accelerated depreciation or additional allowances, as well as grants and subsidised loans). But tax reform and cut-backs in physical investment subsidies have reduced the incentives to physical investment. Most tax incentives and grants for R&D have been retained, and a number of countries are now looking carefully at training incentives and incentives to improve human resource management. In some cases, incentives have been introduced to widen firm-based training. Most other expenditures on intangibles (organisational costs, engineering, marketing) can be deducted from taxable income as they are incurred, and they are now favoured over physical investment. However, as firm strategies give more emphasis to a whole range of intangibles, the question of whether the balance of government policy investment incentives and disincentives is correct must be addressed. Do some intangibles need extra investment incentives? Does investment policy need more careful co-ordination between different components of investment and with overall macroeconomic policy?

Accounting conventions have not kept pace with the rapid development of intangible investment and with the new structures and accounting methods that are required to assess and run computer-controlled and computer-integrated manufacturing and business systems. Managements have poor tools at their disposal to measure the impacts of investment projects, to evaluate financial requirements and seek sources of external finance, and if necessary to articulate demand for government-funded services and investments in the technological and business infrastructure. This is particularly the case for the development of necessary complementary assets in human skills, organisation and management structures.

Complementarities between different kinds of investment are important at firm level and at industrial sector level (between training and R&D, and between training and physical investment) as well. At the micro-economic level, firms which can acquire skills and competences in key areas and build more flexible organisational structures are more competitive and can accumulate greater competence and flexibility. To be successful, investments in AMT also depend crucially on a range of complementary organisational changes and investments in skill development and human resource management. Policies may need to be redirected to ensure equal treatment for a wide range of different investments, or to boost selectively some kinds of intangible and physical investments to achieve the necessary complementarities.

A number of governments have expanded their industrial support systems to supply a greater range of consultancy and advice, particularly to small firms, in order to overcome two distinct market failures. One is the insufficient supply of services, often in areas of advanced technology but also in design, quality, management and information systems. The other is the lack of information available on the demand side: small firms are often unaware of the benefits to be derived from seeking external sources of management and technological advice. Effective government programmes are often shared-cost, market-driven, contractor-supplied; they operate within fixed time limits and provide a relatively wide range of management solutions and technologies. This model of subsidising inputs of selected intangible services into the management and information structures of small firms, more generally applied, may help to improve economic performance.

A large share of foreign direct investment is in technologically advanced manufacturing industries (electronics, pharmaceuticals, chemicals) and in organisationally advanced service industries (transport, communications, financial services). Foreign-owned manufacturing plants use more advanced production technology (robots, computer-integration and operations networks) and are often leaders in training and human resource development. Firms are also increasingly internationalising their technology sourcing and performing more of their R&D abroad, notably in the United States. However, given the growing importance of intangible investments in strategic areas of planning, organisation and technology development, firms still tend to concentrate these assets in their home country, rather than spread them to their foreign production and employment. The policy issue in host countries is to devise strategies to ensure that foreign investment includes a range of key intangibles (R&D, training, information and management assets) and that the technological infrastructure can provide complementary services to support foreign investment.

NOTES

1. The search for a wider and more comprehensive approach to investment initially focused on the notion of "technological investment", grouping technology-related physical and intangible investments. However, the analytical and practical difficulty of separating "technology" components from other kinds of investments has resulted in the simpler course of grouping intangible investments for which information is available or which can be measured and surveyed. This chapter is indebted to exploratory work on conceptualising and defining intangible investment for the OECD by Kaplan (1987); see also, Caspar and Afriat (1988).

2. Previous theory assumed that investment opportunities are created by scientific advance or by inventive activity at a pace independent of investment, or alternatively that investment opportunities exhibit diminishing returns and are thus gradually exhausted as a consequence of undertaking investment. The main proponents of the "new" growth theory are Scott (1989); Romer (1990); Lucas (1988); and Baldwin (1989). Chapter 8 discusses in more detail the implications of these theories for explanations of the total factor productivity slowdown.

3. Discussion of the concept of software and national accounting is drawn from Muller (1990).

4. IFO-Institute results are summarised in Scholz (forthcoming). In other countries surveys have been carried out irregularly, and not always on the same basis. Data for Finland is drawn from Central Statistical Office (1990) and for Italy from Mangano (1990).

5. Available data is drawn from work for the OECD by Kaplan (1987). The data is partial (training is not included) and only partly comparable between countries. Intangible and tangible investment data for Sweden are drawn from *Statistika meddelanden F 13 SM 8802.*

6. Bell and Vickery (1989); Vickery (1986) and (1988). The OECD manual on the technology balance of payments will help improve data collection on international technology transactions (1990*k*).

7. This section draws extensively on OECD (1991*a*). Recent trends in training are also surveyed in OECD, *Employment Outlook,* Chapter 5, "Enterprise-related training" of OECD (1991*j*).

8. For Japan, average training costs across industry are around 5 per cent of total labour costs if training includes employee wages while being trained plus on-the-job training costs. On the same basis, French training costs would be of the order of 4 per cent, with United Kingdom costs somewhat lower.

9. See Schware (1989) and OECD (1988*c*). Data on software R&D in Canada are drawn from the results of a new question on the annual survey of industrial R&D, see Canada, Statistics Canada (1990).

Chapter 6

THE SUPPLY OF SCIENTISTS AND ENGINEERS: CURRENT TRENDS AND CONCERNS

ABSTRACT

Since scientists and engineers play a key role in the innovative process, the possibility of severe shortages has made their future supply a crucial economic issue. However, supply and demand are very difficult to forecast.

The lion's share of the supply of scientists and engineers still originates in each country's native population, educated and trained in national institutions of higher education. In most countries, the fall in the 18 to 24 year-old age cohort that began some ten years ago is likely to continue through the end of the decade. It has so far been compensated by an increase in the overall rate of enrolment in higher education, largely due to increased participation of women. This positive trend is offset by an apparently diminishing student interest in science and technology studies. A variety of factors influence the actual number of scientists and engineers produced, including: the increasing availability and attractiveness of other professions to science and technology degree recipients; the upgrading of technicians; the converting or retraining of graduates with non-scientific degrees; and, of course, the international mobility of scientists and engineers.

Future demand is even more difficult to predict. In addition to demographic factors such as mortality rates and retirement age, it is strongly affected by variables as uncertain as economic growth, changes in the relative weight of different sectors and branches of the economy, changes in their "technology intensity", and changes in government policies. While recognising these uncertainties, studies undertaken in a number of OECD countries stress that recruiting appropriate staff is already difficult in some areas of science and technology. Most countries anticipate a significant increase in the demand for scientists and engineers over the next ten years.

Furthermore, the higher education sector is most likely to experience especially serious difficulties in many countries as a result of the circumstances of its development over the last thirty years (a very rapid expansion in the 1960s and early 1970s abruptly followed by stagnation and sometimes contraction). Massive retirement will lead to intensified recruitment needs from about 1995. The problems in the higher education sector may be even more severe if there is strong parallel demand from industry with a capacity to offer much higher salaries for the same type of personnel.

While recognising the major uncertainties affecting the assessment of the future supply and demand of scientists and engineers, the analyses presented in this chapter suggest that a risk of shortages of scientists and engineers in future years exists and that policies should be adopted to increase supply. The justification for such policies lies in the critical dependence of the economies and societies of OECD countries on science and technology and on the skills of scientists and engineers.

INTRODUCTION

In most OECD countries, although to varying degrees, recent years have been marked by fears of imminent or medium-term shortages of engineers and scientists. In conjunction with certain demographic factors and a relative drop in student enrolments in science and engineering, decreased public R&D spending is seen as potentially weakening the supply of qualified personnel.

Those with a legitimate, although possibly somewhat partisan, interest in this question include the relevant educational and research communities, appropriate public authorities, employers and, not least, organisations representing scientists and engineers. Their sensitivity to signals warning of potential gaps in this particular labour market has been translated into efforts to determine the scope of such an eventuality by surveys of students, educational institutions and employers. These same groups support research on personnel prediction models. The results, while interesting, have not always been conclusive, given the shortage of data and the extreme complexity of the phenomena involved. Yet those who bear responsibility for and exercise authority in the scientific and engineering personnel area continue to support such information and prediction activities, in the expectation that the production of even meagre insights, when joined to other indicators, may provide the kind of evidence which will give rise to the design and implementation of effective personnel policies.

Allusions to possible severe shortages of scientific and engineering personnel by academics, employers and certain ministries are not a completely novel phenomenon. Where markets are used to allocate such resources, shortages and surpluses of varying degrees will always occur. Markets, of course, will clear. However, their putative strength in achieving optimum efficiency of resource allocation may not necessarily satisfy the needs of society. Given the alternatives, avoiding market failure is of high priority. This leads to an emphasis on the supply factor, primarily because of the difficulties of predicting and reducing demand.

1. HIGHER EDUCATION, THE MAIN SOURCE OF SCIENTISTS AND ENGINEERS

The supply of scientists and engineers is easier to determine than the demand. This is due in large part to the demographic data available to many industrial nations. These data usually provide figures for the 18 to 24 year-old age group – in the past and still for the conceivable future the most important source of the skilled scientific and technical personnel required for the functioning of academia, industry and government. Demographic data, however, while precise, are rather coarse as a predictor of numbers of future scientists and engineers, and they are becoming increasingly so.

This is so for several interrelated reasons: the availability of an increased number of study interests and employment opportunities for those with scientific and technical capabilities; diminished participation rates in science and engineering study programmes; postponed entry into higher education; the attraction of non-technical employment for certified scientists and engineers; changing cohort ethnic and gender composition; and the increasing lack of certain facilities and complementary resources to undertake and persist in the recruitment and retention of potentially qualified candidates. As a determinant of future personnel availability, demographic data are also relativised by such factors as immigration and emigration, retraining and upgrading and field mobility.

Overall higher education enrolments

While the supply of scientific and engineering personnel derives from a variety of sources, the lion's share still originates in native-born students educated and trained in national institutions of higher education. The size of this component of future supply is determined by the demographic knowns. For most countries, these reveal, for the 18 to 24 year-old age group, a major downward trend that began about ten years ago and is likely to continue through the middle and perhaps the end of this decade. From the early 1980s to about 1995 the numbers in this age group will have fallen by anywhere from one-quarter to one-third. Figure 16 shows the trends for several countries in Europe and the United States, for the 15-19 year-old age group.

If all other factors remained constant, this demographic downturn would lead to a proportional reduction in the number of students. However, many countries report steady, albeit very modest, increases in enrolments. Thus, Canadian enrolments for first university degree programmes have increased by almost one-third from 1980/81 to 1988/89. In the Netherlands, enrolments in the higher education sector increased by 13 per cent from 1980 to 1988, although absolute numbers declined between 1987 and 1988. University undergraduate enrolments in Japan rose by approximately 9 per cent between 1978 and 1989. And in Norway an increase of almost 10 per cent was noted between 1979 and 1986. The principal reason for increased enrolments, despite reduced numbers of individuals in the higher education age group, is the much stronger presence of women within the student body. To this factor may be added increased numbers of mature (older) and part-time students, and, in certain cases, immigrants. These increases in total enrolment have become noticeably modest over the past few years and indicate the attainment of a plateau which is likely to be maintained with difficulty.

The levelling-off is due to the fact that women now or soon will account for 50 per cent of all enrolments. In some countries they have already become a majority. The combination of unlikely further gains in female participation with noticeable if still slight reductions in the male presence in higher education hardly augurs well for maintaining current enrolment levels in the next few years. For it is exactly in those

Figure 16.   **Supply of 15-19 year olds in Europe and the United States**

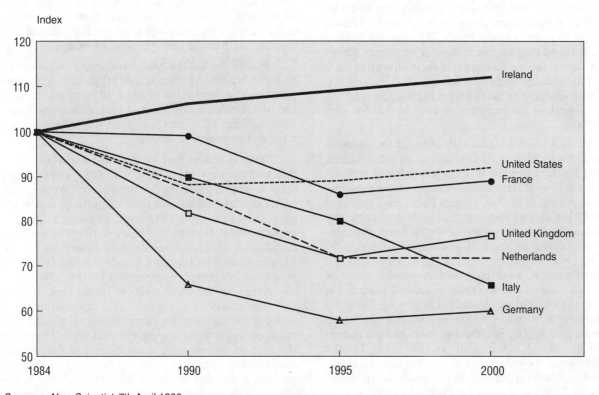

Source: New Scientist, 7th April 1990.

years that the most serious reductions in the pool of potential students will occur.

Enrolments in science and engineering studies

The steady increase in student enrolments despite the smaller numbers of the higher education age group is somewhat offset by an examination of student participation in the science and technology fields, which is generally diminishing. In Canada, first university degree enrolments in the natural sciences and engineering reached a peak in 1985/86 and since then have dropped continuously. In the narrower mathematics and physical sciences areas, this enrolment fell from a high of 31 774 in 1984/85 to 25 062 in 1988/89, provisional. In the United States, total full-time engineering enrolments reached a maximum of 406 144 in 1983 and then descended without interruption to 346 169 in 1988. Data from the Netherlands reveal that student enrolments in science and engineering higher education peaked in 1987 and dropped by almost 4 per cent in 1988. Counter to this trend are still modestly increasing science and engineering enrolments in countries such as Norway, Germany and Japan.

Few data are available on beginning student enrolments in science and engineering. Assuming that the pattern of student withdrawal from these sectors has not radically changed, it seems reasonable to ascribe reduced total enrolments to recently matriculating classes. Some supporting evidence can be found in US surveys of entering university students. Thus, a recently completed study by the Educational Testing Service found that the percentage of secondary school graduates planning to continue their education in a "quantitative science field" (defined as all natural science and engineering fields except the life sciences) fell from a peak of 19 per cent in 1983 to 13 per cent in 1988.

In the United Kingdom, Pearson, of the Institute of Manpower Studies, University of Sussex (Nature, 1989), found that interest in engineering and technology has fallen. Mature students, an increasingly growing component of the UK student body, are disinclined to study the applied sciences. A projection of university graduates in Italy suggests that the numbers of graduates in the natural sciences and engineering in 1992 will be about 81 per cent of 1987 graduates. This can be attributed in large part to demographic developments but cannot exclude shifts in student choices. On the other hand, the percentages of beginning university students in science and technology who do not intend to teach rose slightly in Germany from 30.5 per cent in 1980/81 to 31.5 per cent in 1987/88. In the Netherlands, the percentage of new entrants to higher education choosing science and engineering remained at about 21 per cent between 1980 and 1988.

The participation of women, mature students and minorities in science and engineering studies

While demographic givens place a limit on numbers of individuals of higher education age and thus influence the evolution of future enrolments in science and engineering, several other and more specific factors can be important in determining the distribution of students' curriculum and career choices. They include financial gain, challenging work, the social status of the different professions and individual values. Other determinants include racial and ethnic factors, social class and sex, parental education and student age.

Thus, again according to Pearson (1990), in the United Kingdom, in 1987, managerial and professional social classes, who formed 32 per cent of the population, accounted for 68 per cent of the entrants to the universities and slightly fewer to the public sector. This proportion has hardly changed in the past decade which suggests no widening of the intake into higher education. Many of these entrants contributed to the almost 50 per cent increase in the 1980s of science graduates and the more than one-third growth in engineering degrees. Moreover, since the expected reduction in the number of children in such families is only roughly one-third of that predicted for other socio-professional categories, the fall in higher education enrolments should be smaller than the demographic downturn in general. However, in the 1980s, the composition of the student population began to change because of the entry of large numbers of women and, to a lesser extent, mature students. This growth is expected to continue and be supplemented by students from ethnic minorities whose interest in science and engineering has not been particularly strong.

Within the United States two important minority groups (African-Americans and Hispanics) have been growing as a proportion of university-age students. In the past, neither group has had proportional participation in higher education, and even less within the science and engineering sectors.

Student gender may be the most important factor affecting the numbers of those who will choose to study science and engineering in the future. Women currently display much lower rates of participation in science and engineering than men, despite considerable growth, albeit from a low base, during the past two decades. A review of recent data in certain OECD Member countries reveals a female representation rate in science enrolments – depending on the definition of this term – of the order of 20 + 10 per cent. Neither this percentage nor that of the representation of women among new entrants choosing science has changed significantly during the past several years. In the case of engineering, women account for approximately 10 + 5 per cent of enrolments, a figure that has also not noticeably increased for either total or beginning female students. In fact, recent data from the

United States and Norway reveal a slight reduction in women's representation in this sector.

The relatively low ratio of women to men in science and engineering enrolments is the result of conscious choices by women to opt for other fields of study and careers. Thus, in the academic year 1988/89, 14 per cent of Canadian engineering and applied sciences undergraduate enrolments were women, but they represented only 2.6 per cent of all undergraduate women. In 1989, Japanese women accounted for more than 18 per cent of science undergraduates; however, only a little more than 2 per cent of all undergraduate women chose to enrol in science and only 2.5 per cent in engineering. In 1986, Norwegian data showed that among women enrollees, 16.4 per cent chose natural sciences and engineering; for men, the corresponding percentage was 37.1 and this in a year when the total number of women enrollees was almost equal to that of men. In the Netherlands, from 1980 to 1988 the proportion of women in (university and vocational) higher education choosing science and engineering has remained at roughly 6.5 per cent; in 1988 approximately 13 per cent of all science and engineering enrolments were accounted for by women. In 1987/88, of beginning women students in Germany, almost 19 per cent chose mathematics, natural sciences (including biology, pharmacy and geography) and engineering; these same women accounted for approximately 25 per cent of total mathematics, natural sciences and engineering enrolments.

Just as the ratio of men to women in science and engineering enrolments has appeared to stabilise, so too has the ratio of women's enrolments in science and engineering to total women's enrolments. The situation does not provide much hope of future increases in the total numbers of scientists and engineers. In the United States and Canada, a recent drop in the proportion of women students choosing the physical sciences and engineering compounds the difficulties being encountered in the demographics of the higher education age cohort. While it is theoretically possible to increase the percentage of both men and women in science and engineering programmes, past performance makes this seem unlikely. And yet women surely constitute a large source from which increases in

choices of "quantitative" studies could make a, if not the, most significant contribution in the near future to the scientist and engineer population.

Students are not insensitive to opportunities for employment. In fact, there is some evidence that recent increases in higher education participation rates are due to student perceptions of the economic value of such training. A strong economy in which the production of goods plays an important role has, in the past, been correlated with growth in engineering enrolments and, to a lesser extent, natural science matriculations, e.g. in Germany and Japan. On the other hand, the recent economic relaxation in the United States and Canada may be linked to reduced student interest in "quantitative" curricula.

The quality of pre-higher education

In addition to factors relating to student preferences for different types of study, the moderately long term future availability of science and engineering personnel may be significantly affected by the quality of pre-higher education school systems. Science and mathematics achievement tests given to 13 year olds in 1988 in several countries showed a relatively poor performance for several OECD countries. Students in South Korea, on the other hand, performed relatively well. Within the United States, student mathematics and science achievement, in decline for almost two decades, has stabilised and even shown signs of reversal. A certain minimal proficiency is a prerequisite for those who will later choose to continue in higher education, but the levels of attainment of US children in science and mathematics remain discouraging.

Naturally, the quality of pre-higher education and the level reached by future science and engineering students do not only, or even mainly, affect numbers of future scientists and engineers, but several countries are concerned with the ultimate effect they have on the quality of their future scientists and engineers. In Australia, for example, the marks required from pupils entering university are lower for science studies than for the great majority of other subjects.

2. OTHER FACTORS AFFECTING THE SUPPLY OF SCIENTISTS AND ENGINEERS

Once a pool of scientists and engineers has been formed through education, two important factors influence the available supply in these fields. The first, and most important, is inter-profession mobility. The second is international mobility.

Inter-profession mobility

The seriousness of the likely reductions in the number of scientists and engineers stemming from demographic developments and the apparent stabilisa-

Figure 17. **Terminal degree fields for natural science and engineering (NS&E)**
B.Sc. holders in the United States
(Attained by the high school class of 1972 by 1986)

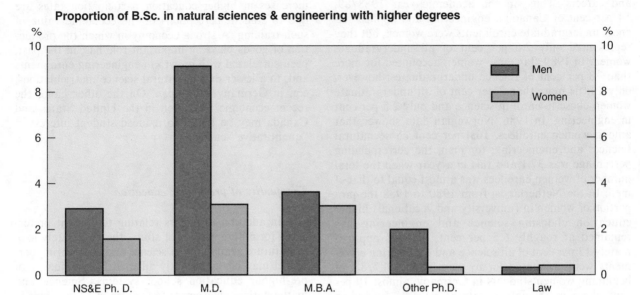

Proportion of B.Sc. in natural sciences & engineering with higher degrees

Source: United States, NSF (1986).

tion in percentages of women enrollees in technical fields is heightened by the increasing availability and attractiveness of professions outside the conventional bounds of science and technology. Many of these occupational opportunities require the knowledge and talents of those who have pursued science and engineering studies. They include such professions as law and medicine, financial analysis, industrial management, policy studies, organisation restructuring, and systems modelling and simulation. Many of the opportunities in such professions offer impressive financial rewards, large degrees of autonomy, and a social dimension that stands in sharp contrast to what is often perceived as the low value accorded to research and related activities.

Unfortunately, little statistical information is available on the extent of this phenomenon. However, a recent analysis by the US National Science Foundation (NSF) of the graduating secondary school class of 1972 in the United States indicates the destination – profession – of a portion of first degree recipients in science and engineering (Figure 17). Only some 3 per cent of men and under 2 per cent of women attained advanced degrees in natural sciences and engineering.

More recent data confirms and extends this distribution. Tables 20 and 21 provide another view of the activities engaged in by US scientists and engineers for the years 1976-1988 (estimates). From these tables one can calculate that in 1976 approximately 9 per cent of scientists and engineers were in non-science and engineering positions. By 1988 this had increased to 15.5 per cent (estimates).

Such personnel are of interest to non-scientific and non-engineering professions partly because of the high reputation enjoyed by scientific and engineering degrees in certain countries, but mostly because of the need for new skills resulting from the increasing application of scientific knowledge and new technologies in more and more professions. In this respect, little hard information is available concerning changes in higher education curricula. Institutions with programmes in business, finance, trade, policy studies and public administration, among others, have introduced quantitative studies, generously supported by computing and communications facilities. Their graduates are not only qualified to participate in their chosen non-scientific and non-engineering professions, but some of them may – with additional training – be able to gain the

Table 20. **Scientists and engineers employed in S&E jobs by field and gender in the United States, 1976-88**

	1976	1978	1980	1982	1984	1986	1988 (est.)
Total scientists and engineers	2 122 100	2 364 400	2 542 700	2 866 700	3 465 100	3 919 900	4 615 500
Men	1 947 200	2 153 000	2 269 900	2 552 500	3 070 400	3 393 700	..
Women	174 900	211 300	272 800	314 200	394 600	526 200	..
Total scientists	843 800	937 500	1 032 800	1 147 500	1 402 900	1 676 400	2 000 000
Men	689 100	753 800	806 200	887 700	1 078 200	1 242 800	..
Women	154 700	183 700	226 600	259 900	324 600	433 600	..
Physical scientists	154 900	168 200	166 300	210 500	234 000	264 900	286 500
Men	143 600	155 700	151 700	190 000	208 000	229 500	..
Women	11 300	12 500	14 500	20 500	26 000	35 400	..
Mathematical scientists	43 800	48 000	57 300	68 300	87 000	103 900	132 500
Men	33 700	36 700	42 100	45 500	68 200	78 900	..
Women	10 000	11 400	15 200	22 800	18 800	25 000	..
Computer specialists	116 000	171 400	196 700	216 100	340 400	437 200	551 800
Men	95 100	131 300	147 600	158 700	251 600	308 700	..
Women	20 900	40 000	49 100	57 400	88 800	128 400	..
Environmental scientists	46 600	56 900	63 100	82 700	89 900	97 300	98 200
Men	44 000	51 600	54 700	71 100	80 800	87 200	..
Women	2 600	5 300	8 400	11 700	9 100	10 100	..
Life scientists	198 200	227 800	267 300	298 000	294 100	340 500	380 800
Men	167 700	191 800	218 400	239 000	226 000	257 100	..
Women	30 500	36 000	48 900	59 000	68 100	83 300	..
Psychologists	103 700	107 400	112 500	105 600	151 900	172 800	229 200
Men	71 600	71 100	70 400	66 400	92 900	99 500	..
Women	32 000	36 300	42 100	39 300	59 000	73 300	..
Social scientists	180 500	157 800	169 700	166 200	205 600	259 800	321 000
Men	133 200	115 700	121 300	117 000	150 800	181 800	..
Women	47 300	42 200	48 300	49 200	54 900	78 000	..
Total engineers	1 278 000	1 426 900	1 509 900	1 719 100	2 062 200	2 243 500	2 615 500
Men	1 258 100	1 399 300	1 463 600	1 664 800	1 992 200	2 150 900	..
Women	20 200	27 700	46 200	54 300	70 000	92 600	..
Astronautical/Aeronautical	55 700	61 100	65 000	77 200	91 800	104 200	112 000
Men	55 100	60 400	63 700	75 100	89 600	100 300	..
Women	600	700	1 300	2 100	2 200	3 900	..
Chemical	76 400	81 900	89 000	101 100	127 500	131 500	132 200
Men	73 700	79 300	84 500	95 300	119 200	121 200	..
Women	2 800	2 500	4 500	5 700	8 300	10 300	..
Civil	182 800	205 200	217 000	243 700	293 000	319 100	311 200
Men	178 100	201 900	211 500	237 900	284 400	307 200	..
Women	4 800	3 300	5 500	5 900	8 500	11 900	..
Electrical/Electronics	267 900	327 000	357 400	413 500	475 000	540 800	601 500
Men	266 500	323 600	350 200	405 400	463 800	523 200	..
Women	1 400	3 500	7 200	8 100	11 200	17 600	..
Mechanical	272 800	296 500	308 800	334 400	414 000	453 700	597 900
Men	270 600	292 300	302 000	327 700	403 300	440 100	..
Women	2 200	4 200	6 800	6 700	10 700	13 600	..
Other engineers	422 700	455 200	472 700	549 200	660 900	694 200	860 700
Men	414 100	441 800	451 800	523 400	631 900	658 900	..
Women	8 400	13 500	20 900	25 800	29 100	35 300	..

Note: Detail may not add to total because of rounding. Total fields for 1988 were estimated based on the 1988 S&E employment rate for that year. Rates were not available by gender.

Sources: United States, NSF (1986) and unpublished data.

Table 21. **Scientists and engineers employed in non-S&E jobs by field and gender in the United States, 1976-88**

	1976	1978	1980	1982	1984	1986	1988 (est.)
Total scientists and engineers	209 100	245 400	317 700	386 400	530 400	706 600	859 100
Men	184 400	214 600	274 900	311 600	412 500	534 100	..
Women	24 800	30 900	42 800	74 700	118 000	172 400	..
Total scientists	115 700	133 500	151 700	258 200	378 500	509 900	624 800
Men	92 200	103 800	111 800	187 400	265 100	343 900	..
Women	23 500	29 700	39 900	70 700	113 400	166 000	..
Physical scientists	34 400	40 100	48 900	16 900	20 100	23 500	24 900
Men	29 100	34 100	42 800	15 100	17 800	20 600	..
Women	4 900	6 000	6 300	1 800	2 300	2 900	..
Mathematical scientists	4 800	5 700	7 000	11 100	13 400	27 100	34 800
Men	3 400	3 800	4 300	8 500	10 300	18 200	..
Women	1 500	1 700	2 800	2 500	3 100	8 900	..
Computer specialists	3 000	5 600	11 100	82 900	96 400	125 400	158 400
Men	3 300	5 500	2 300	61 600	71 100	91 300	..
Women	0	200	8 800	21 300	25 300	34 100	..
Environmental scientists	8 200	12 000	14 500	4 500	8 200	14 000	14 400
Men	6 900	10 100	12 100	3 700	7 000	11 200	..
Women	1 300	1 900	2 300	700	1 200	2 800	..
Life scientists	15 300	16 300	20 200	39 100	59 200	71 300	79 600
Men	11 900	12 700	16 000	29 500	44 700	51 900	..
Women	3 400	3 600	4 200	9 600	14 500	19 500	..
Psychologists	8 800	14 300	15 600	32 800	57 600	80 700	104 900
Men	5 300	8 600	9 000	16 600	28 200	38 900	..
Women	3 600	5 700	6 600	16 100	29 400	41 900	..
Social scientists	41 800	39 600	34 300	71 000	123 600	168 000	207 800
Men	32 500	28 900	25 400	52 300	86 000	112 000	..
Women	9 300	10 600	8 900	18 700	37 500	56 000	..
Total engineers	93 400	111 900	166 000	128 100	151 900	196 600	234 300
Men	92 200	110 700	163 100	124 200	147 400	190 200	..
Women	1 200	1 100	3 000	4 000	4 500	6 400	..
Astronautical/Aeronautical	1 100	900	4 500	3 600	5 400	6 300	6 600
Men	1 300	1 000	4 600	3 600	5 300	5 900	..
Women	0	0	0	0	0	400	..
Chemical	1 100	2 300	5 500	6 600	12 600	17 500	17 400
Men	1 300	2 400	5 500	6 300	12 100	16 600	..
Women	0	0	0	400	500	900	..
Civil	5 400	6 500	15 100	14 500	19 700	27 200	26 700
Men	4 700	6 500	14 800	14 300	19 000	26 200	..
Women	600	0	300	200	800	1 000	..
Electrical/Electronics	15 100	14 500	25 700	24 200	25 700	33 700	37 700
Men	14 900	14 400	25 200	23 200	24 700	32 300	..
Women	200	0	400	1 000	1 000	1 300	..
Mechanical	3 400	2 800	13 800	23 500	31 600	38 900	51 300
Men	3 300	2 900	14 000	23 100	31 300	38 500	..
Women	100	0	0	400	300	400	..
Other engineers	67 300	84 900	101 400	55 700	56 900	73 000	94 600
Men	66 800	83 600	98 900	53 700	55 000	70 700	..
Women	700	1 200	2 600	2 000	1 900	2 400	..

Note: Detail may not add to total because of rounding. Total fields for 1988 were estimated based on the 1988 S&E employment rate for that year. Rates were not available by gender.
Sources: United States, NSF (1986) and unpublished data.

knowledge required for the exercise of certain scientific and engineering tasks. Given the large numbers of students choosing to study business and related disciplines, and the importance placed on the mastery of quantitative skills in the performance of related tasks, the possibility of a transfer or at least an amalgamation of skills should not be dismissed. This changing situation may tend to reduce the opportunity for conventionally trained scientists and engineers to leave their original professional choices.

In most countries, the principal source of annual additions to the science and engineering personnel pool will be recent graduates. However, several other sources of supply deserve mention. One is that of upgraded workers, usually individuals without conventional certification who, by virtue of on-the-job experience, are able to execute tasks normally reserved for recognised professionals. A US estimate of the contribution of this category is about 6 per cent. In France and the United Kingdom, government policies have been introduced which favour promotion from technician and related occupational status to the professional scientist and engineer category.

Another possible, although not as yet very significant supply source is the converting of university graduates. These are individuals who, for the most part, have taken non-science degrees but who have the capacity for quantitative study. A particularly successful example is a one year course to train information technology specialists offered in the United Kingdom.

If the blurring of boundaries among disciplines due to progress in the creation and use of knowledge continues to accelerate, it may lead to situations in which conventional definitions and categories of knowledge sectors, occupations and activities become more fluid. This might in turn relieve actual and potential labour imbalances without disruptive economic and social impacts. The possibility must be kept in mind in any attempt to deal with the issue of the availability of highly skilled personnel, even though experience has shown the difficulties and limits involved in attempting to reconvert scientists in public sector laboratories which have been obliged to re-orientate their activities.

One specific boundary which seems to be breaking down is the one separating training geared specifically to research from training for engineers. It is increasingly recognised that training through research is desirable for engineers even if they undertake no research functions, or do so only temporarily, during their working life.

International mobility

Immigration and emigration may also have an important effect on the supply of scientists and engineers. In most OECD countries, a not insignificant proportion of students in science and engineering is of foreign origin. These students may return to their country of origin to complete their studies or, after obtaining their degree, to look for work. Conversely, supply may be swollen by the return of students who studied abroad. But even when numbers of foreign students are known, it is impossible to say how many of them will remain and find (look for) work in the country of study, and how many will return home.

It is of course not only students and young graduates who immigrate or emigrate, but also scientists and engineers at different stages of their careers. Immigration has traditionally constituted an important source of scientists and engineers for countries like Australia, Canada or the United States. Although the United States expressed its concern some years ago at a certain reduction in the mobility of scientists, notably to and from Europe, 1986 nevertheless saw the arrival in the United States of some 11 000 scientists and engineers, i.e. roughly 7 per cent of the annual increase in this professional category. Moreover, the United States recently adopted measures specifically designed to facilitate the immigration of scientific and engineering personnel. However, despite the obvious short-term benefits, the large-scale immigration of scientists and engineers is not a satisfactory solution to the risk of shortages.

The lack of internationally comparable statistics for international mobility is very keenly felt. In addition to the difficulty mentioned above, it is impossible to distinguish temporary mobility flows, which in the long run benefit the country of origin as much as the host country, from a "brain drain", the spectre of which has recently returned to haunt certain OECD countries. With one or two exceptions, however (notably Greece), OECD countries will probably not experience any significant migratory outflows of scientists and engineers in the coming years. As things stand at present, certain developing countries and the countries of Central and Eastern Europe appear under much greater threat of losing their scientific and engineering graduates through emigration. If this should happen, it could have serious consequences for the development or reconstruction of their economies.

3. THE DEMAND FOR SCIENTISTS AND ENGINEERS

In the preceding sections, an attempt has been made to identify the factors which determine scientist and engineer supply and to comment on the nature of their evolution and likely contribution to existing and future labour supply. Prediction of such stocks has not been attempted for lack of methodologies and necessary data. As a recent OECD study has stated: "In view of the complexity of the human resources supply system and of the difficulty of quantifying many of the variables involved it is not surprising that prediction of the future supply of researchers is fraught with pitfalls and probably about as reliable as weather forecasting" (1989d, p. 24).

Uncertainty concerning future supply is all the greater because it depends to a not insignificant extent on demand. Job prospects are one of the factors which influence decisions taken by students, young graduates or potential immigrants. It is moreover in line with expected future demand that governments, the private sector and educational circles take an interest in supply. In addition, demand is not independent from supply. To cite but one example, the availability of qualified scientific and engineering staff ranks high among the factors determining where companies, and notably multinational companies, choose to locate their laboratories and manufacturing plants. When making an initial estimate, demand is usually considered as an independent variable, yet even so, forecasting demand is still more difficult than forecasting supply.

The fundamental difficulty of forecasting demand

The demand for scientists and engineers is made up of a multitude of factors, some of which have a complex interrelationship. To cite a relatively simple example, in the face of reduced supply, industry and higher education compete with each other to recruit young science and engineering graduates. At the same time, industry is dependent on teaching in the higher education sector and therefore needs a sufficient number of teachers. Nonetheless, the main general factors influencing demand can be described.

As in the case of supply, some are demographic, such as retirement and death. Demand can be affected by changing the age of retirement. Thus, early retirement has been promoted in certain sectors so as not to block the recruitment of young graduates when "natural" demand was low due to the age pyramid and/or a levelling out or reduction of public expenditure. Conversely, the danger of an insufficient supply of scientists and engineers leads some to argue in favour of postponing the age of retirement. Others object that this solution reduces creativity, increases conservatism and thus slows down scientific and technological progress.

A second series of factors, concerning essentially but not exclusively the private sector, relates to the medium and long-term prospects for economic growth, which are notoriously hard to predict, and to changes in the relative importance of different sectors and branches, some of which are more labour-intensive in terms of scientific and engineering personnel than others.

Changes in government policies also have an important influence, both direct and indirect, but are equally difficult to predict. It is well known, for example, that the size of defence-related industries in countries like the United States, France and the United Kingdom means that changes in defence policy have an impact on the volume and structure of demand for scientists and engineers. The changes in East/West relations led some experts to predict an eventual decrease in this component of demand, but developments in 1990 and 1991 may make this change less likely. In general, government expenditures on R&D and in higher education are major factors in determining the demand for scientists. Finally, there are uncertainties: how will any increased and lasting priority given to environmental protection and improvement (see Chapter 9) affect the need for scientists and engineers?

A final series of factors may be grouped under the not altogether accurate heading of the "productivity" of scientists and engineers. It includes a whole series of often controversial issues, qualitative in nature but with potentially important quantitative consequences. For example, to what extent and in what jobs are engineers with post-graduate training and research experience more "productive" than those without such training and experience? Is it possible to increase the vitality and pace of scientific progress by being more selective and retaining only the most "promising" young engineers? Are there ways of raising the student/teacher ratio without at the same time reducing the quality of future scientists and engineers? Can engineers be replaced by staff with a lower degree of initial training (by upgrading technical staff, mentioned in relation to supply)? Can we develop technologies which require fewer scientific and engineering personnel? Would the number of engineers and scientists required to develop and produce such technologies in fact be smaller than the number "saved"?

Government, industry and academia are well aware that demand predictions are circumscribed by great uncertainty. This stems in part from the basic characteristics of democratic societies, in which freedom and the abundance of choice, in large part due to scientific and technical progress, must frustrate any labour force determination efforts. Thus, in some respects, the failure to predict can be viewed as a measure of a society's vitality, initiative and even growth, within a framework that guarantees the respect of basic social and political values. However, this in no way relieves the obligation of governments and others with public responsibilities to secure the uninterrupted functioning of the many activities necessary to meet a wide range of needs and expectations. Planning for an adequate science and engineering labour pool is a form of accountability, regardless of the quality of the results. And in this particular instance, the results may yield insights that, in combination with judgement and intuition, can reduce distortions in personnel markets with potentially very serious national consequences.

The anticipated increase in demand for scientists and engineers

In recent years, several OECD countries have attempted to assess future demand for some or all categories of the scientific labour force. The Netherlands, for example, has attempted to evaluate the future labour market for scientists. Models were developed in which past trends were combined with likely future supply and demand and in which the annual turnover rate arising from retirement and job changes was fixed at 5 per cent of existing jobs. The results are summarised in Table 22. On the basis of the calculations, it appears that significant shortages of scientific and engineering personnel will take place between 1985 and 2000. The estimates set out in these tables are based solely on an extrapolation of trends.

Any changes in the socio-economic situation or policies adopted could lead to very different results.

A joint effort between the UK Council for Industry and Higher Education and the Institute of Manpower Studies at Sussex reveals that the United Kingdom will need a considerable increase in the number of scientists and engineers in the next ten years. At present, recruiting in electronics, computing, engineering and R&D is difficult. Since 1982, industry has increased the stock of engineers by 25 per cent. Within the next five years it is expected that professional and related occupations will see growth of the order of more than 17 per cent.

In France, the *Bureau d'Informations et de Prévisions Économiques* (BIPE) has estimated that between 1987 and 2000 some 250 000 additional engineers and technical staff will be needed – a number quite in excess of current production rates.

On the other hand, a study issued by the Centre of Planning and Economic Research in Athens indicates that for the period 1984-95 Greece will have a surplus of university trained personnel in all areas (Glytsos, forthcoming). But since the range of fields covered by the study is very wide, this overall surplus does not exclude difficulties or shortages for certain specific skills.

In the United States, the 1989 NSF *Science Indicators Report* concluded that, on the basis of the employment indicators used, apart from a shortage of engineering graduates, the present labour market situation for scientists is more or less satisfactory. Looking

Table 22. **Annual demand and supply shortages of graduates entering R&D occupations in the Dutch labour market**
1985-2000

	Demand	Supply	Shortage	Average shortage (% of supply)
Agriculture	110 - 190	20 - 80	90 - 110	(200)
Nature	540 - 960	140 - 430	400 - 530	(165)
Technology	640 - 1 130	70 - 220	570 - 910	(510)
Health	160 - 310	90 - 280	30 - 70	(30)
Economics	100 - 140	20 - 70	70 - 80	(165)
Law	60 - 90	30 - 80	10 - 30	(35)
Behaviour and society	250 - 420	100 - 240	120 - 210	(95)
Language and culture	110 - 180	40 - 140	40 - 70	(60)
Total	1 980 -3 420	520 -1 600	1 460 -1 820	(145)
(average)	(2 700)	(1 100)	(1 600)	

Note: The estimated annual shortages - i.e., averaged over the period 1985-2000 - have been calculated on the basis of the difference between the lowest and highest variant for supply and demand. A shortage is expected in all higher education and research plan (HOOP) sectors. The relative shortage per sector (arithmetic mean) is given in brackets as percentage. For instance, the estimated shortage of agricultural researchers - 100 on average - is twice as great (200 per cent) as the average annual supply (50). The estimated shortage in the technology sector is over five times the average annual supply.

The table should be read as follows. The high and low variants indicate the highest and lowest value produced by the two calculation methods for both supply and demand. The total shortage in the year 2000 is estimated at 1500 to 2000 researchers.
Source: Netherlands, Ministry of Education and Science (1989).

to the future, the same report states that over the 1988-2000 period, US private industry is expected to create more than 600 000 additional jobs for scientists and engineers. S/E employment is projected to increase by over 33 per cent, nearly four times that of the overall industrial labour force. Despite the substantial growth, the projected gains in S/E requirements should not match past increases, due to the overall slowdown expected in the 1990s of growth in the labour force, total employment, and GNP (United States, NSF, annual).

The report also notes that in a vigorous economy, temporary imbalances appear unavoidable (e.g. recent shortages in some engineering sub-fields and computer specialties). However, corrective action may be required in the 1990s, because of the decline in the 18 to 24 year-old college-age cohorts and increased retirements of experienced science and engineering workers. The report speculates that some of the gap may be filled by increased enrolments of 18 to 24 year-olds, older students and foreigners who remain to work in the United States. Small shifts in the percentages of students choosing to train in and enter S/E fields could provide an adequate supply of new entrants to the S/E workforce. However, adjustments in enrolments patterns in response to a growing demand for S/E graduates may not prove sufficient. Further adjustments may entail substantial costs and possibly affect the quality of the S/E work force.

Special higher education problems

Despite uncertainties about the future evolution of global demand for scientists and engineers, it would seem that in many OECD countries the higher education sector is likely to experience very serious difficulties in the coming years, due to the particular way in which this sector has evolved. With few exceptions, OECD country university systems expanded extremely rapidly during the 1960s and early 1970s in response to post-war demographic growth, the greatly increased demand for higher education and also a demand for research. In many countries, this explosion, made possible by the prevailing economic climate, suddenly gave way, from 1975, to a period of stagnation or even contraction, due to the combined impact of the economic crisis, the demographic downturn which followed the baby-boom, and the fact that demand for research and higher education could not continue to grow at the same pace indefinitely.

Given this pattern, the age pyramid for university staff has a special structure: it is narrow at the base and the summit, with a considerable bulge in the middle for the 45-60 year age group which will gravitate towards the top roughly until 1995 before giving rise to very sizeable recruitment needs. This general problem is particularly well illustrated by the situation in Germany. After having weathered growth rates of more than 10 per cent per year from 1960 to 1975 in senior academic posts, increases dropped to 0.1 per cent per annum between 1980 and 1985. In that period, the rate of staff retirements was 1.8 per cent per annum. During the next twenty years, however, this is expected to increase to 5 per cent per annum. Other nations in which large proportions of research activity are conducted in academic and public sector facilities, and which will face the same difficulties, include Norway, France, Australia, Canada and the Netherlands.

The same concern regarding academic science and engineering faculty has been voiced in the United Kingdom. Pearson (1990) notes that unusually large numbers of academics will retire during the next twenty years. Consequently, staff recruitment by universities could rise by as much as 50 per cent between 1995 and 2000, the period in which the demographics of the university age population becomes positive. Perhaps as many as 400 scientists and 200 engineers would be needed each year. In numerical or even in percentage terms, this is not really very large. However, the concern expressed underscores the importance of attracting sufficient numbers of highly qualified staff to the academic sector. In the present situation, where academic salaries are considered unattractive, shortages are likely to skew the quality balance of science and engineering staff to the disadvantage of higher education. In the United Kingdom, current indications of inadequacies in fields such as information technology, physics and the biological sciences would only be exacerbated by increased recruitment difficulties.

In the United States, an NSF report on future scarcities of scientists and engineers (1990) gives close attention to scientists and engineers earning a Ph.D. degree. Figure 18 presents "most likely" estimates of demand in the academic, private and government sectors from 1988 to 2006. While at present, 5 000 Ph.D. positions become available annually as a result of death and retirement, the number is expected to increase to 8 000 by the end of the century and to 10 000 by 2006. More than half of this replacement demand derives from the academic sector. New position demand is primarily ascribed to growth in research spending and, within higher education, to enrolment increases. On the basis of model calculations, demand is expected to increase precipitously in the mid-1990s (see Figure 19), causing severe imbalances which cannot be easily or quickly rectified, given the relatively long time spans needed to produce Ph.Ds.

Figure 18. **Potential jobs for new Ph.D.s at 1988 salary levels, expected enrolments and continuing national R&D trends in the United States**

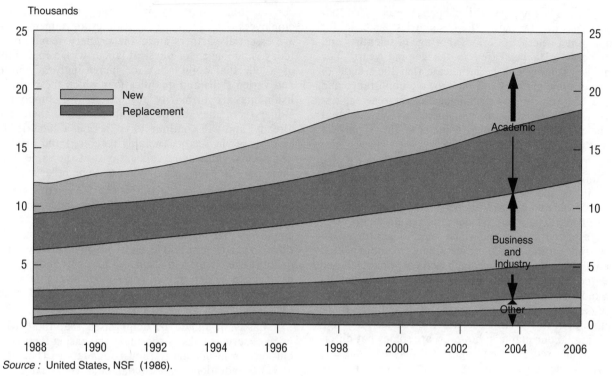

Source : United States, NSF (1986).

Figure 19. **Market adjustment to shortages of new NS&E Ph.D.s compared to demand for and supply of Ph.D.s at constant salary in the United States**

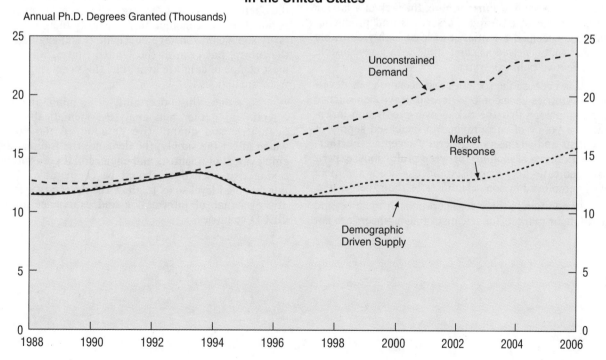

Source : United States, NSF (1986).

4. SOME POLICY CONCLUSIONS

At the beginning of this chapter, it was stated that a key question for most OECD countries has been, for some time, whether a serious shortage of scientists and engineers is to be feared in the coming years. The analyses summarised here indicate that the weight of demographic changes and the growing importance and pace of scientific and technological progress suggest that the situation must be monitored carefully and that policies must be adopted to increase the supply of scientists and engineers.

The justification for an active government policy in this field lies not only in the risk of shortages but also in the special characteristics of the labour market in scientific and engineering personnel. In conventional labour markets, where efficiency of resource allocation is considered a valid and viable mechanism for assuring the public good, adjustments to unexpected demand are usually accomplished via financial and related measures. This usually occurs as a result of abundant human resource reserves, short-term training, and the fairly high degree of fungibility to be found even in the craft sectors. The science and engineering labour markets, however, are somewhat different. While scientist and engineer supply and demand may balance on the basis of economic considerations, the allocation of resources would not be socially adequate if, for example, the key tasks in control, monitoring, inspection and regulation which underlie the operation of modern societies were not performed or not properly carried out. Furthermore, the lack of a labour force capable of ensuring the services and producing the goods deemed essential or at least highly desirable can hardly be viewed as compatible with a responsible and effective political system.

It is because the societies and economies of developed countries have come to depend heavily on science and technology that the risk of over-production inherent in a policy of encouraging an increased supply of scientists and engineers is generally accepted. Furthermore, in terms of mobility between professions, experience indicates that scientists and engineers are more easily employed in non-scientific jobs than the reverse.

In addition to these considerations, it must be said that labour markets for scientists and engineers are not conventional markets involving employers and employees only. Higher education establishments also are essential partners since the primary requirement today is no longer for physical labour with a minimum of intellectual ability but for increasingly advanced and varied skills and qualifications which these establishments are primarily responsible for supplying. While there may be no overall shortage, there are already, and will continue to be, serious difficulties in certain specific sectors (notably teaching) and for certain special skills and qualifications; at the same time there may be excess production of some other skills.

That is why any labour policy for scientists and engineers must, whatever else it does, endeavour to promote permanent and close co-operation between employers and higher education and training institutions, in particular at all levels at which qualitative and quantitative changes in the needs for skills and qualifications can be determined with sufficient accuracy. In this way, the educational system can actually adapt to these changes.

Lastly, it should be emphasised that those who warn of serious risks of shortages are basing their conclusions on insufficiently robust foundations (Fechter, 1991). In addition to the fundamental complexity and uncertainty of the supply of, and even more the demand for, scientific and engineering personnel, there is an almost universal shortage of data and information – especially internationally comparable data – about the factors involved (Zinberg, 1991). A number of national studies conclude that generally there will be a fairly significant shortage of scientists and engineers in the future, but even in the United States, where the most complete data are available, this conclusion is far from unanimous.

To reduce this uncertainty to a minimum, it is essential to gather and analyse systematically both qualitative and quantitative data on all the factors which affect the supply, the demand, the training and employment of scientists and engineers. It goes without saying that such data should be comparable at the international level so as to allow the necessary systematic exchange of information and experience among OECD countries.

Chapter 7

HUMAN RESOURCES AND NEW TECHNOLOGIES IN THE PRODUCTION SYSTEM

ABSTRACT

The importance of "human capital" to national wealth was recognised during the post-war period. The 1980s saw renewed government interest in human resource development, and questions have been raised concerning the suitability of existing education and training systems for meeting the needs of international competition and the particular challenges posed by new technologies.

Changes in demand for human resources in the production system are much more rapid and congruent across enterprises and nations than changes in supply. At the macro-economic level, a major feature of the more knowledge-intensive economy is rapid growth of tertiarisation. At the micro-economic level, a new model integrating new technology, work organisation and skill formation, with particular emphasis upon flexibility and the quality of human resources is now gradually replacing the previous paradigm of corporate management and organisation of the workplace (discussed in Chapter 4). New technologies may make available new options for work organisation and human resources development. The latter is becoming a new factor in competitiveness, and newly created jobs tend to demand higher levels of skills. The main features of the emerging skill structures are increased emphasis upon multi-skilling, customer-oriented communication skills, problem-solving and entrepreneurial skills.

The capacity of a critical number of firms in a given country to move towards post-tayloristic forms of work organisation appears to be strongly influenced by that country's education and labour market institutions. Institutions of formal education and training seem slow to respond to changes in demand. Lifelong learning has become the essential prerequisite for the emerging knowledge-intensive economy, with its rising minimum competence threshold. Curricula and school career paths must be organised to provide every school leaver with fundamental competence in literacy and numeracy, reasoning, communication skills, positive work values and attitudes, and the ability to learn new skills.

The implementation of a system designed to achieve these goals and to ensure lifelong learning requires close co-ordination at the policy level. It involves the building of new networks and partnerships between different public and private providers of education and training for both youth and adults. Greater transparency in programmes must also be ensured. Enterprise-based education and training have increased, and a great variety of programmes to combat unemployment have been launched by public authorities. The public sector should concentrate on providing education and training after initial education where the private sector cannot and will not do so. In the emerging knowledge-intensive economy, an efficient partnership between different providers of education and training would enable a more effective use of limited resources and also facilitate the development of adequate supplies of human resources by including the adult learner, who is a crucial resource for enterprises and nations during the decade ahead.

INTRODUCTION

In addressing issues of the relationship between new technologies and human resources, different methodological approaches can be applied. A large body of micro-economic studies of enterprises and/or sectors focuses, in particular, on the qualitative changes taking place. Fewer macro-economic studies have been made, because of the complexity of the subject and the lack of adequate data. As a result, while both perspectives will be considered in the following pages, micro-economic findings will dominate the discussion of developments and changes in demand and supply and their interaction. The final section identifies some policy issues related to demand and supply measures and to the need to strengthen the data base[1].

The importance of the "human capital" component of national wealth was recognised and analysed by many economists during the post-war period. During the 1980s there was increased government interest in human resource development. Many national governments have begun to address the issue of whether existing systems of education and training are adequately equipped to meet increasing international competition and the particular challenges posed by new technologies. In some quarters, fears that new technologies may increase unemployment have been reinforced by an apparent paradox: in many OECD countries, economic growth has been accompanied over a long period by very high unemployment.

One school of thought has tried to explain this paradox in terms of an emerging "jobless growth society", where a growing economy is unable to create full employment as traditionally defined. Another explanation is that it is a temporary phenomenon that increased deregulation and freer market forces would solve over time.

Recently, a third school of thought and analysis has emerged which argues that many OECD countries are rapidly moving towards a more knowledge-intensive economy based to a large extent upon a new techno-economic paradigm (Freeman and Soete, 1985). In these new structures, major clusters of new technologies will significantly change the way economies operate; they have considerable potential for productivity increases and thus lead to rising growth and employment. If these new technologies have so far had relatively little impact upon productivity increases, it is because society has not succeeded in matching these advances with the necessary innovation in the workplace. This interpretation of the paradox of economic growth and high unemployment postulates that the crucial bottleneck is insufficient human resources development at both the micro-economic and macro-economic levels. As Eliasson and Ryan (1986) noted, when human knowledge replaces machines and plants as the moving factor behind productivity change, and when the required knowledge is missing, output suffers however much finance and hardware are supplied. A recent study on the competitiveness of the United States economy (Dertouzos et al., 1989) argues that the first of five principal imperatives for a national effort to regain competitiveness is the need to invest more heavily in the future and, above all, in human capital. An OECD report (1988b) reached a similar conclusion, stressing that neither the technical nor the economic potential of major new technologies can be fully realised without concomitant, even anticipatory, social and institutional changes, particularly in education and training systems.

In addressing the issue of human resources development, the changing demographic situation must also be taken into account (see Chapter 6). At the micro-economic level, many companies may well be forced to look for workers among groups they once ignored. They are likely to find themselves caught up in such issues as education, child care and literacy. In some countries, competition for young people between companies and the formal education systems could develop. Companies might find it advantageous to employ young people and give them education and training before they have finished formal education in order to secure the renewal of their work force. At the macro-economic and government policy level, the demographic changes may lead to greater flexibility of retirement age. From this perspective, government policy towards adult education and training and retraining will gain considerable importance during the decades ahead.

1. CHANGES IN THE DEMAND FOR HUMAN RESOURCES

Overall changes: the increasing importance of the service sector

Today, the crucial role of human resources in the full utilisation and diffusion of new technology is generally recognised, as is the contribution of new technologies to the development of a more knowledge-intensive economy (see Box 33). It can be argued that, at the macro-economic level, one of the major features of more knowledge-intensive economy has been the rapid growth of the tertiary or service sector of the economy where new technology often plays an important role. The overall proportion of employment in the service sector rose from 49 per cent in 1970 and 57 per cent in 1980 to 62 per cent in 1988 (see Table 23). While in

1970, only the North American countries had recorded levels of over 60 per cent, eight countries reached this range by 1980 and twelve by 1988. The United States proportion was then over 70 per cent, while in Germany and Japan it remained under 60 per cent. To some extent, this growth reflected the practice of contracting out service-type activity from firms in the manufacturing sector to firms in business services, which saw a particularly rapid expansion over the period.

The rising level and diffusion of information and information technology and its integration into every aspect of productive work, in conjunction with the overall movement towards knowledge-based products and industries in most advanced countries (and

---

Box 33. Human capital and knowledge intensity

The importance of the "human capital" component of national wealth has been widely recognised in recent decades. It has grown rapidly in advanced economies, not only in absolute terms but also relative to physical capital. Kendrick (1976) estimated that the stock of intangible human wealth (comprising the education, training and health care embodied in the labour force) grew from 48 per cent of the stock of physical capital in 1929 to 75 per cent in 1969. Education and training accounted for the bulk (89 per cent) of the stock of human wealth in the latter year.

As advanced economies become more knowledge-intensive, human resources become more central to economic progress. According to Drucker (1983), "knowledge work" has become the economy's growth area in the mass conversion to new technology; this means that productivity will increasingly be determined by the knowledge and skills workers put into their tasks. Eliasson and Ryan (1986) notes that human knowledge has replaced machines and plant as the moving factor behind productivity change, and that when the requisite knowledge is missing, output suffers – however much finance and hardware are supplied. The requisite human resource development involves investment principally in education and training, with secondary contributions from health and labour mobility.

Table 23. **Share of services in total civilian employment, 1966-89**

Percentages

	1966	1970	1975	1980	1985	1989
Canada	57.4	61.4	64.6	66.0	69.5	70.1
United States	58.3	61.1	65.3	65.9	68.8	70.5
North America	58.2	61.1	65.2	65.9	68.9	70.4
Japan	45.1	46.9	51.5	54.2	56.4	58.2
Australia	53.3	55.0	59.4	62.4	66.2	68.0
New Zealand	48.1	48.6	53.5	55.4	56.5	64.3
Austria	41.1	44.0	46.6	49.3	52.9	55.1
Belgium	49.2	52.5	56.5	62.1	66.7	68.7
Denmark	48.6	50.7	58.8	62.4	65.2	66.9
Finland	38.3	42.8	49.0	51.8	56.5	60.2
France	43.9	47.2	51.1	55.4	60.4	63.5
Germany	41.3	42.9	47.6	51.0	54.5	56.5
Greece	30.4	34.2	36.8	39.5	43.7	47.1
Iceland	45.5	46.9	49.5	51.4	55.8	59.2
Ireland	40.7	43.1	45.8	49.2	55.2	56.5
Italy	37.8	40.3	44.1	47.8	55.2	58.2
Luxembourg	41.6	46.3	49.6	56.5	61.9	65.4
Netherlands	52.1	54.9	59.4	63.6	67.0	68.8
Norway	46.5	48.8	56.1	61.9	65.4	68.1
Portugal	30.9	37.1	32.3	36.1	42.2	45.7
Spain	35.7	37.4	39.6	44.6	49.9	54.0
Sweden	48.3	53.5	57.1	62.2	65.3	67.0
Switzerland	42.1	45.4	50.2	55.0	58.3	59.3
Turkey	17.5	17.9	21.3	26.2	27.7	29.5
United Kingdom	50.0	52.0	56.8	59.7	65.8	68.4
OECD - Europe	40.5	42.9	46.6	50.6	55.0	57.6
EC	42.6	45.0	48.9	52.7	57.8	60.7
OECD - Total	46.6	49.3	53.5	56.6	60.3	62.5

Source: OECD Labour Force statistics.

increasingly in newly industrialised economies), have reinforced the process of "tertiarisation". This evolution has also created a considerable amount of blurring, notably between what might otherwise be considered as pure manufacturing or as pure service activity. The blending of service and manufacturing activities has intensified in three distinct areas: *i)* the research, design and planning of new products; *ii)* the development and use of software for manufacturing processes; *iii)* the adaptation, customisation and marketing of these new products.

Another important aspect of "tertiarisation" is the productivity/employment nexus[2]. Knowledge of past, present and expected productivity developments in the various service sector industries should therefore be of considerable help in assessing and analysing employment trends. However, the severe problems in measuring service sector productivity (notably biases on account of choice of base year, secular changes in the self-service component of consumption and changes in the complexity and heterogeneity of service sector products) must be kept in mind (see Chapter 8). Examination of total productivity figures from OECD countries (Englander and Mittelstadt, 1988) and relevant empirical literature (Kendrick, 1988) nonetheless reveals a number of general characteristics of the relationship between productivity and employment:

i) Service sectors with slow productivity growth in the past tend to be those which have also recorded high levels of employment growth.

ii) However, this relationship appears to be unstable over longer periods of time. For example, distribution services (retail and wholesale trade, accommodation and food) recorded among the highest absolute and relative employment gains in the early stages of service sector growth but showed a considerable deceleration of employment growth in a number of countries during the latter part of the 1980s.

iii) Inter-industry variability in the productivity/ employment nexus appears functionally related to variations in the structure and conduct of particular service industry markets (local monopoly characteristics, size and size distribution of firms, tradeability and international competition, technology content of services).

iv) In general, the more capital and information technology (IT) intensive communication and transportation sectors have shown high productivity and low employment growth. At the other extreme, traditional tertiary activities such as personal services, trade and accommodation and food have, in several countries, shown a quite dramatic rise in employment but a poor record in productivity growth. The extent to which this sector has been able to absorb an abundant supply of cheap and unskilled labour has largely been contingent on wage flexibility and the erosion of certain labour

relations institutions. This can been seen in international differences in job growth and employment creation in distribution services. (Compare, for example, Australia, the United States, Italy and Japan to Sweden, the Netherlands, France, Germany and Canada.) Productivity developments and employment creation in the social services, a major service sector in terms of employment, have been largely contingent upon government policies.

Changes in skill requirements at the level of the firm

From a variety of case studies at the enterprise level (OECD, 1989*o*) it is possible to identify, in broad terms, how modern enterprises are reacting to the challenge of the knowledge-intensive economy. The principal features (see Box 34) are shared by most enterprises, although they are being addressed in different ways, largely as a consequence of different traditions and trends in the different national labour markets, industrial relations and skill formation systems. Changes resulting from the new knowledge-intensive economy significantly affect skill structure within firms; they are occurring notably in the areas of work organisation, occupational structure and the pace of change.

It is extremely difficult to define the emerging skill structures in any detail. Activities and industries within the economy vary widely, and the skill structure itself is greatly influenced by a number of factors which, by themselves, are difficult to capture. For instance, work organisation is crucial to the kinds of skill structures that develop in a given enterprise. Broadly speaking, two opposing trends exist: one still follows a taylorist approach; the other, more effective in this context, involves employees in more responsible tasks, team-work and human resource development. The second type is increasing rapidly, although the former still predominates in many work settings.

Another factor affecting skill structures and related to changes in work organisation is changes in occupational structure. The adoption of information technologies is changing them in two ways (OECD, 1988*b*). The first reflects the potential and actual use of information technologies to automate many routine and less complex jobs. This phenomenon has been clearly observed in the manufacturing industries. Thus, the short-term labour and labour-saving effects of information technology are slanted towards less skilled jobs. Similar compositional changes in occupational structure can be expected in white collar areas where the productivity potential of information technology will lead to reduced employment of less skilled clerical employees.

The second aspect relates to changes in the size and distribution of workplaces. Here one observes a move towards small-scale units of production of both goods and services. Often this smaller unit has a less hierarchical structure and work organisation, allowing for greater responsibility and participation of individual workers. As a result, the shape and structure of occupations is likely to change from the conventional pyramid, with large numbers of unskilled and semi-skilled workers at the bottom, to something resembling an orange or an onion in which a bulge occurs at the middle level of skills.

Still another factor influencing the skill structure is the accelerating pace of change in products and services. Increasingly, enterprises are being obliged to meet two principal conditions in order to stay in business: they must produce high-quality products and services and they must shorten the life cycles of these products and services. To handle this situation, enterprises are forced to develop more flexible work organisation, to emphasise multi-skilling and to build up a skill reserve prepared to respond to unforeseen changes in markets and final demand.

Phases in the training of the work force

Given these developments in work organisation, occupational structure and the general pace of change, it becomes clear that any specific and clear-cut picture of emerging skill structures is more or less impossible to establish. However, some of the main features of a broader picture can be discerned. It is important to begin by identifying the different forces that lie behind a decision at enterprise level to invest in the education and training of the work force (on this topic, see also Chapter 5). Three different phases can be identified (see Figure 20), and each has different implications for the skill structure and competences needed.

In *a first phase,* the enterprise decides to invest in the training of its work force in relation to its tangible investment and often in conjunction with the introduction of a new technology or the launching of new products or services. This training often tends to be product-specific. Once the new equipment and products have been mastered, these investments tend to flatten out, particularly if the products are expected to have a rather long life cycle. This phase could be called a *product-driven* training strategy, and it is clearly slanted towards specific skills.

A second phase emerges when enterprises must respond to market changes in the life cycles of products and services by producing a greater variety of products of higher quality. In this phase, enterprises tend to look for more flexible work organisation, and they increasingly focus on developing multi-skilled workers capable of performing many different tasks within a flexible work organisation. Job rotation and on-the-job training become the rule. The most advanced enterprises will also educate and train in order to build up a skill reserve to meet unforeseen changes.

In this phase, education and training efforts lose the *ad hoc* characteristics of the first and become a permanent feature of the enterprise's business strategy. The skill structure that develops is shaped by the multi-skilled worker, who masters and integrates formerly isolated and fragmented domains. This could be called the *market-driven* phase in the education and training strategy of an enterprise.

Figure 20. **Interactions between education and training investment**

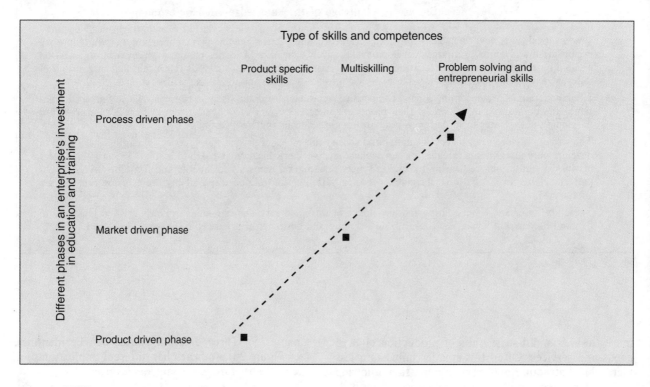

Source: OECD.

A third phase may emerge, particularly in small and knowledge-intensive enterprises. Many of these are in the service sector, either as independent enterprises or closely related to the manufacturing industries as business service firms. The main raw material of these enterprises is, increasingly, knowledge and information. Their education and training strategy tends to concentrate on recruitment at the university level and the creation of a positive learning environment in the workplace. In these enterprises, learning and working increasingly become an integrated process. The skill structure tends to centre on problem solving and entrepreneurial skills. In this *process-driven* phase, most tasks demand the acquisition of new knowledge, and learning is an essential part of successful execution.

In reality, many firms must have an education and training strategy which addresses all three phases, while for others, phase three is less relevant. But the general trend of a critical mass of lead enterprises is to move from the first to the second phase; the fastest employment growth in the service sector, although not in absolute terms, is generally in enterprises in the third phase (e.g. business service firms). However, the assessment of this investment is very complex. Recently, the need for more accurate measures has been felt in both the policy and research areas. Assessment of investments for the first phase are not very difficult, but assessment of investments in the second and third phases poses serious problems. Yet, these are the phases which an increasing number of enterprises are entering, and they are those most likely to set the pace in this area in the coming decade.

Management and skill structures

It is obvious that the role of management is crucial in this process. Case study material from a number of countries gives the overall impression of a shift from fordist-taylorist practices of human resource management to the new "toyotist" model with its greater potential for coping with a more knowledge-intensive economy. Despite apparently converging views on the advisability of developing and expanding the new model, enterprise practice remains mixed (see Chapter 4 and Table 11). In the wake of the two oil crises, enterprises, sectors and nations tended to adopt conservative strategies at the margin of fordism. Conventional fordist methods have been used by some firms in new geographical areas (e.g. credit and direct investment in Latin American countries in the 1970s). Others have kept previous management devices but introduced defensive flexibility, lowering wages in order to preserve obsolete fordist jobs or new tertiary jobs. Still others are using the opportunities offered by the new information technologies in order to maintain fordist principles (strong division between conception and execution, strengthening of monitoring of labour, control of labour intensity by new computerised machine tools). Finally, the rise of the service sector and the deepening cleavage in the labour market have been widely used in order to alleviate the crisis within the fordist manufacturing sector (M. Boyer, 1988).

Nevertheless, international comparisons suggest that these strategies do not resolve the crisis in the previous management model, even if they can help it to pass from one system to another. The continuing strong presence of strategies along very traditional fordist lines in North America seems to have delivered poorer results than the more innovative management style worked out in Japanese firms (see Chapter 4). Similarly, the very sluggish adaptation of British manufacturing has created direct investment opportunities for Japanese firms in some key sectors, such as the automobile and consumer electronics industries. The rather defensive strategy used by English managers has been outperformed by the surprisingly successful introduction of new and different principles.

Systematic statistical analysis of firms' trajectories during the 1970s/1980s confirms that the new strategies have provided better performances. In the case of France, it has been convincingly shown that maintaining taylorist principles of de-skilling has usually led to very disappointing results (Choffel *et al.*, 1988). Small or medium-size firms with high technical knowledge and sufficient worker skill have succeeded in obtaining access to new external markets which have replaced national ones. Large firms that have decreased or abandoned the de-skilling of blue collar workers inherent to fordism have succeeded in limiting job destruction, while a large majority of those still following taylorist strategies have encountered serious difficulties.

2. EDUCATIONAL ATTAINMENT AND THE PROVIDERS OF EDUCATION AND TRAINING

Competence levels of the work force

Looking more closely at education and training, two basic issues have to be addressed. The first relates to target groups for education and training and the second to the providers of education and training. Given the current demographic situation in many OECD countries and the rising minimum competence threshold at the workplace, issues concerning the adult learner have become quite as important as those relating to the young learner. An overall view of educational attainment among the working-age population in OECD countries (Figure 21) presents a pattern of broad dispersion in a state of flux. The largest share is at Level A (with less than an upper secondary education), but it is shrinking, particularly among younger age groups, as a result of the expansion of educational opportunities in the 1960s. This change is especially marked among women, for whom the proportion at Level A has decreased faster than for men, with a corresponding more rapid increase at Level E (with at least one university degree completed). However, more than half of the working age population in eight of the sixteen countries for which data are available has failed to complete upper secondary education.

Table 24 presents the annual percentage changes in level of attainment for total working-age population for countries for which data for multiple years are available. It suggests wide variations across countries with respect to the pace of change in the educational

Table 24. **Annual percentage changes in educational attainment levels of working age population**

	Level of education				
	A	B	C	D	E
Australia 1983-88	−2.02	0.54	3.16	1.97	3.32
Austria 1981-87	−5.83	1.54	7.32	..	12.30
Belgium 1970-87	−1.28	5.77	..	..	3.07
Canada 1975-87	−2.50	1.89	..	0.55	3.97
Germany 1978-87	−2.16	−3.75	0.90	4.20	2.75
Italy 1980-87	−1.32	−4.91	..	..	3.60
Japan 1974-87	−2.76	1.54	..	..	3.55[1]
Norway 1972-87	−5.73	6.32	..	..	5.80
Sweden 1971-87	−2.71	1.86	5.31	7.08	..
United States 1972-88	−2.83	0.51	..	2.23	3.34

1. Since levels D and E are separated in 1987, but not in 1974, they have been combined under level E for the purpose of this analysis.
Source: OECD.

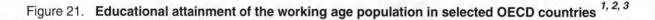

Figure 21. **Educational attainment of the working age population in selected OECD countries** [1, 2, 3]

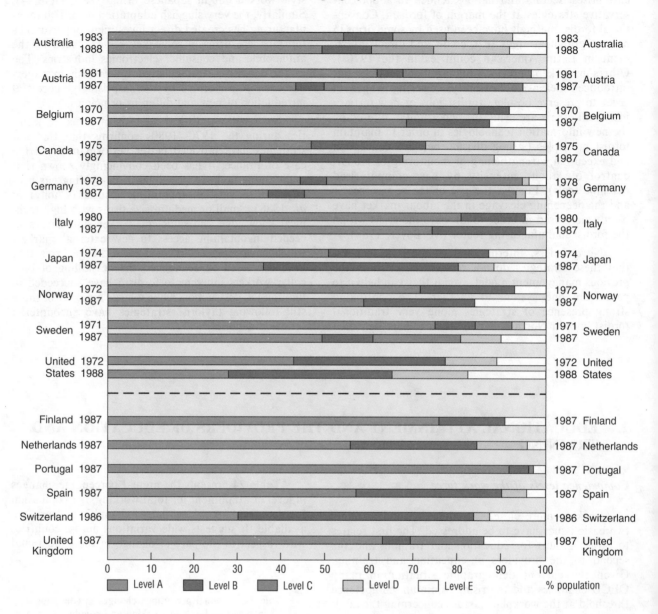

Level A Level B Level C Level D Level E % population

1. Educational attainment is classified on the basis of data submitted by national authorities. The classifications presented here may not reflect the full diversity of education and training systems in cases where data provided are not sufficiently detailed.

2. A number of countries include in their data persons whose level of attainment cannot be determined. For the purpose of the analysis, those persons have been omitted. They comprise the following percentage of working age population: Australia (1983) 3.7%, (1988) 4.6% (students); Japan (1974) 9%, (1987) 10% (students); the Netherlands (1987) 13.5% (students); Portugal (1987) 0.3% (other qualification); Sweden (1971) 1.1%, (1987) 1.9% (other qualification or no answer); United Kingdom (1987) 5.6% (other qualification or no answer).

3. Data refer to persons 20 and other, with the following exceptions: 15 and over in Australia; 15-74 in Finland; 20-64 in Italy, the Netherlands, Sweden (1987) and the United Kingdom (males); 20-60 in the United Kingdom (females); 20-74 in Norway (1971) and Sweden (1971).

Source: OECD (1989j).

attainment profile. Austria, Japan, Norway, Sweden and the United States have seen substantial rises in attainment levels. These are all the more impressive in light of the fact that the young people who, at the margin, are more highly educated than their elders comprise a relatively small and declining share of total population in all of those countries except the United States.

Box 35. Education and employment

Educational attainment affects labour market outcomes. In general, educational attainment has a weak but consistent influence on the employment experience of men during their prime-age working years, with higher levels of attainment being associated with lower levels of unemployment. The relationship appears to be weaker for women than for men.

Available evidence from some OECD countries indicates that educational attainment appears to re-emerge as a more influential factor explaining differences in unemployment experience for older workers. This may be linked to structural change, and the difficulty that those with low levels of attainment encounter when they try to leave occupations and jobs in declining industries, and seek to update their qualifications or requalify for new jobs. Educational attainment seems to be growing in importance over time, particularly for those with the lowest levels of attainment, as differences in unemployment rates by level of attainment widen.

The relationship between educational attainment and unemployment is strongest for those with the lowest level of attainment; they consistently face the highest incidence of unemployment. Furthermore, the risk they face seems to be growing. While the evidence does not support the "up-skilling hypothesis", which suggests an overall secular increase in skill requirements, it is consistent with a rising "minimum threshold" of competences required of those in the labour market (OECD, 1989*j*).

In contrast, the rise has been minimal in a number of countries such as Australia, Belgium and Italy and is especially worrisome in countries where overall attainment levels are low. Educational reforms in Australia and Belgium have led to increased attendance; more generally, high rates of youth unemployment have effectively reduced the alternatives to (and

opportunity cost of) education. But large proportions of older people are still at low educational attainment levels. Their levels of competence may prove increasingly inadequate in the face of rising qualifications requirements, and they will encounter greater difficulty than more educated workers in updating their qualifications through retraining, because of limited basic educational competences. The relationship between educational attainment and unemployment is strongest for these adults; they consistently face the highest incidence of unemployment, and the risk seems to be growing (see Box 35).

Public and private providers of education and training

The second major issue concerns providers of education and training. During the 1980s, dramatic changes occurred in the provision of education and training after initial education, which itself has changed very little. First, there has been a marked increase in enterprise-based education and training. Second, a great variety of education and training programmes have been launched by public authorities to combat unemployment, both for youth and adults. As a consequence, there has been increasing demand for more co-ordinated strategies among different providers, both within government agencies and between government and the private sector.

The "division of labour" between providers, not least in terms of the content of education and training, is a significant issue. Among different private providers, it can be expected that the market will sort things out, but the situation is less clear for the "division of labour" between the private and the public sector. However, if one assumes that the private sector will continue to increase its investment in education and training because it is profitable, increasing pressure will be placed on the public sector to prepare an educable work force in which the private sector can further invest.

This has profound implications for a core curriculum in initial education. The debate on the core curriculum often postulates an end product that will remain valid for the rest of an individual's life. But if lifelong learning is the goal, the core curriculum in initial education must provide, not an end product, but a stepping stone for further education and training. This implies emphasising transferable basic skills (OECD, 1989*g*, including problem solving and entrepreneurial skills, and the development of positive attitudes towards later education and training in different settings (see Box 36). Such a core curriculum, it could be argued, would then come close to the aims of "liberal education", although it would have to cater to all and not just to an elite, as has often been the case. Thus, the

Box 36. Educational goals for an evolving work force

The skills which are increasingly required of the work force strongly resemble those which secondary and post-secondary education has, theoretically at least, attempted to teach for centuries. They are basically "higher order thinking skills", and they involve a cluster of elaborate mental activities requiring nuanced judgement and analysis of complex situations according to multiple criteria. Higher order thinking requires effort and depends on self-regulation. The path of action or correct answer is not fully specified in advance. According to Resnick (1987), the thinker's task is to construct meaning and impose structure on situations rather than expect to find them already apparent. Those who reflect on the evolution of education also point to the need to recognise the existence of several types of intelligence, and they emphasise that education should not be excessively geared towards one type – logical/mathematical at present – to the detriment of others (Gardner, 1985; Bourdieu, 1985).

It has been been argued (Perelman, 1984) that the emerging knowledge-intensive economy demands a new kind of "learning enterprise", focused on adults rather than children, on learning rather than education, on technology rather than institutions and on private competition rather than public administration. According to the same observer, the adult learning crisis will really put the United States economy at risk until the beginning of the next century. To this provocative analysis must be added a heightened political awareness of unexpectedly high levels of adult functional illiteracy, which makes it in fact relevant to identify an adult learning crisis that is amplified by demographic and economic changes (Chisman, 1989a and 1989b).

The crucial importance of reforming children's education should not be overlooked, although its influence upon the economy is more distant, given that the labour force's annual renewal, based upon individuals leaving the education system, is in the range of 2 to 4 per cent.

---

public sector's "liberal education" would increasingly provide a truly vocational education.

However, the public sector's role does not stop at initial education. The first to be educated and trained in the private sector are those who are already best educated and trained in initial education. For obvious reasons of costs, the individuals more difficult to educate and train will only be drawn into private training programmes in situations of acute labour shortages. Although there are signs that enterprises have started dipping deeper into the pool of trainable workers because of fears of labour shortages, it is nevertheless obvious that market forces act principally at the middle and upper end of the scale. It is thus at the lower end that the public sector would have a particular role to play.

The private sector is nonetheless likely to address the issue of the low end of the scale in one of two ways.

It may invest, but to do so effectively it would have to develop a partnership with the public sector. Alternatively, driven by the need to reduce the costs of a less competent and trainable labour force, the private sector may increasingly substitute physical capital for human capital, invest "off shore" where qualified labour is cheaper, increase outsourcing, etc. The jury is still out as to which of these two strategies will dominate during the 1990s; OECD countries differ considerably with respect to public legislation and traditions in the private sector that may favour one approach or the other. In any case, the lesson for public policy seems clear: the public sector should concentrate on providing education and training after initial education where the private sector cannot or will not. This does not mean, however, that public policy should not also have a part in "setting the rules of the game" for private providers, particularly in terms of guaranteeing transparency and equity of access for the individual.

3. RESPONSES FROM THE EDUCATIONAL SECTOR

It is clear that increased educational attainments are needed for the whole population and particularly for those who did not complete secondary school. The expansion of education and training opportunities and the improvement of standards have in fact become major goals of education reform strategists in many OECD countries. The solutions put forward vary, but most aim at increasing the numbers of young people at the upper secondary school level at least. This would amplify the rise in educational attainment which has particularly benefited the service sector. Further changes, including changes in higher education, will require major adaptation of curricula, as well as reorganisation of the initial public and private education and training systems. The reform could equal in scope the changes induced by commitments which made first cycle secondary level education available to all in the 1950s and 1960s. It can be compared to the introduction of vocational education – in another period of social, technical and economic change – at the beginning of the twentieth century (OECD, 1989*t*). The question is whether the systems which are at present blamed for falling or insufficient standards and which have traditionally been structured to cater to elites can meet these new challenges.

Schooling as preparation for work

Both the educational system and the workplace provide skills and knowledge, as well as culture and values. With respect to the latter, much is made of the fundamentally different orientations of business and schools. Education is seen to focus primarily on the individual and his or her personal development as a citizen, a member of a family and a community and only then as a "worker". Business is seen to adopt a more instrumental view. Whatever the relevance of these arguments for explaining past policy approaches and their apparent limitations, the present situation can no longer support such a simple dualism, as the personal qualities and social and cognitive skills required by new technologies bring the value domains of work and education closer together.

At the same time, objective discontinuities between in-school and out-of-school settings cannot be discounted. Time frames, constraints and "managerial control" differ. In many respects, such dualism is unavoidable and even desirable. Education must always keep a certain distance from its environment if it is to fulfil its immediate purposes and its long-term objectives effectively. Nevertheless, schools should not be hermetically isolated from the world around them.

In fact, "situated" learning and closer relations between schools and the environment – and notably the work environment – have always been advocated by educational thinkers as an integrating process. This may take the form of simple site visits or more formalised and ambitious partnerships, although these are difficult to organise because the industrial infrastructure is lacking. This is unfortunate, as perilous levels of youth unemployment would make such partnerships particularly useful. In the United States, many such partnerships have been criticised as too fragmented, unco-ordinated or sporadic to bring about significant changes (United States, The Conference Board, 1988). The need for such partnerships is not questioned, but if they are to be effective, evaluation, co-ordination and replication of successful innovations must be institutionalised.

Policies designed to facilitate the transition from the educational system to the labour market through work experience are often advocated. Existing evidence on the precise effects of such work experience is contradictory, and more appraisal is needed. In the United States, where work experience is usually unrelated to school programmes, it generally eases short-term insertion into the labour market but sometimes to the detriment of educational attainments and later employment and career prospects (Stern, 1989). In other countries, where policies with stronger links to the educational system are implemented, results tend to be more positive. It would seem that these initiatives or programmes would be more beneficial if attention were given to type and quality of jobs involved and if built-in links with schools ensured supervision and assessment so as to maximise learning and transfer effects.

The curriculum and the teaching/learning process

The broad skill requirements identified for the service sector, in particular, would seem to be best served by general education, with vocational courses being of less relevance than to manufacturing. In many countries, however, technical or vocational degrees with higher standards and greater "liberal arts" content are being created, sometimes in upper secondary schools traditionally devoted to general education, thus bringing these courses of study closer to the needs of the service sector. Much, of course, depends on how the reforms are perceived and received in the industrial and business world, but in the long run, they may have far-reaching effects on social attitudes towards technical learning and training.

Research carried out by the OECD clearly indicates that curriculum and school career paths should be organised so as to ensure the acquisition of threshold skills, or core competences, that are prerequisites to insertion into working and social life (OECD, 1989*h*).

159

These include basic literacy and numeracy, reasoning and communications skills, positive work values and attitudes, as well as the ability to learn new skills. These are minimum requirements, even for the lowest level service sector jobs in a complex society.

In a growing number of school systems, technology and/or computer science are being included in the curriculum at secondary level, for both general and vocational education (OECD, 1986a). This may help bring about the "computer literacy" many enterprises feel they require. However, computer literacy must be integrated in a broader technological perspective, including history, alternative strategies for use, and social and economic consequences, including internationalisation and cross-cultural transfers. A merely instrumental approach which excludes an understanding of the wider social and economic implications of the use of technology should be avoided.

This broader conception of technological literacy should blend into a notion of liberal education expanded to reflect scientific and technological issues in a critical perspective. The aim should not be to teach a new set of values and ethics, but rather to enable students to integrate scientific information and derive ethical judgements.

In terms of values and attitudes – an essential component of skill requirements – the curriculum should aim to reinforce the attitudes sought and valued by service sector employers. This is often seen more as a matter of teaching practices and patterns of knowledge transmission than of content, although the relation between teaching methods and attitude is anything but clear-cut, for in spite of very traditional teaching methods, Japanese society is not uncreative. On the whole, however, schools remain ill-prepared for changes in this area. Recent reforms in the State of California show that it is far easier to upgrade curriculum and testing than to change the nature of pedagogy (Kirst, 1988).

Pedagogical reform is tied to deep-rooted social practices, and generalised changes in school practices will only take place when tied to similar changes in social practices. The influence of social values should not be overlooked, as they appear to play an important part in forming teaching practices and teachers' expectations and in motivating students to learn.

The attitudes and personal qualities discussed above are influenced and shaped by the tools, processes and content of education. While they cannot be taught, they can be reinforced and developed, notably through the organisation, delivery and evaluation of learning. Learning patterns are important in their own right, and attention should be given to them in the same way as it is given to outcomes. This may have bearing on the debate about quality in education: improving the quality of outcomes might be facilitated by improving the quality of processes.

With regard to teachers, the relevant issue may be less technological change *per se* than more general concern about reform and the quality of education and teaching. The status of the teacher as a professional, as well as his or her role and position in schools, is complicated by the creation of intermediate levels which blur the distinction between teaching and other school duties. Recruiting the "right" teachers is a major policy issue in many OECD countries, but achieving a

---

Box 37. Policy issues for teacher training

Most debate on teacher training has opposed the pedagogical and substantive aspects of the teacher's role, but arguments have also been advanced in favour of greater exposure to "real world" situations and a less exclusively academic training. This would certainly help bring education and the world of work closer together; it would also facilitate changes in teaching practices. This general concern points to specific and pressing areas for policy development:

- upgrading the technological and other broad-based knowledge of existing teachers;
- promoting the systematic development of the qualities sought through in-service education;
- changing the training and career patterns of teachers so as to provide for significant non-teaching work experience at some point;
- finding and validating alternative entrance paths to teaching in order to ensure diversity of teacher supply;
- using teaching resources from the service and the economic sectors to make learning more relevant to the world of work.

These issues, which have been taken up in OECD work on the conditions of teaching (Doyle, 1989; OECD, 1989h) have massive resource implications. Policy development will need to take these implications fully into account.

160

satisfactory combination of the two elements of appeal of teaching jobs – real and psychic income – is proving a complex exercise (see Box 37).

The involvement of higher education

As stressed in the Intergovernmental Conference on Education and the Economy in a Changing Society (OECD, 1989h), few today question the importance of involving higher education in efforts to cope with the complex, interrelated and changing problems with which OECD countries are confronted. Higher education systems, with intellectual resources spanning the whole range of scientific and technical achievement, have a unique role to play. An important question, however, is how they can respond more effectively to the needs of society without endangering their longer-term responsibilities to science, their role as repositories of national culture and their function as the essential critics of much that is taken for granted in the social order.

In all OECD countries, albeit to varying extents, the higher education system is a major component of national research capability. Between 13 and 30 per cent of R&D is carried out there (OECD, 1981a). The irony of the last few years has been that just as the importance of the need to "mobilise" the university research effort was coming to be recognised, the universities reached a low point in their level of resources (see Chapter 1). Lack of resources and "new blood", a consequence both of declining public spending and the research policies of the 1970s, meant, in some cases, obsolescent equipment and demotivated staff. Higher education institutions were less able to respond to the

demands being placed upon them than they had been for years. Still more recently, steps have been taken in a number of Member countries to remedy this situation: "new blood" schemes, new investment in apparatus and carefully targeted additional resources have certainly helped this mobilisation process.

Furthermore, in the last few years, the governments of almost all OECD countries have come to see science and technology as a source of new strategic opportunities (see Chapter 1). Many reports attest to the importance which governments have come to attach to new science-based industries as a source of growth and employment. This point of view is supported by industry, despite differences of opinion as to how governments should intervene. It is nonetheless widely agreed that as industry moves into new areas of advanced technology, it develops a growing dependence on the "science and engineering base". This has led corporations to develop stronger links with universities (see Chapter 3).

These initiatives have certainly played a role in stimulating the development of the new high-technology sectors. It is here that the greatest benefits have been reaped from the measures taken to bring new blood into the universities and facilitate the movement of ideas out of them. Certain areas of professional training, including post-graduate business studies, have also grown rapidly as graduates have assessed and sought to maximise their employment opportunities in difficult times. In addition, there can be little doubt that if much remains to be done, resources are being used more efficiently and that universities are increasingly being managed according to best practice (OECD, 1989h).

4. MISMATCHES AND THE INTERACTION BETWEEN DEMAND AND SUPPLY

There is growing concern in many OECD countries about an emerging "skill gap" due to rapid and convergent changes in demand, coupled with insufficient response on the supply side, both domestically and in the context of growing global economic interdependence.

The nature and scale of mismatches

National concerns about skill gaps have begun to rank high on the policy agenda. This was brought out forcefully at the Dutch Government/OECD TEP Conference held in Utrecht in November 1989, which

identified three areas of skill shortages. The first relates to the widespread under-preparation of the work force in most OECD countries, particularly for the requirements of the service sector, discussed above. Generally, this shortage is referred to as the "skill supply problem", and it is acutely felt with respect to companies' managerial labour force in most OECD countries.

The second area is "occupationally specific skill shortages". Nursing shortages in the health care sector as well as shortages of scientific, engineering and computer software professionals currently exist in a large number of OECD countries. Occupationally specific shortages tend to be country-specific, since they reflect different market conditions and the differing ability of national educational and skill preparation systems to respond to different needs. Such shortages appear in national employment statistics in the "job vacancy" rates. It can be argued that such shortages are often a reflection of the dynamism of an economy. They need not be worrisome so long as their magnitude does not interfere with a sector's ability to continue growing and/or serving its market, but this has already happened in some sectors in some countries.

The third area is the job shortage for workers with inadequate marketable skills, or underskilled workers. This type of shortage cannot be remedied exclusively through policies attempting to raise the skills of these workers, but must also be dealt with through government policies emphasising job creation. It is not simply a question of matching unemployed workers with existing job vacancies, if only because these are often few in number and at skill levels superior to those for which many of the unemployed can reasonably be trained or retrained; jobs must also be developed at skill levels which the unemployed have a reasonable chance of attaining. In the Netherlands, for example, there are approximately 80 000 to 100 000 job vacancies, mostly at relatively high skill levels, but nearly 600 000 unemployed.

In terms of remedies, the skill supply problem will necessitate both different educational and training preparation for a new generation of managers, and increased numbers of qualified personnel. Extensive retraining of employed managers by firms themselves is also a pressing concern.

It can be argued that for occupationally specific shortages, firms and industry associations must take the lead in designing solutions. It should also be pointed out that these shortages need not always result from insufficiencies or deficiencies in the training system. In the nursing profession, for example, several countries have very high turnover rates because of work-related stress. In the United States, nurses remain in their profession only ten years on average; improvement in work organisation might help bring these rates down and thus help solve part of the problem. In addition, shortages may be relieved by increasing use of technology rather than by seeking new labour pools, although such strategies might have a negative impact on efforts to deal with job shortages.

The issue of job shortages for the least skilled requires public sector intervention and macro-economic policies aimed at fostering job creation. Efforts have been undertaken in several OECD countries to build retraining programmes into unemployment assistance. Note should also be taken of the success in several countries of the Passive Income Support programmes whereby particular groups – typically single parent women on welfare – are provided with additional income support, so that it becomes economically feasible for them to re-enter the working labour force. Apprenticeship programmes must also be rejuvenated as part of an employment strategy aimed at disadvantaged groups.

Transformations in the school and early post-school skill preparatory system will require particular attention, although a distinction must be made between countries with a strong apprenticeship tradition and those with a weak or non-existent one. In the first group, many countries continue to maintain extensive apprenticeship systems for disappearing occupations but do not devote sufficient resources to developing programmes for new occupations. In addition, today's apprenticeship systems should be quite different from those of the past.

The German system, for example, has become much less occupationally specific over the 1980s; to the extent that it prepares individuals for occupations, it does so much more in terms of broad groups of trade skills. In addition, the German "dual" system places much greater emphasis on general education. However, important distributional problems occur in the existing apprenticeship system. Apprentices placed with Siemens or Deutsche Bank, for instance, are likely to receive much better skill preparation than those placed with small firms, where they are likely to be employed primarily as a source of cheap labour.

In the United States, with a weak tradition of apprenticeship but a long history of school-based vocational/technical education, emphasis is currently being placed on reforming the existing vocational/technical education track to better suit the new demands of the labour market. Here, too, the growing need for broader and more general education and the need for providing stronger preparation for the world of work have been recognised. One proposed solution would retard the occupational focus of the vocational/educational track and create an occupational track spanning the last two years of high school (11th and 12th grades) and the first two years of Community College.

Interaction between supply and demand

Although current changes in demand reflect a general trend that cuts across enterprises and countries, enterprises dispose of a great variety of human resource strategies to cope with new market conditions. In many cases, the strategies pursued by a given enterprise in a given country often have a close relation to the way that country's education, labour market and training institutions function. Three principal strategies have been identified in OECD work:

i) The human resources-intensive strategy
The enterprise recruits people with a good and broad educational background and then complements it with intensive on-the-job training and education, coupled with a flexible work organisation to allow for job rotation.

ii) The polarisation strategy

The enterprise focuses its development of human resources on a core group of employees through innovative measures in skill formation and work organisation. Beyond this group of core employees lies a buffer group of employees with less formal ties to the human resource development of the enterprise. This group is often highly vulnerable to shifts in the business cycle of the enterprise.

iii) The mobility strategy

The enterprise recruits at a high level of education, usually university level. However, very little structured in-service education and training exists. Learning and earning are part of the same work process, and inter-firm mobility is very high. Where learning is part of the job and high mobility the norm, the focus of the enterprise's human resource development will be on recruitment and permanent learning enhancement.

The human resources-intensive strategy can be developed in countries where the formal education system takes the majority of 16-19 year olds through upper-secondary education with either a good general or a good vocational education. If this situation is coupled with a tradition in the internal labour market of great flexibility and a co-operative culture of education and training at the enterprise level, then most conditions for developing this strategy are met.

The polarisation strategy is often found in countries with a rather weak vocational education system and/or a relatively high degree of failures and drop-outs from general secondary education. If this situation is coupled with a relatively weak internal labour market flexibility and a weak tradition of investment in training at the enterprise level, then the polarisation strategy will often develop.

The mobility strategy is found in countries with a very competitive high technology and private service sector. If a tradition of strong external labour market mobility exists together with a very responsive system of higher education to meet the needs of sophisticated and specialised training, then this strategy will be effective.

Obviously, there are variations and combinations of these strategies, but this schematic presentation nevertheless highlights the complex relationships existing between human resource development at enterprise level and the policies pursued relative to labour market and education and training systems. Often these policies set the framework for the kind of human resource strategies that it is possible and feasible to develop at enterprise level.

5. SOME POLICY CONCLUSIONS

From the preceding analysis of changes in demand and supply and their interaction, it seems possible to identify at least three principal messages. First, changes in demand are much more rapid and congruent across enterprises and nations than changes in supply. A new model – integrating new technology, work organisation and skill formation, and with emphasis upon flexibility and human resources – seems gradually to be replacing tayloristic principles of organisation. Second, sluggishness of changes in supply seems due in part to slow response to demand from institutions of formal education and training, a situation which has often led informal and private institutions of education, training and skill formation to respond in a manner suggesting the beginnings of a market. Third, it seems that the capacity for a critical number of enterprises in a given country to create a more efficient, post-taylorist work organisation is strongly influenced by education and labour market institutions.

Emerging policy issues on the supply side

At the OECD Intergovernmental Conference on Education and the Economy in a Changing Society it was stressed that a commitment to invest in human resources can only be effective when educational policies are pursued in tandem with other policies embracing different public authorities as well as the social partners. This will obviously not be an easy task (OECD, 1989*h*). Just when education and training are under pressure to contribute more to the effective economic performance of individuals, enterprises and economies, the policy issues are becoming harder to manage. The discussion above shows that the distinctions between education and training are blurring and the notion of an educational and training endpoint is evaporating. Moreover, education and training are less and less the concern of education ministries alone; they are increasingly provided by labour market authorities, employers and others whose main mission lies elsewhere. Finally, although responsibility for scientific and technological research and the preparation of a highly qualified work force have never been confined solely to institutions of higher learning, these activities are increasingly taking place outside the education community.

In this changed context, education policymakers are likely to find that the customary levers for policy change are ineffective, or at least insufficient and, by themselves, not always appropriate. Box 38 lists the orientations and objectives identified by the OECD Education Committee as desirable and necessary changes to enhance, in socially and economically acceptable structural ways, the systems of education and training. These goals are considerably broader and more complex than in earlier periods. Selecting strategies for change is now more complicated than specifying objectives and choosing directions within a largely hierarchical authority structure. That approach may have its place within the formal education enterprise. But, more and more, it must be complemented by a strategy of consensus building and partnerships, based on a both new structure of incentives and rewards internal to conventional education arrangements and on the joint pursuit by education authorities and other *de facto* education providers of mutually agreed objectives. To succeed, the varied goals and behaviour of institutional players and individuals and the dynamics by which they make and carry out decisions must be taken into account.

The result is not likely to be neat. The multiplicity of goals pursued, combined with the abundance of ways in which education is provided, means that strate-

gies for change will be neither simple nor coherent: what changes curriculum in one setting may not change it in another, what increases training by large enterprises may not do it for small enterprises. Multiple overlapping mechanisms may be needed to achieve the mix most appropriate to given national circumstances.

Governments will have to guide and improve an expanding market for education and training. Many OECD countries already have a fairly well-developed market for education and training, particularly at the post-compulsory level. A multitude of education and training providers give consumers – whether individuals pursuing their own development or employers trying to upgrade the skills and qualifications of their workers – choices with respect to field of study, quality of instruction, location, training content and cost. But the market for further education and training of adults assumes increasing importance, as abrupt and dramatic changes in skills and qualification requirements, shifts in the occupational structures and uncertainty about future developments have made the notion of initial education as adequate preparation for a lifetime career obsolete.

Unfortunately, markets for further education and training of adults have been far from perfect. There are obvious cases of market failure: distortions in supply and demand have led to sub-optimal allocation of resources and a lack of transparency concerning available opportunities. Some market outcomes have been unacceptable from a social point of view. Under-representation of low-skilled workers in further education and training or under-training by enterprises in declining sectors may be efficient by certain narrow economic criteria, but they may violate norms of social equity and impose unacceptable social and economic costs in the longer term. Indeed, these kinds of market failure and market outcomes compel governmental intervention.

Policy strategies for lifelong learning and recurrent education – as opposed to the constant extension of the "front end model" of education in combating educational inequalities – must be revitalised. Such strategies were developed some 15 years ago (OECD, 1973) but were largely abandoned in the wake of the oil crises and the recession at the end of the 1970s. Renewed interest in education and its quality and standards during the mid-1980s tended, on the whole, to ignore the policy debate about educational structures which was so vital to the earlier strategies. Today, when the *quality of the total work force* is at stake, a lifelong learning system becomes necessary to ensuring attainment and maintenance of the rising minimum competence threshold.

For OECD governments, structural adjustment has recently become a guiding principle in many policy areas, and it can be hoped that needed adjustments in education will find inspiration in the earlier policy debate about recurrent education, in which changes in educational structures were a prime objective. However, the proposals which received universal approval at that time were too peripheral or too general to be truly significant. The extension of individual choice and the improvement of adult learning opportunities were wholly unobjectionable aims, but once concrete policy was formulated, those who discerned practical consequences inimical to their own vested interests put up strong resistance. However, to interpret lifelong learning simply as a sizeable expansion of adult education and training not affecting the structure, content and methods of initial education is to ignore issues as crucial as tradeoffs between youth and adult education.

To implement a system of lifelong learning would require careful policy co-ordination aimed at stimulating the building of new networks and partnerships among different public and private providers of education and training, and among providers for youth and adults. This is a formidable task. It requires breaking down administrative and psychological barriers: formal education generally responds with difficulty to external events and those outside the formal educational system are often prevented from profiting from existing facilities. For the emerging knowledge-intensive economy, partnerships and networks make sense in two respects: they would allow more effective use of limited resources and, at the same time, the development of the adult learner – a crucial resource for enterprises and nations during the coming decade.

Policy for the demand side

Public policy has a limited influence on the demand for human resources. However, public policy might be further developed in a number of areas, such as:

i) The review of tax systems and incentives in order to promote greater propensity for firms to invest in tangible and intangible resources;

ii) Specific incentives and measures to stimulate small and medium-sized enterprises to invest in education and training of their work force. Such measures should take into account the need for innovation in delivering education and training, for instance by using new technology-based equipment;

iii) Stimulation and guidance for the evolution of industrial relations systems and collective bargaining towards the new post-fordist model;

iv) Development of labour market policies to give priority to active programmes leading to insertion into working life rather than passive subsidy programmes.

The pressing need for further data

Policy formulation is hampered by a scarcity of aggregate data. Case studies and sectoral studies have generated considerable amounts of data, and evidence from the research community suggests that these will continue, particularly because of mounting interest in the analysis of changes on the demand side and at the enterprise level. Micro-economic data has furnished the information for most of the above discussion. There is an urgent need to strengthen the data base at the aggregate and macro-economic level. At least five areas need to be considered, namely:

i) the scope and direction of enterprise-based education and training;

ii) the educational and training attainment of the adult labour force;

iii) the ratio between tangible and intangible investment at enterprise level;

iv) educational indicators of input, process and output and their relationships to the formal education system;

v) the scope and direction of education and training provided by the market.

NOTES

1. This chapter is much indebted to recent and ongoing work in the OECD as well as the TEP colloquia held at Utrecht (1989) and Helsinki (1990).
2. In the short run, low levels of productivity or lagging productivity growth can have an employment-stimulating effect (Petit, 1986; Weiermair, 1988). In the long run, however, cost/price pressures as well as the forces of increased domestic and/or global competition are likely to lead to opposite effects.

Chapter 8

TECHNOLOGY AND ECONOMIC GROWTH

ABSTRACT

This chapter examines the main macro-economic relationships between technology and economic growth through the predominant analytical approaches. The first section discusses the links between investment, productivity and growth. Greater investment can increase total factor productivity (TFP) and therefore economic welfare in the long run in two principal ways. First, it may lead to the more rapid diffusion and adoption of new production methods and techniques. A key aspect of the "quality" of the capital stock is its technical sophistication, and because new inventions and techniques are largely embodied in new machinery, newer capital should be more productive than older capital. By increasing the rate of substitution of new for old capital, higher rates of gross investment would raise the rate of growth of productivity.

Second, the so-called "new" growth theories argue that greater investment in both physical and human capital creates externalities and aggregate economies-of-scale effects. The new theories emphasise the role of economy-wide returns to scale, expenditure on R&D, human capital formation and the mediating role of investment in the diffusion and promotion of technical change. At present, the available empirical proof of the validity of the new models is still weak and, conceptually, these models remain quite simple. While this is not necessarily a drawback, further insight could probably be gained by incorporating into the theoretical models some of the complexities inherent in innovation, investment and technological diffusion.

Productivity growth in most OECD countries began to slow in the second half of the 1960s. However, seen in the longer historical perspective, the excellent performance of productivity during the 1950s and the 1960s is more striking than its subsequent weakening. There was no single cause for the productivity slowdown, a number of factors being at work simultaneously. TFP recovered slightly during the late 1980s, and the rate of decline of capital productivity in the business sector in particular has slowed in some countries. The improvement of capital productivity in the 1980s is probably related to increases in the real and the relative price of capital which stimulated a more economical use of this factor. Labour productivity in some manufacturing sectors has also risen substantially. This trend has been associated with a variety of influences, in particular exposure of enterprises to increased international competition.

Regarding future productivity trends, a position of cautious optimism seems appropriate, based upon the hypothesis that the structural reforms of the 1980s have unleashed dynamic forces by increasing competition. This will sooner or later lead to higher TFP growth by i) *stimulating the search for process and product innovation;* ii) *diffusing technical change more rapidly within and across economies; and* iii) *accelerating complementary institutional and organisational innovations.*

INTRODUCTION

The notion of economic growth as it is used throughout this chapter refers to a sustained expansion of the productive potential of the economy which – in the long run – converges with the growth of aggregate output. The conventional view explains long-term growth as a weighted average of the growth of aggregate inputs. Under the assumption of constant returns to scale, doubling of inputs entails a doubling of output. In most OECD countries labour accounts for about two-thirds of factor input and capital for one-third[1].

However, the growth in factor inputs typically falls short of explaining the growth in output. Indeed, over the longer term, the unexplained part – the so-called Solow residual – accounts for as much as one-half of observed output growth. According to the standard interpretation, this residual represents disembodied technical progress, usually referred to as total factor productivity (TFP).

One of the main tools used in the measurement and analysis of TFP is the growth-accounting framework. In general, such analyses assume the existence of an aggregate production function, constant returns to scale, cost minimisation and competitive input and output markets. TFP must be interpreted carefully, because it almost certainly reflects many factors other

Table 25. **Productivity (business sector)**

Average percentage change at annual rate

	Total factor productivity[1]			Labour productivity[2]			Capital productivity		
	1960-73	1973-79	1979-88	1960-73	1973-79	1979-88	1960-73	1973-79	1979-88
United States	1.6	−0.4	0.4	2.2	0.0	0.8	0.2	−1.1	−0.4
Japan	6.0	1.5	2.0	8.6	3.0	3.2	−2.5	−3.1	−1.7
Germany	2.6	1.7	0.7	4.5	3.1	1.6	−1.4	−1.1	−1.1
France	4.0	1.7	1.6	5.4	3.0	2.6	0.9	−1.0	−0.5
Italy	4.6	2.2	1.0	6.3	3.0	1.6	0.3	0.3	−0.6
United Kingdom	2.3	0.6	1.8	3.6	1.5	2.4	−0.6	−1.5	0.4
Canada	2.0	0.7	0.3	2.8	1.5	1.5	0.5	−0.7	−2.0
Austria	3.4	1.3	0.9	5.8	3.2	2.0	−2.3	−3.2	−1.8
Belgium	3.8	1.4	1.2	5.3	2.8	2.3	0.4	−1.9	−1.3
Denmark	2.8	1.2	1.0	4.3	2.6	1.7	−1.0	−2.2	−0.9
Finland	3.4	1.7	2.3	5.0	3.4	3.2	0.1	−1.8	0.3
Greece	5.8	1.5	−0.7	8.8	3.4	0.1	−1.0	−2.7	−2.6
Ireland	4.1	2.6	2.5	5.0	3.6	3.9	2.0	0.2	−0.9
Netherlands	3.0	1.4	0.8	4.8	2.7	1.6	−0.6	−1.3	−0.9
Norway	3.6	−0.4	1.4	4.1	0.1	2.0	0.9	−2.9	−1.6
Spain	4.1	1.5	2.1	6.0	3.5	3.4	−1.8	−4.9	−1.9
Sweden	2.9	0.5	1.2	4.1	1.5	1.8	0.0	−1.9	−0.6
Switzerland	1.9	−0.4	0.8	3.2	0.8	1.6	−1.9	−3.8	−1.2
Australia	1.7	0.8	0.5	2.7	2.2	1.1	−0.4	−1.8	−0.6
New Zealand	1.1	−2.1	0.7	1.8	−1.5	1.4	−0.5	−3.4	−0.7
OECD Europe	3.3	1.4	1.2	5.0	2.6	2.1	−0.4	−1.4	−0.7
OECD	2.9	0.6	0.9	4.1	1.4	1.6	−0.4	−1.5	−0.8

1. TFP growth is equal to a weighted average of the growth in labour and capital productivity. In both cases, the sample-period averages for capital and labour shares are used as weights.
2. Output per employed person.
Source: OECD.

than technical progress. Examples of such factors are: the contribution of production inputs other than capital and labour (such as natural resources); improvements in the quality of capital and labour; deviations from perfect competition; and errors in the measurement of output, labour and (especially) capital. These factors, among others, have hindered the study of the determinants of technical change.

Nevertheless, if it is assumed that the residual represents "technical change", and if technical change is assumed to augment only the productivity of labour, the standard model generates several long-term predictions which are approximately consistent with known stylised facts[2]. While this predictive ability accounts for the popularity of the conventional view in both theoretical and applied work, the fact remains that the most important single determinant of long-term output growth remains entirely unexplained[3]. In the absence of an operational theory of technical change, technical change is simply assumed to be exogenous[4]. This weakness was brought into sharp relief by the decline in productivity growth experienced by OECD countries in the late 1960s and 1970s (Table 25). In a nutshell,

since the model cannot explain why rates of technological change and productivity growth were large in the first half of the post-war period, it sheds no light on why these growth rates are now lower.

For many macro-economic policy purposes it does not matter whether movements in TFP come from changes in worker attitudes, management quality or pure technological change. The path of potential output growth, the trajectory of wages and the return to capital compatible with price level stability are all equally affected. On the other hand, it is important to understand the nature of the Solow residual – especially because of its importance to long-term economic growth – in order to model, or endogenise, technical change in growth models. The ultimate aim of such research would be to analyse policies that might raise, in a sustainable way, productivity and output growth.

In the discussion which follows, the link between investment, productivity and growth is treated first. The chapter goes on to examine productivity trends in the 1970s and 1980s and to elaborate productivity measurement issues. A final section deals with the prospects for productivity growth.

1. INVESTMENT AND PRODUCTIVITY GROWTH

Greater investment could increase TFP and therefore economic welfare in the longer run through two main channels. First, it can lead to the more rapid diffusion and adoption of new production methods and techniques. Second, according to the so-called "new" growth theories, greater investment in both physical and human capital may create externalities and aggregate economies-of-scale effects.

Embodiment effects

A key aspect of the "quality" of the capital stock lies in its technical sophistication. If new inventions and techniques are embodied in new machinery, newer capital should be more productive than older capital. Higher rates of gross investment could raise the rate of growth of productivity by increasing the rate of substitution of new for old capital.

Simple observation provides evidence that new technology is embodied in new types of capital – computers embody the technology of fast electronic computing, just as adding machines embodied the older technology of mechanical computing. Despite the common-sense appeal of the embodiment hypothesis, however, supporting evidence has often been difficult to find in aggregate economic data. There is, however,

indirect evidence consistent with the hypothesis. For example, capital growth tends to contribute more to output growth than its share in factor income would suggest[5]. Also, many "vintage" models – which assume embodied technical change – provide further indirect support. Finally, there is substantial literature establishing a relationship between R&D and productivity, although this work has not established that the fruits of R&D are embodied in capital.

A simple and direct proxy for the amount of embodiment is the "capital replacement rate", or the ratio of gross investment to capital stock. A positive relationship between the capital replacement rate and the growth rate of TFP would indicate that embodiment effects could explain changes in TFP. Figure 22 plots the rate of change of TFP and the capital replacement rate in selected major OECD countries[6]. Ignoring the cyclical "noise"[7], trend relationships can hardly be detected in the examples given here – the TFP slowdown in the 1970s is roughly coincident with the fall in the capital replacement rate only in Japan and France. For other countries, either the relationship is less clear or inexistent, or movements in TFP growth appear to lead those in the capital replacement rate. In the OECD countries, over the period 1960-88, the two sub-periods 1960-73 and 1974-88 correspond roughly to before and after the productivity slowdown that

Figure 22. **Embodiment, productivity and capital replacement**

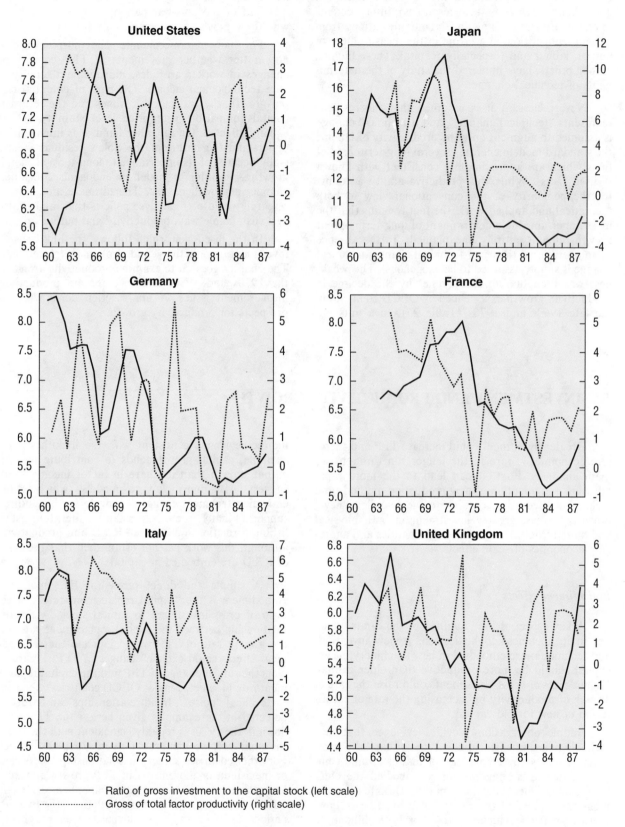

Ratio of gross investment to the capital stock (left scale)

Gross of total factor productivity (right scale)

Source: Ford *et al. (1990).*

Figure 23. TFP growth and average age of the capital stock

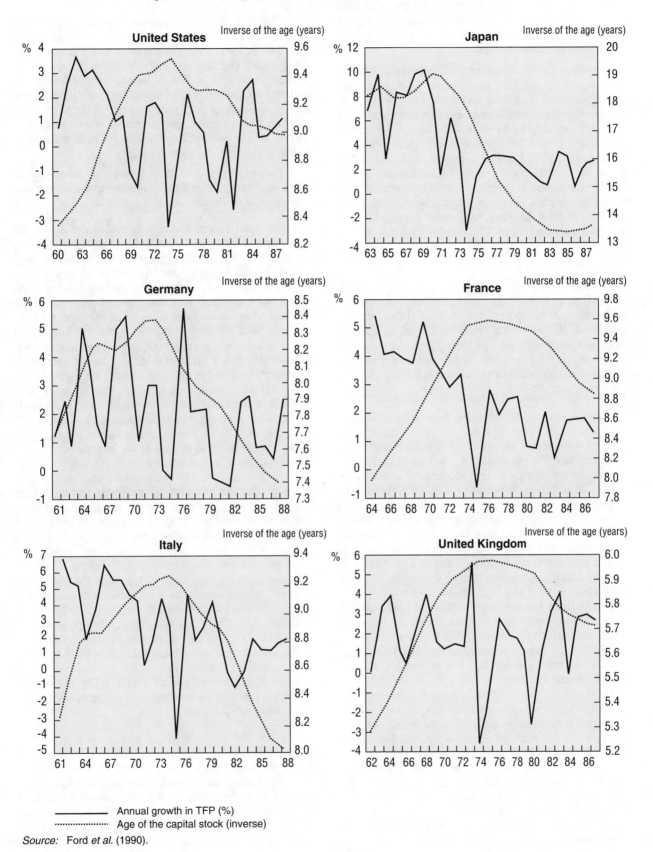

Annual growth in TFP (%)
Age of the capital stock (inverse)

Source: Ford *et al.* (1990).

171

occurred in virtually all OECD countries in the early 1970s. Data for the first sample period seem to lend some support to the embodiment hypothesis, but the relationship vanished in the post-1974 period, except for the singular case of Japan.

Another proxy for the rate of embodiment is the average age of capital, since newer capital should embody superior technology and therefore should be more productive. Given investment flows and scrapping rates, it is possible to estimate the average age of the capital stock by the perpetual inventory method. Growth rates of TFP and the inverse of the estimated age of the capital stock are shown for selected major OECD countries in Figure 23[8]. The embodiment hypothesis implies that the two curves should move together, at least in terms of trends. Again, there is considerable variation in TFP due to cyclical factors, but there appears to be no strong trend correlation for most countries.

Perhaps the most dramatic case of the embodiment of new technology is the rapidly increasing importance of computer and related (e.g. robots) investment. But no strong productivity pickup appears to be associated with all this investment. Although as yet this puzzle (the so-called "Solow paradox") has not been resolved satisfactorily, several possible solutions have been suggested:

i) There is no puzzle, since the accumulated investment in computers has not yet raised the capital stock sufficiently to have had much effect in the aggregate. For example, a back-of-the-envelope calculation suggests that computer investment may have raised output growth by only about one-twentieth of one per cent per annum, which is far too small to be detected in aggregate data[9]. Moreover, in those countries using the Bureau of Economic Analysis (BEA) method of quality correction (i.e. the United States, Canada and Australia – see below), computers are, by definition, no more productive than any other piece of capital.

ii) Productivity has not risen because of inadequacies in training, applications software or organisational restructuring (Freeman, 1987b; and Kimbel, 1987). This hypothesis comes very close to arguing that technical change is, at least in part, disembodied (see Chapter 2).

iii) Productivity has indeed risen, but the major impact of computers has been in sectors where output is notoriously difficult to measure – primarily what has been labelled the information-intensive industries such as finance, insurance, business services, communication, real estate, transportation, storage, wholesale and retail trade and the government sector. However, there is little evidence that output measurement errors in the service sectors could explain a major part of the TFP slowdown.

"New" growth theories

The so-called "new growth theories" offer more radical explanations of the Solow residual and, therefore, of the long-term growth of productive potential. They emphasise the role of economy-wide returns to scale, expenditure on R&D, human capital formation, and the mediating role of investment in the diffusion and promotion of technical change[10]. Each of these mechanisms extends the standard model described above, with important implications for both micro-economic and macro-economic policies.

Two examples provide a flavour of the nature of these new theories and the relationship between them and the conventional view:

– It has been argued that the hypothesis of aggregate increasing returns to scale is consistent with the observed long-run aggregate relationships and allows one to eliminate the Solow residual altogether. Specifically, capital (including human capital) accumulation may have two effects: the direct effect of increasing the stock of capital and an indirect effect of increasing the stock of knowledge available for future production and investment. Thus, the stock of knowledge can be thought of as proportional to the stock of capital. As a result, the coefficient of capital is, in effect, much higher than the one-third implied by standard growth accounting. If it were unity, an increase in the investment rate (the ratio of investment to output) could raise the long-term growth rate of the economy.

– Alternatively, one can think of the economy as being composed of two distinct economic activities: first, the production of goods using capital and labour, as in in the standard model; and, second, the production of knowledge (i.e. R&D), also using capital and labour[11]. A key difference between goods and knowledge is that the former are "rivalrous" – only one person can eat a particular piece of meat – but the latter is "non-rivalrous" – the fact that someone knows something in no way prevents someone else from knowing it too. Put differently, R&D generates "external" effects (i.e. spillovers), since people other than the inventor can share, and perhaps profit from, the knowledge (see Chapter 2). A second key feature of R&D production is that an increase in the level of resources (capital and labour) devoted to R&D results in an increase in the rate at which new knowledge is created.

From the perspective of economic policy, both these models posit that a once-and-for-all change in economic conditions can lead to a permanent change in the rate of economic growth. This stands in sharp contrast to the conventional view, which assumes that such a change can have only a temporary effect on economic growth, or, equivalently, can in the long run only affect the level, not the growth rate, of productive capacities.

To see the distinction, consider the effect on long-term growth of a once-and-for-all increase in gross investment (or savings), as seen from the conventional viewpoint and then from the perspective of the new growth theories. In the conventional view, only the level of output will rise in the long run because increasing capital intensity reduces the marginal product of capital and, at the same time, increases the resources that must be devoted to replacing depreciated capital. The reduction of the marginal product of capital translates into a decline of the rate of return on capital, thus reducing the incentives to invest. As a result, the accumulation of new capital gets squeezed and, eventually, comes to a halt, leaving the long-run growth rate unaffected.

In the view of the new growth theories, a 1 per cent increase in the stock of capital increases output by 1 per cent, and will not entail a decline of the marginal product of capital because of the external effects on productivity growth. As a result, an increase in gross investment can sustain a permanent increase in the growth rate of output. In terms of the second example, the growth rate of output will rise permanently if at least some of the new investment takes place in the R&D sector, thereby raising the rate of productivity growth in the economy as a whole through external, or spillover, effects.

Thus, the new growth theories predict that policy actions that raise investment or R&D should have dramatically larger effects on aggregate output than they would in the conventional view. Owing to the "miracle of compound interest", even a small, but sustained, increase in the growth rate will eventually dwarf any once-off gain in the level of real output. If the new theories are correct, the benefits from a range of policy reforms – removing disincentives to save, lowering trade barriers, reducing distortionary taxation, increasing support for R&D and increasing human capital accumulation – would be much larger than those calculated by using standard models of economic growth.

An example of how much larger these benefits might be is provided by an analysis of the output effects of the 1992 Single European Market Programme, using ideas drawn from the new growth theories and positing the presence of aggregate increasing returns to scale (Baldwin, 1989). The standard analysis, as presented in the Cecchini Report (Cecchini, 1989), focuses on the static effects of trade liberalisation, which by their nature give rise to only once-and-for-all improvements in output. The report concluded that the *level* of EC income could rise by 2.5 to 6.5 per cent in the long run. The new growth models, in contrast, predict that the *growth rate* of EC income could increase by between 0.3 and 0.9 per cent per annum. This implies that the level of EC income could be 3 to 9.4 per cent higher after ten years, 6 to 19.6 per cent higher after twenty years, and so forth.

In view of the size of these hypothesised effects, it is important to examine the evidence in support of the new growth models. This evidence is of two sorts. The first examines aggregate returns to scale directly, and the second tests the theories indirectly by examining cross-country correlations.

To illustrate the nature of the evidence for increasing returns to scale, aggregate Cobb-Douglas production functions for the business sector were estimated for the seven largest OECD economies (Table 26). No constant-returns-to-scale restriction was imposed. The data are in levels rather than growth rates because use of growth rates emphasises cyclical

Table 26. **The "new" growth theory: unconstrained production function estimates**

	Capital	Labour	R^2
United States			
1955-1989	0.9	−0.08	0.98
	(0.11)	(0.02)	
1974-1989	0.05	1.2	0.98
	(10.0)	(0.14)	
Japan			
1962-1989	0.64	0.8	0.99
	(0.05)	(0.4)	
1974-1989	0.6	0.8	0.99
	(0.06)	(0.3)	
Germany			
1960-1988	0.8	0.75	0.99
	(0.02)	(0.15)	
1974-1988	0.73	0.63	0.99
	(0.03)	(0.22)	
France			
1963-1988	0.9	1.2	0.99
	(0.01)	(0.18)	
1974-1988	0.83	0.57	0.99
	(0.03)	(0.17)	
Italy			
1960-1988	1.0	−0.14	0.99
	(0.01)	(0.1)	
1974-1988	0.34	2.5	0.98
	(0.11)	(0.5)	
United Kingdom			
1961-1988	0.85	1.14	0.98
	(0.03)	(0.16)	
1974-1988	1.1	1.1	0.95
	(0.05)	(0.16)	
Canada			
1960-1989	0.07	1.7	0.99
	(0.07)	(0.08)	
1974-1989	0.32	0.88	0.98
	(0.06)	(0.16)	

Note: Dependent variable: real business sector value added (log). Standard errors in parentheses. Constant terms are not reported.
Source: OECD.

factors while the hypothesis relates to long-term growth. The estimated coefficients are interpreted as "elasticities" of output with respect to the input. For example, the estimated elasticity for capital for the United States of 0.9 implies that an increase in the capital stock of 1 per cent will raise output by 0.9 per cent. The standard errors in parentheses below the coefficient estimates are rough indications of the precision of the estimate. As a rule of thumb, a range of two standard errors is considered to be "statistically insignificant". Thus, the estimate of 0.9 for the United States is probably consistent with a "true" elasticity of one, but is inconsistent with one of zero. The R^2 statistic is a rough measure of how well the explanatory variables (capital and labour) explain movements in output. For regressions done in levels, values near one are not unusual. For most countries (Canada is a notable exception), the estimated coefficient on capital is close to unity. However, the equations for the United States, Italy and Canada exhibit considerable instability. Also, for many countries the coefficient on labour is much higher than the conventional estimate of two-thirds.

The second sort of evidence, from cross-country regressions, attempts to show that countries with higher output growth have also had higher levels of the variables that are hypothesised to affect growth (investment, for example, or education) (Romer, 1989; and Easterly and Wetzel, 1989). Positive correlations between investment and output and between TFP and investment exist in cross-country data, but are weaker than might be expected if the new growth models were correct.

Cross-section regressions for all 24 OECD countries have attempted to correlate output growth with some aggregate variables suggested in the literature: investment, exports, imports (all as ratios of output), and their growth rates. They give very mixed results (Table 27). For example, although the coefficient of the investment-output ratio is uniformly high in all regressions, the low value of the t-statistics suggests that little confidence can be accorded to the result (as a rule of thumb, a t-statistic less than two suggests the coefficient is not statistically distinguishable from

zero). The low R^2 values also imply that the explanatory variables explain little of the variation in output growth from country to country.

While a sympathetic reading of these results, and of results produced by other researchers, suggests some support for the new theories, the evidence presented to date is, in fact, quite weak. However, the new models are still in their infancy and considerable work remains to be done before the empirical relevance of the issues raised by this literature can be judged. In theoretical terms, the models remain quite simple. While this is not necessarily a drawback, invention, innovation and diffusion are complex processes, and further insight could probably be gained by incorporating some of these processes into the theoretical models. In addition, the predictions generated by these models so far have proved difficult to test empirically, with the result that it is very difficult to distinguish the new theories from the conventional view. Further work will no doubt sharpen the predictive aspects of the new growth models, thereby permitting more refined tests.

Table 27. **The "new" growth theory Cross-country regressions**

Equation	Investment	(growth)	Exports	(growth)	Imports	(growth)	R^2
1.	13.2 (1.6)	−0.35 (−1.2)	0.86 (0.1)	0.62 (2.0)	−2.5 (−0.3)	−0.3 (−1.3)	0.18
2.	10.4 (1.2)	−0.21 (−0.7)					0.00
3.			−3.0 (−1.7)	0.47 (1.9)			0.25
4.	16.2 (2.0)	−0.4 (−1.4)	−1.5 (−1.4)	0.4 (1.8)			0.19
5.	14.1 (1.7)			0.38 (1.8)			0.11

Note: Dependent variable: Average growth rate of real GDP. Independent variables are ratios of real GDP. T-statistics in parentheses. Constants not reported. Sample period: 1960-88.
Source: OECD.

2. PRODUCTIVITY PERFORMANCE IN OECD ECONOMIES

The productivity slowdown of the 1970s

Despite widely-held perceptions of 1973 as a watershed year, TFP growth in some countries actually began to slow in the second half of the 1960s, notably in the United States and Japan (Figure 24). In some smaller countries, a significant weakening of TFP gains only became apparent after the first oil price shock[12]. Seen in the longer historical perspective of the century as a whole, what stands out is the excellent performance of productivity during the 1950s and 1960s rather than its subsequent weakening (Table 28)[13]. OECD output per person-hour grew by an annual 4.5 per cent in the period 1950-73 compared with an average rise of around 2 per cent in the first half of this century and just over 2 per cent since 1973.

It is generally agreed that there is no single cause of the productivity slowdown. Prominent contenders are:

– reduced opportunities for catching up with the technological leader, the United States;

– a fall in the effectiveness of R&D expenditure, as measured by the fall in the number of patents per unit of R&D spending; and

– the prolonged and large under-utilisation of resources in the 1970s and 1980s, following the two oil price shocks. A sustained departure of actual output from its trend lowers potential output growth, reduces fixed capital formation and, through embodiment effects, lowers TFP growth. These feedback effects have received rising attention in the 1980s, as traditional approaches proved unable to identify the main source of the TFP slowdown (Abramovitz, 1991).

The large rise in energy prices in 1973 probably did not have major direct impacts on TFP. This conclusion is based upon the timing of the TFP slowdown in some countries, the weight of energy in OECD economies and estimated results of aggregate production functions, with energy and other raw materials as additional inputs of production. In addition, no important

Figure 24. **Total factor productivity[1] trends**
Percentage change

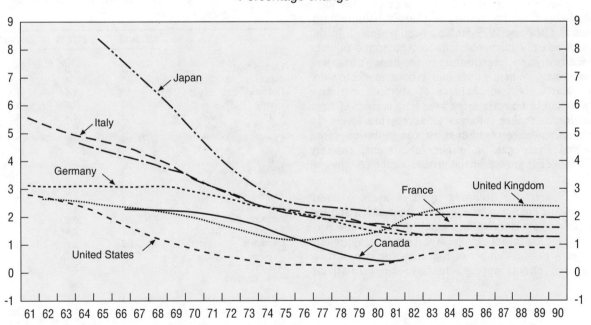

1. The growth in total factor productivity is equal to the weighted average of the growth in labour and capital productivity using the period averages for factor shares as weights. The "trend" has been derived using "centered" moving averages.

Source: OEDC (1989*f*).

Table 28. OECD productivity growth[1], 1900-86

Annual average compound growth rates

	1900-13	1913-50	1950-73	1973-86
United States	1.7	2.4	2.4	1.2
Japan	2.3	1.7	7.6	3.1
Germany	1.5	1.0	6.0	3.0
France	1.6	2.2	5.0	3.4
Italy	2.4	1.7	5.5	2.1
United Kingdom	0.9	1.6	3.2	2.5
Canada	3.5	2.4	2.9	1.5
Australia	1.1	1.6	2.7	1.8
Austria	1.5	0.9	5.9	2.8
Belgium	0.9	1.4	4.4.	1.7
Denmark	2.2	1.6	4.1	1.5
Finland	2.1	2.3	5.2	2.5
Netherlands	1.1	1.7	4.3	1.8
Norway	2.1	2.5	4.3	3.3
Sweden	1.6	2.8	4.4	1.6
Switzerland	1.6	2.7	3.3	1.6
OECD average	1.8	1.9	4.5	2.2

1. GDP per hour worked.
Source: Maddison (1989).

reversal of the TFP slowdown has yet occurred in response to the sharp decline in real energy prices in the second half of the 1980s.

On the other hand, it is undeniable that the indirect effects from two rounds of sharply rising energy prices in 1974 and 1979-80 have been powerful. In the face of sharply increased inflation, economic policies turned restrictive, contributing to the under-utilisation of resources. Where goods and labour markets were particularly rigid in the face of shocks, restrictive polices had to be maintained over long periods of time in order to reduce inflation to acceptable levels. In addition, policies of budgetary consolidation often involved sharp cuts in public investment, possibly reducing TFP growth in the private sector (Aschauer, 1989).

Although the declining trend in average hours worked, the changing labour force composition and the increasing importance of some industries with low productivity growth appear to have curtailed aggregate TFP growth in a number of countries, the size of their combined effects appears to have been relatively modest.

Productivity performance in the 1980s

After falling sharply in the late 1960s and 1970s, TFP has recovered a little over the current cycle, although it remains disappointing when judged against the record of the 1960s and early 1970s, when TFP in the OECD area increased by nearly 3 per cent a year. Moreover, the slight recovery in TFP growth in the 1980s has not occurred in all OECD countries. It is confined to less than one half of the countries for which data are available (data for the United States, United Kingdom, Finland, Sweden, Switzerland and New Zealand are included in Figure 24).

With respect to *partial factor productivity*[14] two features stand out: one is the behaviour of capital productivity in the business sector; the other the revival of labour productivity in manufacturing.

While the level of *capital productivity* (the ratio of output to capital) kept on sinking in the current cycle in most countries (including Japan), the rate of decline has lost momentum in ten countries (Table 25). In the OECD area, capital productivity declined at an annual rate of 1 per cent in the period 1979-88 compared with a decline of 1.5 per cent in 1973-79 period. Given the rising weight of investment for environmental protection in most OECD countries – which, if added in, would contribute to an even sharper decline of capital productivity – the outcome is all the more remarkable. Reversing earlier trends, capital productivity even began to rise in the United Kingdom, Finland and Australia.

Table 29. Labour productivity[1] in manufacturing

Average percentage change at annual rate

	1960-73	1973-79	1979-87
United States	3.7	1.0	3.3
Japan	10.5	5.4	5.8
Germany	5.0	3.4	1.6
France	6.5	3.8	1.9
Italy	6.5	4.4	4.3
United Kingdom	3.8	0.6	3.6
Canada	4.1	1.0	2.2
Austria	5.3	3.5	3.0
Belgium	6.2	4.9	4.6
Denmark	4.7	3.8	1.2
Finland	4.2	2.7	4.7
Netherlands	6.6	4.3	1.5
Norway	3.4	0.6	2.5
Spain	8.7	3.9	4.3
Sweden	5.2	1.1	2.8
Turkey	6.5	2.9	3.6
Australia	2.2	2.6	2.7
New Zealand	..	..	2.0
Standard deviation	2.1	1.5	1.3
OECD Europe	5.6	3.2	2.9
OECD	5.2	2.4	3.7

Note: For the period 1960-87, except for the United Kingdom, Belgium and Norway, 1960-86; for Canada, 1961-87; for Spain, 1964-86; for Sweden, 1963-87; for Australia, 1969-87; for New Zealand, 1977-86.
1. Value added per employed person.
Source: OECD.

176

The improvement of capital productivity in the 1980s most probably is related to increases in both the real and the relative price of capital, stimulating a more economical use of capital as a factor of production (OECD, 1979c). Among the factors tending to raise the real and relative price of capital have been the removal of interest rate ceilings, the withdrawal of tax incentives to fixed asset formation and nominal wage moderation, a consequence of high unemployment (Atkinson and Chouraqui, 1985). To the extent that the comparatively favourable performance of capital productivity reflects reduced distortions in financial markets, it has a positive effect on TFP. In addition, abstracting from the influence of changing relative factor prices, technological progress may have become more capital-saving, an example being novel methods of monitoring inventories (see Chapter 4, Section 1).

In contrast to developments in the business sector as a whole, *labour productivity in manufacturing* has risen substantially with output per employed person in 18 countries, growing at an annual 3.7 per cent in 1979-88, or 1.3 percentage points higher than in 1973-79 (Table 29). The recovery of productivity in manufacturing was also more broadly based, with only seven out of 18 countries suffering from a further weakening of labour productivity in the 1980s. The most striking improvement took place in the United States, the United Kingdom and Finland where gains in labour productivity in the 1979-88 period exceeded even those observed in the 1960s.

The rising trend of labour productivity in manufacturing has been associated with a variety of influences. In a setting of low profitability and keener international competition, employment was reduced in the early 1980s. Services were contracted out, unprofitable capital vintages were removed and new technologies came on stream. In the case of the United States and the United Kingdom, a sustained rise in the real exchange rate in the early 1980s also played a part in inducing stronger efficiency gains in the open (i.e. mainly manufacturing) sector. In most OECD countries progress in structural reform gathered momentum in the late 1980s with, in many instances, privatisation and deregulation further stimulating economic dynamism.

There has also been a tendency for the aggregate level of labour productivity among OECD countries to converge, reflecting a process of catch-up at the industry level rather than shifts among industries with different average productivities. A comparative study of 28 manufacturing industries covering 13 countries found that productivity levels converged in all but three industries between 1963 and 1982 (Dollar and Wolff, 1988). Unlike 1963, when the United States was the leader in virtually every industry, eight different countries held the lead in at least one industry in 1982. The driving force behind this process of "diffusion" is competition between firms in different coun-

tries. If one firm benefits from a significant innovation, firms which produce competing products are forced to imitate in order to stay competitive.

Although available only for 13 countries up to 1985, sectoral data suggest that much of the rise in labour productivity growth in manufacturing reflects developments in only a few sectors, particularly machinery and equipment (including computers) and primary metals industry. In contrast, efficiency gains weakened further in the food, textile and construction industries (Figure 25).

In addition, given perceptions of rising rates of technical progress, the vigorous revival of labour productivity in manufacturing seems, at first glance, consistent with the proposition that new technologies are mainly created and absorbed by manufacturing (Englander et al., 1988). However, a different picture emerges when recent productivity trends in manufacturing and gross fixed asset formation are examined. In the United States and the United Kingdom, the level of gross fixed investment in manufacturing declined in the 1980s, while productivity accelerated. This points to a restructuring process, in the course of which unprofitable firms and vintage capital were withdrawn. In contrast, in Finland and Sweden, labour productivity and gross fixed investment were both accelerating in the 1980s, perhaps pointing to a wider diffusion of innovations ("embodiment" effects).

The special case of the service sector

For virtually all countries during the past three decades, productivity growth in manufacturing, a sector relatively open to international trade, has been higher than in the business sector as a whole, which includes non-tradeables like construction and most services. In the 1960s and early 1970s, the average gain in OECD output per employed person in manufacturing exceeded that in the service sector by large margins (Table 30). After narrowing in the 1970s, this differential widened during the 1980s as measured efficiency gains increased in manufacturing, while they weakened further in private services and construction. This widening of the gap between rates of sectoral labour productivity growth appeared in about two-thirds of the countries for which data are available[15].

Disaggregated data for 13 countries (Meyer-zu-Schlochtern, 1988) show a further synchronised weakening of TFP trends in most service sub-sectors after OPEC II, the only exception being the retail and wholesale trade sector where TFP began to rise again, albeit by a small amount. In three service sub-sectors (real estate and business services; community, social and personal services; and financial institutions and insurance) combined inputs grew more rapidly than output, reducing the level of TFP.

Figure 25. **Total factor productivity by sector**[1]
average percentage change at annual rate

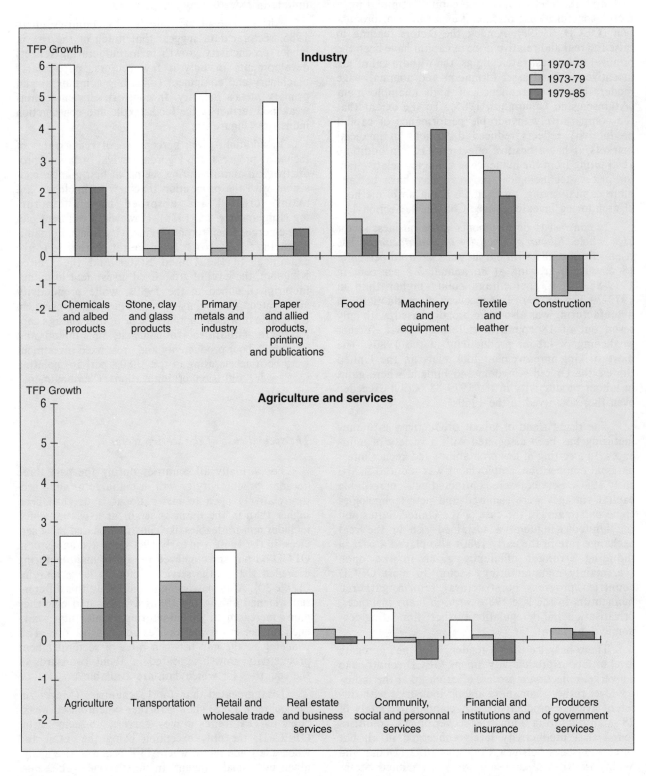

1. 13 countries: United States, Japan, Germany, France, Italy, United Kingdom, Canada, Belgium, Denmark, Finland, Norway, Sweden and Australia.

Source: OECD.

Table 30. **Labour productivity[1] in the service sector,**
1960-86

	1960-73	1973-79	1979-86	1960-86
United States	2.2	0.1	0.3	1.2
Japan	7.4	2.1	2.6	4.8
Germany	3.6	3.2	2.1	3.1
France	4.0	1.4	1.8	2.1
United Kingdom	2.3	0.7	1.1	1.6
Italy	4.5	1.3	−0.7	2.3
Canada	2.0	1.1	1.1	1.5
Belgium	2.8	1.0	0.8	1.3
Denmark	3.3	1.5	0.8	1.9
Finland	3.0	2.7	2.8	2.9
Netherlands	2.4	2.7	1.3	2.0
Norway	1.2	2.8	1.1	1.6
Sweden	2.4	1.3	1.0	1.7

1. Value added per employed person.
Source: OECD.

The data for individual countries reveal a few important exceptions to this generalisation. In the United States, Japan, the United Kingdom, Canada, Belgium, Denmark and Finland – mostly countries with a significant recovery in manufacturing productivity – TFP performance after 1979 improved in at least two of the six main service sub-sectors. In the case of Japan, the United Kingdom and Belgium, the improvement was broadly based within the service sector. Even so, compared with the acceleration of TFP in industries such as basic metals and machinery and equipment, the revival in service efficiency was on the whole modest.

The reason for the difference in performance between manufacturing and the service sector lies both in the statistical conventions for measuring service productivity and in the lower potential of services to absorb new technologies. Furthermore, the service sector has traditionally been sheltered against the full pressure of international and – in many instances – domestic competitive forces. To some extent, the appropriate technology to overcome high information and transaction costs did not exist, thus providing a "natural" shield against competition; but government regulation has also played a role in this context. New technologies have promoted rapid development in trade in services and, in several cases, have undermined the sustainability and even the justification of regulations that historically have restricted competitive forces in this sector. From this it might follow that the scope for using modern technologies in the service sector has increased and that competition will compel the service sector firms to perform more efficiently. The reason why this has not yet appeared in the statistics will be dealt with below.

3. MEASUREMENT ISSUES

Attempts to explain the "Solow paradox" have revived interest in the way in which output and inputs are measured. Is it true that measurement errors have increased over time, overstating the productivity slowdown in the 1970s and understating the revival in the 1980s? In other words, have official statistics portrayed an unduly pessimistic picture of overall efficiency gains?

From the numerous studies which have examined this issue, two principal conclusions emerge:

- First, mismeasurement has increased over time, largely on the output side rather than on the input side, exaggerating the productivity slowdown in the 1970s and understating the scale of the subsequent recovery in the 1980s.
- Second, the extent of rising mismeasurement has been too small to explain more than a minor part of the TFP slowdown in the 1970s. Similarly, the rise in mismeasurement is too small to explain why TFP growth in the 1980s has failed to return to its long-term trend[16].

There are two main sources of increased mismeasurement. The first is the difficulty of measuring service sector output, coupled with the increasing share of that sector. In sectors such as government, social, community and personal services and private non-profit institutions, labour input data are typically used to derive output estimates. This constrains labour productivity to zero. Moreover, new technology has been appearing in those service sub-sectors where mismeasurement is notorious, e.g. the health industry, financial institutions and insurance, space exploration and defence. As a result, measurement errors have tended to rise.

A second source of increased mismeasurement relates to more rapid changes in the quality of goods. Productivity measurement is difficult for new or rapidly changing products, because conventional price indices fail to capture fully the quality changes. A prominent example is the increase in the power of computers and related equipment. One response, pioneered by the US Bureau of Economic Analysis (BEA) and

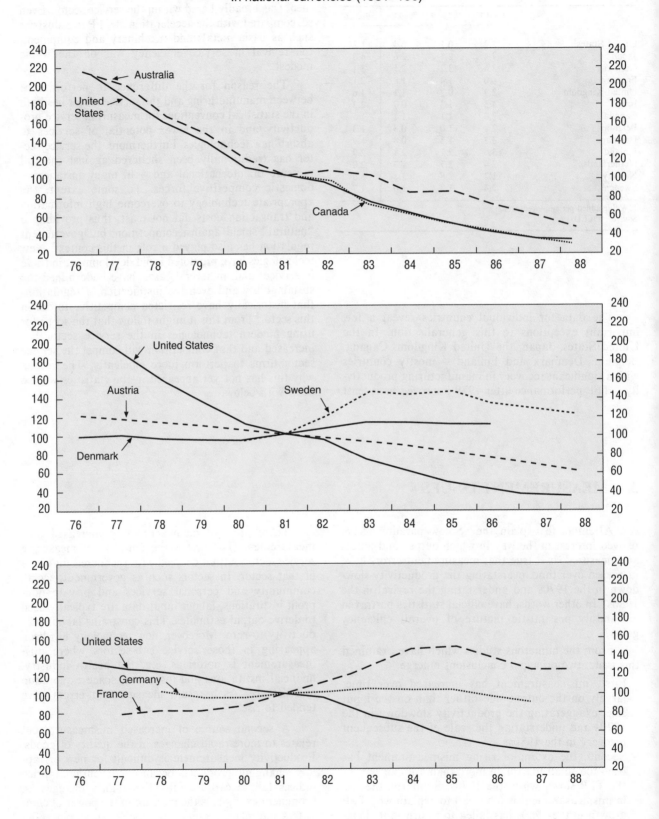

Figure 26. **Prices for computing machinery**
In national currencies (1981=100)

Source: Ford *et al.* (1990).

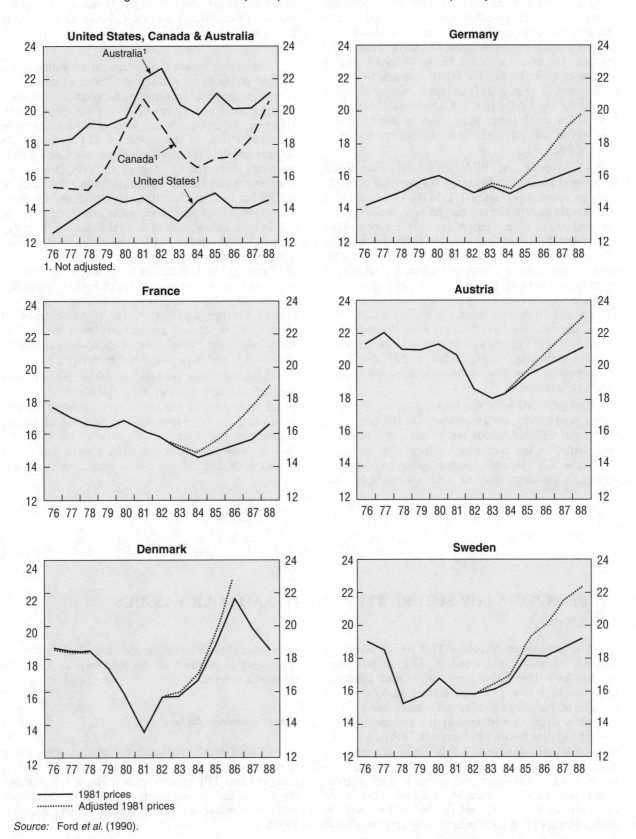

Figure 27. **Original and adjusted business investment-output ratios (%)**
using BEA series for computer prices to make investment-output adjustment

United States, Canada & Australia

Australia[1]

Canada[1]

United States[1]

1. Not adjusted.

Germany

France

Austria

Denmark

Sweden

—— 1981 prices
········ Adjusted 1981 prices

Source: Ford *et al.* (1990).

181

since adopted by Canada and Australia, has been to adjust computer prices explicitly for quality changes using a hedonic price index. Under this approach, prices of computers include the values assigned to their characteristics (processor speed, memory and so forth). Several other countries – Japan, Denmark, France and Sweden, for example – are likely to make similar adjustments in the future. What emerges from this adjustment is that quality-adjusted computer prices – those for the United States, Canada and Australia – have fallen much faster than prices in countries that use standard national accounting procedures (Figure 26).

As a result, the BEA procedure yields much higher real investment figures, given nominal expenditures on computers (Figure 27). In the three countries that have already adopted the hedonic methodology, the real share of computers in total business fixed investment has risen much faster than the nominal share, reflecting the measured price decline. In contrast, in countries that have not adopted a BEA-type procedure, real and nominal investment shares for computers do not deviate very much from each other. The use of a BEA-style index also implies a significant rise in the share of total business fixed investment in total output. Typically, the share was between 2 and 4 percentage points higher by 1987, with the effect growing over time, along with the share of computers in investment.

The implications of the BEA procedure for measuring productivity are ambiguous. On the one hand, traditional national accounting procedures understate real output in the computer industry. On the other hand, the US National Income Accounts currently aggregate the components of GNP according to their 1982 prices, which gives too large a weight to the computer-producing industry, where productivity advances are far above average. On balance, it appears that the productivity growth in US manufacturing over the past few years has been overstated (OECD, 1989k).

Measuring inputs is also fraught with difficulties. Labour productivity is often measured as output per employed person, ignoring reductions in standard working time and increased part-time employment. However, adjusting for this kind of mismeasurement has little effect on the profile of TFP growth[17]. As regards capital, the equipment mortality or survival functions employed by most of the OECD countries (variants of bell-shaped distribution functions) may be faulty. But it is not clear that this introduces a systematic error into capital stock measures. Similarly, the rise in mismeasurement of capital linked to the surge in energy prices in the 1970s appears too small to have a significant effect on the TFP growth (Olson, 1988). On the other hand, under the conventional approach to measuring capital, embodied technological knowledge is assumed to be homogeneous in different vintages of capital. On this assumption, the effect of technical change on output and productivity growth is understated when the "true" rate of technological change embodied in new equipment has been rising[18].

All in all, there are indications that measurement problems have risen over time. However, there is a broad consensus in the discussion – and it became apparent in the 1989 International Seminar on Science Technology and Economic Growth at the OECD – that the slowdown in productivity growth was a real phenomenon and not just due to measurement errors.

4. PROSPECTS FOR FUTURE TFP GROWTH AND POLICY ISSUES

The assessment of future TFP trends is one of cautious optimism: the trend of TFP growth may increase over the medium run, with most countries likely to share in a TFP acceleration, at least in the business sector and particularly in manufacturing. This outlook is largely conditioned by the disappearance of several negative forces which eroded TFP gains in the past. New positive influences have appeared, including more assertive supply-side policies (with ensuing market expansion and increased domestic and international competitive pressures)[19] and new technologies (with further decreases in information and transaction costs and a speeding up of international transmission of knowledge). Both of these developments, mutually reinforcing each other, should tend to raise TFP over the medium term.

Macro-economic factors

"Tomorrow's" productive potential may depend in part upon the intensity of factor use "today". Thus, it is argued that TFP growth tends to rise or fall, depending on whether realised levels of profitability have been repeatedly higher or lower than expected (Abramovitz, 1991).

Figure 28. Change in TFP growth and share of gross fixed capital formation in GDP[1]
(1979-88/1973-79)

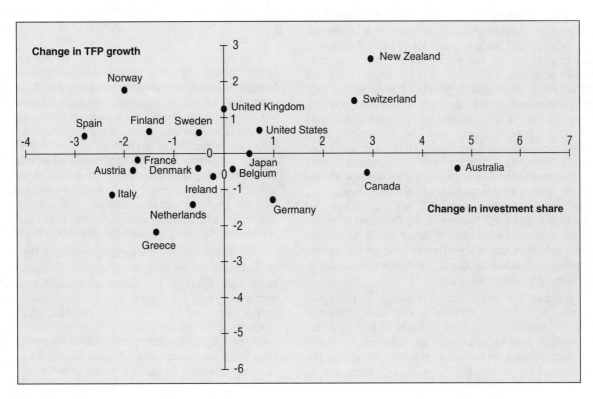

1. Business sector.
Source: Ford *et al.* (1990).

Rates of capacity utilisation are projected to remain relatively high over the next couple of years – in the spectrum of medium-term prospects, an economic crisis does not appear very probable. If there are dynamic economies of scale, strong and sustained output growth should translate into increased efficiency gains. Factors such as learning-by-doing, the expansion of markets and the associated increase in the division of labour are relevant in this context. Empirical evidence points to a strong influence running from changes in average (medium-run) rates of capacity use to changes in medium-run TFP growth (Englander and Mittelstadt, 1988).

In addition, gross fixed asset formation, stimulated by high capacity use, has been buoyant over the past few years, and its share in real GDP is projected to rise. Since investment is the vehicle by which new technologies enter the production process, sustained investment growth could lead to a more rapid diffusion of innovation than in the period of sluggish investment

activity during the 1970s and early 1980s. Some weak evidence for embodiment effects is provided by Figure 28, which plots changes in the average TFP growth rates between 1973-79 and 1979-88 against corresponding changes in the share of gross investment in output. Generally, increases in TFP growth appear to have been associated with increases in the investment share.

Structural adjustments

Technological knowledge is often viewed as generated largely through innovative activities undertaken by firms to develop and improve products and production processes rather than through research activities (see Chapter 1). Clearly, the pace at which such knowledge is created depends in part on competitive pressures and expectations about profitability, and some contribution to increased TFP growth over the medium term may thus come from the progress towards structural reform achieved in the OECD countries.

In major policy areas – such as financial markets, foreign direct investment, taxation and regulatory policy – where markets have been opened up or economic incentives redressed and strengthened, increased competitive pressures will stimulate the positive sum game of "creative destruction". Market liberalisation (e.g. the completion of the common European market), along the lines of institutional competition rather than *ex ante* harmonisation, can be regarded as a "Schumpeterian event" (Siebert, 1989) that will entail mainly dynamic efficiency gains. Moreover, many OECD countries – and among them some with the highest shares of the public sector – are bringing a broad range of issues concerned with the scope and efficiency of the public sector to the forefront of political discussion. An attempt is being made to change management structures and incentives and to either introduce or simulate market processes in order to increase operational efficiency.

As has already been mentioned, capital has become more mobile, increasing allocative efficiency. This is exemplified by the recent surge in foreign direct investment, leading to transfer of "enterprise" strategies and related increases in firm-specific productivity (Kanawaty *et al.*, 1989). Also, US firms are no longer exercising technological leadership in all domains; Japanese and European firms have also been competitive along a highly differentiated technological frontier. Because there are more actively competing players in the game, the "potential at the frontier" may be more effectively pushed outward.

However, policy areas relating to international trade, agriculture and industrial subsidies remain problematic. The international trading system, because of increasing emphasis on result-oriented trade-regulating measures, continues to be under strain. Growing concern about international competitiveness and specific characteristics of market structures in high-technology industries have led to widespread development of industrial policies focusing on so-called strategic sectors and technologies. These policies can contribute to trade frictions and may lead to a build-up of excess capacities and protectionist pressures, thus restraining further productivity growth (see OECD, 1991*h).

A major area of uncertainty concerns the impact of increased investment for protection of the environment (see Chapter 9). Given traditional accounting methods, such investment typically reduces capital productivity and TFP. On the other hand, as the environment deteriorates and as economic development risks becoming unsustainable, environmental inputs may need to be reflected in gross output in a multifactor production function ("green GDP").

5. CONCLUSION

The proposition that investment in R&D and technological progress are essential for future growth has not yet been conclusively empirically demonstrated. Nevertheless, economists generally agree that R&D and technical progress do indeed play a crucial role in economic growth. Despite present limitations, the "new growth theory" brings to the forefront the interaction between growth and technology-related tangible and intangible investment. If the "increasing returns" associated with the features of technological change discussed in earlier chapters can be successfully introduced into macro-economic growth analysis and modelling, the results may show more satisfactorily than they do at present that policies for technology have significant impact on growth and cannot simply be equated with micro-economic policies that improve static efficiency.

Structural adjustment reforms carried out in the 1980s have helped open up markets and strengthen competitive pressures, thereby reducing the inefficiencies resulting from protectionism and uncompetitive behaviour. This should translate sooner or later into higher TFP growth by i) stimulating innovation; ii) diffusing technical change; and iii) speeding up complementary institutional and organisational innovations. However, as national economies become increasingly interlinked, a parallel growth of concentration (see Chapter 10) may bring cartelisation in some areas. This may imply the need for international co-operation in pursuing effective competition policies to preserve the momentum for long-run efficiency increases.

NOTES

1. The list of inputs is often expanded to include different types of capital (structures versus machinery and equipment, for example), different types of labour (skilled versus unskilled) and other inputs (energy and intermediate inputs), although, from the point of view of explaining long-term growth, such modifications have not proved crucial.

2. These are: i) real wage rates rise more or less at the rate of technological change; ii) the real return to capital is roughly constant; and iii) the capital-output ratio is also stable.

3. TFP, measured as a residual, has sometimes been referred to as "a measure of our ignorance" (Nelson, 1981a, p. 1032).

4. Of course, technical change is not perceived as being literally independent of economic processes. On the contrary, as the vast literature surveyed in this report testifies, a major theme in studies of growth economies has been explanations of productivity change.

5. See Englander and Mittelstädt (1988); this theme is taken up below in the context of the "new" growth theories.

6. The evidence for all Member countries is contained in Ford et al. (1990).

7. While it would be possible to filter out the cyclical information in order to highlight any underlying trends in the two series, such a procedure could also destroy the timing between changes in the two series. Timing is obviously of some importance, since, according to the embodiment hypothesis, changes in the capital replacement rate should precede those in TFP growth.

8. The evidence for all Member countries is contained in Ford et al. (1990).

9. These calculations were made by Romer (1988).

10. Pioneers of the "new" growth theory include Romer (1987), Lucas (1988), Scott (1989), and Baldwin (1989).

11. Romer (1989) works out the implications of such a model.

12. The growth rates of both labour and capital productivity fell harply in the 1973-79 period compared with the 1960s. But labour productivity growth fell more sharply than capital productivity or TFP growth rates, a consequence of the slower rise in the capital-labour ratio.

13. According to estimates by Maddison (1989, Table 1.5), for the five major OECD countries TFP growth was higher in 1973-84 than in 1913-50 for Japan, Germany, France and the United Kingdom, whereas it was markedly lower in the United States.

14. TFP growth is usually interpreted as "technical progress" or as a shift of the production function towards higher levels of production at given input levels. Increases in the partial (capital, labour) productivities may reflect both technical progress of factor substitution due to changes in relative factor prices.

15. For example, the share of service employment in total employment has been rising in OECD countries. Given the normal differential between growth rates of productivity in industry and services, this compositional effect has restrained the recovery of labour productivity growth, but by only a small margin.

16. Evidence on the effects of measurement error is provided by Gordon and Baily (1991), Griliches (1989), Kendrick (1991) and Englander (1991).

17. Adjusting the recent data on labour productivity for changes in the number of hours worked yields no stronger profile for the revival of overall efficiency gains in the 1980s. In many countries, the pace of reduction in hours has been slowing in the 1980s.

18. See van Zon (1991). Conventional measures may also seriously understate the capital stock, if allowance is not made for intellectual capital as a complement to physical capital. Under a broad definition, intellectual investment includes outlays on research and development, patents, licenses, advertising and software development.

19. In the United Kingdom, Finland and Spain, for example, all countries with assertive supply-side policies, TFP growth increased in the 1980s, even though the investment share remained unchanged or fell (OECD, 1989f, p. 31).

Chapter 9

TECHNOLOGY AND THE ENVIRONMENT

ABSTRACT

Over the past two decades, environmental challenges have become both broader and more acute. The notion of "sustainable development" links the more traditional concerns about physical pollution to wider concerns about the conditions under which the geosphere/biosphere can sustain equitable long-term growth and welfare for the world population as a whole. Sustainable development concerns are likely to become increasingly important for policies in the 1990s.

The nature and extent of truly global environmental problems are still insufficiently known. For instance, the magnitude and the timing of predicted temperature changes are controversial, and the potential consequences of global warming on health and environment conditions are even more uncertain. The economic analysis of environmental problems is still sketchy, and existing price signals are unable to take into account most of the environmental consequences of economic activity. The new environmental challenge poses policy problems in the area, already familiar in environment discussions, of public goods and externalities. The OECD has promoted the "Polluter Pays Principle" and other economic measures as means to internalise the external costs of pollution. With much broader and longer-term concerns, the principle will remain important. However, in the area of international policy discussions, while the use of economic instruments should become more important, the scope for inter-governmental agreement seems at present modest.

Environmental policies in OECD countries have generally sought to regulate new products, processes and facilities more than existing ones. Reliance on command and control style regulation has resulted in the diffusion of end-of-pipe environmental control technologies. It has not provided a strong enough incentive for fundamental redesigns of products and processes. Moreover, although this approach has proved to be relatively effective in controlling large industrial sources of pollution, it has not succeeded in controlling more diffuse forms of pollution, in particular by urban consumers and by farmers. Finally, because regulatory policies have focused on the most obvious problems and the easiest means of addressing them, future risk-reduction and clean-up efforts will require far larger financial outlays.

The results of two decades of regulatory policies have been mixed. On the whole, regulatory policies have promoted the diffusion of "best available technologies" but have had less success with innovation specifically aimed at environmental objectives. Analysis suggests that there remains considerable scope for diffusion of "best available technologies", especially outside the OECD area, where issues of cost-sharing and technology transfer have international policy implications. The present challenge is to develop prevention policies and the requisite technologies. Innovation will entail a major science and technology effort. To fund it, environment-related programmes will have to receive a much higher order of priority in the R&D budgets of governments and firms. Greater international co-operation in environmental research is a priority area.

This experience confirms the need to develop broader-based and longer-term policies for pollution prevention and reduction of the sources of pollution. In the long run, the only affordable and effective way to diminish pollution of the scale and type societies face today is to work "upstream" to change the products, processes, policies and pressures that generate waste and give rise to pollution. But this shift of policy will probably prove to be much more difficult than earlier dependency on end-of-pipe solutions. It will involve major changes in the behaviour of large firms as well as consumers, important paradigm shifts and changed technology trajectories. It will involve major shifts in economic policies, taking environmental concerns into market prices and through a major shift in fiscal structures and policies.

INTRODUCTION

For most of the post-war period, technology was viewed almost exclusively from the standpoint of its contribution to innovation and economic growth. It was considered to be an essential condition of economic progress and the competitive efforts of enterprises and nations. Technology was also regarded as a means of realising social objectives or improving the quality of life. Other dimensions were considered to be of secondary importance.

In the late 1960s, critics began to question the value of technical change and economic growth. Environmental interest groups began to perceive technology as a demon that fouled the air and poisoned the earth. They asked whether, in the light of its exhaustion of materials and energy resources, economic growth could be sustained and whether environmental problems might not endanger the very existence of the human race. A first major study, with a world focus, took up some of these issues (Meadows *et al.*, 1972).

The concept of sustainable development, elaborated by the World Commission on Environment and Development (WCED), is based on the premise that economic and ecological systems are now completely interlinked (WCED, 1987). It implies that global goals to maintain economic growth, eliminate poverty and deprivation, conserve the environment and enhance its resource base must be defined simultaneously (see Box 39). A workable strategy for the elimination of the interrelated problems of poverty, inequality and environmental degradation requires a new approach to economic growth and environmental management. These issues will be addressed notably at the 1992 UN Conference on Environment and Development (UNCED) in Brazil.

The experience of OECD countries, reviewed below, suggests that it has not been straightforward to integrate (even relatively specific) environmental concerns into efficient policies. There is community of interest around some central environmental concerns, and policy mechanisms (with the prospect of "clean" technologies) to implement environmentally favourable outcomes. For much of the "sustainable development" agenda, there are neither criteria to judge between alternatives, nor – much more important – societal processes to implement effective policies. The more sophisticated view of technology associated with this approach recognises in technology a source both of economic progress and of serious social and environmental problems. Technology is not an autonomous force with its own internal trajectory; it reflects the social, economic, and organisational priorities of major interest groups in industrial societies.

This suggests that, properly channelled, with the right incentives and appropriate analysis, technology can be used to help solve some of the problems that it has been accused of creating. As one scholar has suggested, technical advance is "critical for the long-term conservation of resources and the improvement of the environment. The prevention of most forms of pollution and economic recycling of waste products are alike dependent on technological advance" (Freeman, 1982). To meet the global challenge, steps must also be taken to ensure that, as technological processes and products are applied to help solve the problems created by poverty and inequality, they do not create new environmental problems. In particular, ways must be found for new, cleaner and safer technologies to be diffused internationally as quickly and as inexpensively as possible.

Most of the analysis in this chapter deals with attempts to find means of improving the use of technology and resources within present trajectories and to better monitor the relations, at the world level, between industrial society and its natural environment. Most past environmental policies have sought to achieve the (better) diffusion of existing "best available technologies". These can still offer considerable benefits, especially in the short term. Thus, energy saving, which carries a very high priority, will, in the short term, primarily involve better diffusion of existing technologies. However, recent scientific assessments made with regard to certain key global issues, notably global climate change as accentuated by "greenhouse effects", suggest that the previous approach can no longer suffice. "Sustainable development" now requires the search for totally new technological trajectories (see Chapters 1 and 2) and, subsequently, the major investments required for establishing productive activity on foundations which would no longer threaten essential ecological and climatic parameters.

The pace of economic growth as currently measured will necessarily be determined by the scale of these investments as well as by the extent to which presently known methods for the reduction of chemical emission, waste and noise and the saving of energy are applied effectively by countries (see below Section 5) and so affect the growth of GDP. This is why this chapter cannot be separated from the preceding one. Time has been lost in financing the R&D programmes required for reducing the OECD's countries collective degree of *ignorance* and so of *uncertainty* about the scale and the effects of major on-going environmental changes. As environmental problems and challenges come to be understood and measured with greater precision as a consequence of scientific and sociological research they will almost certainly influence the entire manner in which the technology-economy relationship is considered in the future.

1. THE LEGACY OF POLLUTION-PRONE TECHNOLOGIES AND SOME ECONOMIC ISSUES

Several areas which have been affected by the pollution-prone technologies of this century now have a bearing on the current environmental situation at the global level (Speth, 1990). In all, the trends need to be halted or reversed if sustainable development is to be a reality.

First is the trend from modest quantities of environmental pollutants to the huge quantities associated with the explosive growth in economic activity and human population. Since the beginning of this century, the gross world product is some 20 times larger than in 1900 and world population has tripled. Pollutants have increased dramatically: fossil fuels have been the principal energy source for the technologies of this century's economic growth, for example, and their use has increased by more than ten-fold since 1900. As a result, over 200 million tons of sulphur and nitrogen pollutants are being added to the global atmosphere each year. In addition, burning of fossil fuels has contributed to the 25 per cent increase in carbon dioxide – the principal greenhouse gas – from the beginning of the century.

Similarly, the vast quantities of solid and liquid waste materials being generated by the technologies used in industrial operations and human settlements are exceeding the ability to manage them effectively and safely. Throughout OECD countries, waste management has become a critical challenge for governments. Public opposition to waste incineration is increasing, due to environmental concerns, and space for burying refuse is being exhausted. Shipments of wastes across state and national borders for disposal are becoming politically unacceptable. These problems will continue to increase if the trend is not reversed, because by the middle of the next century, the global economy is likely to be four or five times as large as it is today and the world's human population of 5 billion will probably double.

The *second* major issue concerns the chemical and nuclear industries and the technologies which substitute synthetic chemicals for natural products. Many of these synthetic products have been found to be toxic even in small quantities, and certain human-made chemicals persist in the environment for long periods, accumulating in soil, water supplies and biological systems. Millions of tons of pesticides, for example, are used throughout the world each year. Only a fraction of this ever actually reaches the pests (US Council on Environmental Quality, 1981); if the pesticides are not biodegradable, the major part of these inherently toxic chemicals will remain in the environment for a long time.

Even chemicals which were originally considered to be safe are bringing unpleasant surprises. The case of chlorofluorocarbons (CFCs), a non-toxic mainstay of modern society, is probably the best known. They were thoroughly tested for toxicity to man and to classic aspects of the environment, and no thought was given, as they were developed and marketed, to the effects they might eventually have on a remote environment. Yet it was later recognised that, as a result of a complex set of chemical reactions, they were the major source of the destruction of the ozone layer of the upper atmosphere.

The *third* trend is the broadening geographical scope of environmental issues. What once was seen largely as a preoccupation of the industrialised nations of the OECD area is becoming more acute in many developing and Central and Eastern European countries, as cities in these parts of the world are now experiencing levels of air pollution that exceed those found elsewhere. Environmental pollution by technologies and inefficient use or mismanagement of natural resources not only diminish the quality of life for the citizens of these countries but also erode prospects for the national economic growth needed to upgrade living standards over the longer term.

The *final* issue is the broadening scope of the environmental threats themselves. The focus has gradually moved from air and water pollution near sources and resource degradation at local and national levels, to problems of a regional, transboundary nature, such as acid rain, shared water resources, desertification and deforestation. Since the middle of the 1980s, global environmental problems have become a major concern, triggered largely by the threat to the stratospheric ozone layer and the prospects of global climate change (see Box 40). These latest additions to the international agenda for action have introduced not only an elevated sense of public awareness and concern but also a host of complex new issues related to technology (Kemp and Soete, 1990).

The potential effects of greenhouse gases (GHG) on global climate change is emerging as the central environmental problem. It represents a limiting case of environmental problems, because: *i)* the potential environmental damages are very large, and, perhaps, irreversible; *ii)* the uncertainties about what will happen are very large; *iii)* the time horizon for these damages is very long: very few people living today are likely to suffer the consequences in their lifetime; *iv)* the costs of meeting the problems are potentially very large; *v)* the problem is truly "global". Given these propositions, the whole environmental agenda is changed. The implications of global climate change represent a qualitative jump in regard to the other more immediate "classical" environmental and sustainable development concerns.

Despite widespread uncertainties about the nature and extent of global environmental threats, certain general characteristics of these concerns, as they relate to science and technology[1], can be defined:

i) No single coordinating concept, such as "avoiding pollution", permits the establishment of overall international priorities for science and technology environmental policies. Nevertheless, the environmental interest groups and S&T communities do not see a conflict among most of the existing priorities. The view that "pollution avoidance" corresponds in general to "minimising energy and raw material use" might serve as a guide.

ii) Recognised concerns are increasingly long-term, thus implying issues of inter-temporal welfare.

iii) They threaten accepted notions of aggregate equilibria and smooth adjustment, by introducing massive uncertainties and risks, including that of irreversibilities.

iv) They raise questions about attitudes to technology: "optimism" versus "pessimism".
v) They pose institutional problems, and in particular, the need for new transfer sciences and their organisation.

Some economic issues

Environmental issues and their policy implications are now recognised as relevant to economic analysis and decision making. However, while environmental concerns can be related to important strands in traditional economic analysis, much remains to be done. First, much economic discussion about the environment has been based on the well-developed strand of economic analysis relating to free "externalities", reflecting the "public goods" nature of environmental resources. With public goods (e.g. air and water), decisions based on market prices may not reflect social costs and benefits. It is widely argued that, in the absence of government intervention to reflect the externalities in market prices, this has seriously disfavoured a proper management of the environment. Second, there is a growing body of economic analysis based on the "non-renewability" of environment resources, with spatial and inter-temporal (e.g. inter-generational) implications[2].

Some of the many approaches to "sustainable development" can be taken into this economic analysis through the notion of two types of capital (stock and flow) – "environmental" and "human-made" – which determine in traditional growth theory, future consumption. By identifying a separate component – environmental capital – which is only partially reproducible and substitutable by human-made capital, the analysis can focus on the conditions for ensuring sustainable development across countries and generations (Nicolaisen *et al.,* 1991). The "sustainability constraint" is that the total stock of capital should not decline. In general, the shadow price of environmental capital is likely to be higher than its market price (because of externalities) and rising over time (because human-made capital is more easily reproducible and increasing its share of the total capital stock). Conversely, the shadow discount rate for environmental investments should be lower. It has been argued (e.g. Cline, 1991) that the resulting overall discount rate should be as low as 2 per cent. From a technology point of view, this focuses attention on the extent of substitutability between environmental and human-made capital, and, given the inter-temporal issues, on technology optimism/pessimism and technological and ecological risks.

Some observers argue, on grounds of low substitutability and/or technology pessimism and caution, that overall GDP growth should be constrained to rates several percentage points below those of the past three decades. Most, however, argue (implicitly) that continued economic growth (with appropriate environmental policies) can generate technology innovation and diffusion that is compatible with the "sustainability constraint". The implicit assumptions are about substitutability, the shadow discount rate for environmental investment and the longer term risks.

A different, but related, discussion starts from empirical analysis of the costs of "environmental abatement" and considers the implications for economic growth of projecting these forward on different scenarios of environmental objectives (see e.g. Nicolaisen *et al.*, 1991, for a discussion of the evidence and issues in the OECD countries). Some illustrations are given in Section 5 b), which suggest that, with all the data and analysis uncertainties, the economic cost of meeting most environmental objectives should not be unacceptable with appropriate policies. However, most of the analysis does not relate to the "sustainability" problems of non-OECD countries, while the global climate issues present much more intractable problems.

In both cases, there is a problem about the valuation of economic growth, consumption and welfare. In economic analysis terms, there is now considerable interest in using environmentally related shadow prices to reassess the familiar national accounting concepts, and to develop indicators. If social accounting based on sustainable development considerations is taken into the familiar national accounting frameworks, the analysis of aggregate output and income growth might be affected. Plausibly, environment-related expenditures which are treated as costs of production in the conventional accounts have slowed the pace of conventionally-defined GDP growth. In policy terms, the difficult questions are about how far the implementation of shadow prices would influence producer and consumer behaviour.

All the economic analysis makes more or less explicit assumptions about technology innovation and diffusion. The environment discussion necessarily involves relatively long term projections with policy scenarios going beyond much recent experience. It has proved very difficult (even e.g. in energy models) satisfactorily to bring together sufficiently disaggregated economic data while articulating specific technology choices (Nicolaisen *et al.*, 1991). Yet, one strand in the argument of this chapter and the whole report is that while new opportunities and indeed whole new paradigms can be opened up by technology, technological trajectories will be shaped by the decisions which determine future costs of production and patterns of economic activity. To the extent that environmental concerns take a central place in policymaking, environmentally driven technologies (affecting innovation, diffusion, and technological paradigms) may shift industrial structures and competitive advantages, both nationally and internationally.

At the same time, there is widespread evidence of market failure with existing price signals: much environmental investment would be profitable but is not undertaken, for reasons that seem to involve market structures, corporate cultures and investment time horizons, given present price signals and rates of reward to environment friendly measures by firms. Finally, environmental concerns enter the question of product quality and safety, which have been increasingly important for OECD countries in determining competition, investment and industrial processes in recent years. These issues will be dealt with in greater detail in Section 5.

2. NATIONAL ENVIRONMENTAL POLICIES IN OECD COUNTRIES

Environmental policies, developed and implemented in OECD countries since the mid-1960s, have responded to a concern for the same broad issues – air, water, waste and toxic chemicals – in the same general way. The primary focus has been on regulating large industrial sources of pollution, with overwhelming reliance on command and control regulation to force development and use of end-of-pipe pollution abatement technologies. Beneath these similarities, however, lie major national differences in social priorities, in the degree to which regulatory policy has forced technological development, and in the procedures used to implement environmental policy. This section attempts to examine the policy context for both technology and environmental issues in various OECD countries.

In the United States, stringent emissions standards based on human health considerations have

often been adopted, sometimes without consideration of the economic costs or of the monitoring and implementation burdens they impose on industry. The philosophy underlying these emissions standards involves "technology forcing", that is, requiring industry to devise and adopt better pollution control technologies and, in the best cases, environmentally sound technological processes and products.

Environmental quality standards, like those in the US Clean Air Act for stationary sources, were intended to encourage technology development and application by threatening plants with closure if they proved unable to meet strict deadlines for emission reductions. Technology-based standards (e.g. emissions standards for automobiles) seem to force technology by requiring controls based on the future state of the technological art (Stewart, 1981). According to most assessments, this approach has not, in general, achieved the desired result, because of problems encountered in implementation. Government regulators have often found it difficult to determine the "appropriate" state of the technological art. Pressures from industry and concerns about economic or technical feasibility have often resulted in modifications which weaken stringent standards.

Where implemented, the technology-forcing approach has given firms strong incentives to adopt *existing* pollution control technologies that allow them to meet standards in a cost-effective manner in the short term (Rehbinder and Stewart, 1985). By thus discouraging firms from investing in the development of new technologies, the approach may not have achieved its full potential.

United States policy has been designed in principle to give firms as much technical flexibility as possible in the methods of complying with standards. In practice, United States regulators have often written standards in a way that encourages the use of particular control technologies. One example is the detailed technical requirements associated with the Clean Water Act. Another is the source performance standard for coal-fired power plants. Concern about the competitive interests of coal producers producing high sulphur coal led the Clean Air Act in 1977 to be modified to require, in effect, the use of scrubbers regardless of the sulphur content of coal used (Ackerman and Hassler, 1981).

The emphasis on the "polluter pays principle" in the Clean Air Act, the Resource Conservation and Recovery Acts and other programmes has been one of the major reasons for the rising costs of pollution control to US industry. However, political reluctance to impose costly measures on industry has also led to situations which subject new industrial facilities to more stringent regulation that existing ones. It has been argued that differences in the treatment of existing and new facilities diminish incentives to invest in new plants, thus hindering innovation and modernisation of existing plants (Stewart, 1981).

In Europe, as environmental concerns emerged in the 1970s, different countries assumed a leadership role on different issues. Germany pioneered in the adoption of lead and automobile emissions standards; the Netherlands instituted an ambitious waste management policy. European Community policy has introduced a complex intervening variable affecting environmental regulation. It has been influenced by the need to eliminate obstacles to trade and distortions in competition. The philosophy of the EC involves the establishment of an open, competitive system in which no nation or firm is placed at a disadvantage because of a single member state's legal or political action.

The result of this emphasis on harmonisation has been a cautious and economically-orientated pattern of standards adoption at the Community level, with environmental policy usually not designed primarily to force technological change (Rehbinder and Stewart, 1985). EC standards only mandate changes that are technically feasible. Standards of automobile emissions are a prominent example of this approach. European emissions standards started at a relatively low level of control and were subsequently upgraded several times in the wake of technological progress (Heaton and Maxwell, 1984). Similarly, the product standards on the biodegradability of detergents represented the current state of technological development when they were written and did not impose undue burden on industry. Although the EC standards policy has not concentrated directly on forcing development of new pollution control technologies or environmentally less harmful products and processes, this has been partially compensated by the sponsorship of research and development where it is believed that technological progress should be fostered for the improvement of environmental conditions (Rehbinder and Stewart, 1985).

Japan has adopted a "growth and planning orientated approach" to environmental policy, which integrates regulatory and industry promotion policies (Heaton and Maxwell, 1984, p. 16). Japanese economic policy has been characterised by a unique national consensus among industry, banks and government on rapid economic growth and the means to achieve it. A critical factor in environmental policy has been government's efforts *vis-à-vis* private economic activity to integrate environmental and industry promotion policies. This reflects the Japanese viewpoint of a "community of interest" between environmental and economic objectives. One of the principal mechanisms in this policy is the Ministry of International Trade and Industry's (MITI) sponsorship of research and development programmes on an industry-wide basis, such as the programme on flue gas desulphurisation. Another technology research programme instituted in the early 1970s enabled Japanese automobile manufacturers to meet clean air standards. One result of this project was the development of micro-electronic systems that gave Japanese producers a competitive advantage over foreign manufacturers. MITI has recently initiated a programme to assist the microelectronics industry in developing cleaner substitutes for CFCs. This is an important element in a new initiative for technological breakthrough, focusing on research and development of environmentally sound technologies. More generally, Japan has out-performed other major industrial nations in the area of energy efficiency (OECD, 1985*d*).

Relationships between regulation and technological changes depend largely on interactions among environmental problems, subject technologies, industrial sectors and firms. While generalisations should be avoided, two explanatory variables stand out: the "rigidity" of the technology in question and the degree to which regulation looks beyond existing technology.

It should be noted that the view of technology implicit in regulatory policy has often led to a situation of frustrated ambitions, as in the case of technology forcing, and to more control over new products, processes and facilities than over existing ones. In combination, this has often reinforced a technological *status quo*. In process control regimes, information-gathering and standards are largely based only on currently available technology. If such standards were continually revised in view of long-term environmental objectives, an incentive to innovate would follow (Heaton, 1990). It is likely that longer term environmental objectives based more expressly on technological potential and on more stringent standards and implementation would also act as powerful financial incentives to firms which manufacture pollution control equipment to develop superior control technologies.

3. ISSUES RELATING TO THE LEVEL AND STRUCTURES OF ENVIRONMENTAL R&D

An accurate assessment of R&D efforts directed towards the environment is greatly hindered by the deficiencies of available statistics. Two main reasons – partly interdependent – account for this situation:

i) Practices differ from country to country with respect to the integration of environment-related activities, and notably R&D, into administrative and business organisations. The specific approaches adopted in each case make the R&D effort more or less visible and therefore more or less easy to capture in statistical enquiries.

ii) The definition of environmental R&D appears to require re-thinking: the definition adopted in the standard OECD classification may be too restrictive.

Independently of measurement problems, R&D directed to environment issues remains extremely low, with negative consequences on the capacity of the OECD and its Member countries to assess risk correctly and to define appropriate policies. Global warming offers a perfect illustration of this.

Trends in government funding of environmental R&D

Despite these limitations, some salient facts emerge. Government environmental R&D expenditures only began to respond to heightened attention to these issues during the 1980s. The available OECD data on government expenditures on environmental R&D suggest that in most OECD countries, the increase largely occurred in the latter part of the 1980s. Table 31 indicates government R&D appropriations for environmental protection in 1985 dollars, and Table 32 presents the percentages of total government R&D expenditures for this area. Moreover, since the early 1980s, a number of countries have shown a decrease in R&D efforts for environmental protection as a share of total government R&D expenditures.

Only five countries allotted a share above 3 per cent at the end of the 1980s. It would appear either that environmental R&D has not yet become significant in most OECD countries, in spite of heightened awareness of environmental problems, or that some relevant R&D is not easily identified as "environmental".

Table 31. **Government R&D appropriations for environmental protection**

1985 US$ millions

	1980	1983	1985	1986	1989[1]	1990[1]
Australia	31.1[1]	28.4[2,5]	23.3[2,5]	16.3[1,2]	34.0[2]	36.0[2]
Austria[2]	1.8	1.7	3.9[5]	4.6	4.8	4.8
Belgium	16.7	16.5	16.1	14.3	8.5[1,5]	8.3
Canada[2]	29.6	35.6	47.5	50.4	40.3	44.4
Denmark	5.6[3]	3.8[5]	5.4	4.1	15.8	15.0
Finland	2.2	3.2	5.6	6.5	8.4	7.5
France	73.0	29.7	44.0	39.2	70.7	..
Germany	159.1	225.2[5]	267.0	270.1	296.1	370.5
Greece	2.4	3.5[5]	5.2	4.0	6.4	..
Iceland[1]	..	0.0	0.0	0.1	..	..
Ireland	0.6	0.8	0.8	1.0	1.0	0.9
Italy	21.9	72.6	42.4	43.3	101.0	..
Japan	..	..	38.4[1]	36.8[1]	34.1	..
Netherlands	..	..	49.1	49.7	64.1[5]	67.5
New Zealand	..	..	..	..	..	6.8
Norway	11.6	10.8	10.7	10.3	15.4	17.4
Portugal	..	..	..	4.9	6.7	..
Spain	3.8	6.5	4.1	3.8	36.0	..
Sweden	19.5	20.6	19.8	21.1	45.7	45.6
Switzerland[2]	..	..	..	16.3	5.1	..
Turkey	..	..	..	..	..	..
United Kingdom	51.8	85.7	94.1	83.2	85.8	97.7
United States[2,4]	306.3	221.3	258.0	262.4	293.1	333.9
Yugoslavia	..	..	..	..	..	..

1. Provisional Secretariat estimates or projection based on national sources.
2. Federal central government only.
3. Over-estimated.
4. Excludes most or all capital expenditures.
5. Discontinuity in series from preceding year.
Source: OECD, STIID database (September 1991).

Table 32. **Share of environmental protection in total government R&D appropriations**

(percentage of total)

	1980	1983	1986	1989[1]	1990[1]
Australia	2.9[1]	2.6[2,5]	1.3[1,2]	2.8	2.8
Austria[2]	0.5	0.4	1.0	0.9	0.9
Belgium	2.8	2.6	2.3	1.2[5]	1.2
Canada[2]	1.6	1.6	2.0	1.6	1.8
Denmark	2.1[3]	1.2[5]	1.0	3.0	3.0
Finland	0.8	1.0	1.6	1.7	1.5
France	1.1	0.3	0.4	0.7	..
Germany	2.0	2.8[5]	3.2	3.4	4.2
Greece	2.6	3.2[5]	2.7	3.1	..
Iceland[1]	..	0.1	0.6	..	..
Ireland	0.6	0.9	0.9	0.9	0.8
Italy	1.0	2.1	0.9	1.9	..
Japan	..	..	0.5[1]	0.4	..
Netherlands	..	..	2.9	3.8[5]	4.0
New Zealand	..	..	..	..	3.6
Norway	3.0	2.7	2.3	2.7	2.9
Portugal	..	..	2.9	..	..
Spain	0.6	0.9	0.4	2.1	..
Sweden	1.7	1.6	1.6	3.3	3.2
Switzerland[2]	..	..	4.5	1.4	..
Turkey	..	..	..	..	..
United Kingdom	0.7	1.1	1.1	1.3	1.4
United States[2,4]	0.8	0.5	0.5	0.5	0.6
Yugoslavia	..	..	..	..	..

1. Provisional Secretariat estimates based on national sources.
2. Federal central government only.
3. Over-estimated.
4. Excludes most or all capital expenditures.
5. Discontinuity in series from preceding year.
Source: OECD, STIID database (September 1991).

A recent EC study leads to certain conclusions about overall and national priorities for environmental concerns. Traditional environmental areas (largely associated with "command and control" regulation) accounted for 53 per cent of total EC public environmental R&D budgets. "Ecosystem" research received almost 20 per cent, essentially due to the massive German commitment. Longer term climate concerns represented almost 11 per cent of commitments, very much concentrated in the United Kingdom. The relatively new theme of "technologies for environmental protection" received some 8 per cent of total commitments, with a large share from the Netherlands. Other, more specific, concerns were important for some countries: "protection of the cultural heritage" in Belgium and the Netherlands and "seismic hazards" for France and Greece (IEEP, 1990).

Common trends in environment research policies

Several features seem to characterise the evolution of R&D environment research policies in most countries, as seen notably in the OECD study (OECDc, forthcoming):

- From an institutional viewpoint, there is a marked tendency to establish mechanisms for rationalising and coordinating efforts hitherto scattered in different organisations or under different interdepartmental committees or programmes set up to address specific issues of national or global interest.
- The nature of research has gradually evolved from earlier, largely descriptive forms towards more analytical and experimental forms directed to the understanding of the inner mechanisms of ecological, climate and other processes affected by human activities. At the same time, there is an intensification of basic research; in the 1970s and early 1980s, technology-oriented research, notably for anti-pollution actions, received the greatest emphasis. Finally, the size of individual research projects is increasing, and projects are becoming more capital-intensive.
- International co-operation is increasing rapidly, in view of the growing attention to transnational environment issues. Available financial data seem to indicate growth in the share of government funding allocated to international programmes, and a number of international collaborative research activities have been set up in recent years. Some are intergovernmental (especially in the framework of the European Community) and some are sponsored without direct formal government support (for instance, under the auspices of the International Council of Scientific Unions). The R&D involved ranges from basic to pre-competitive research, and most is conducted in national institutions. The involvement of the enterprise sector in international collaborative research is beginning to materialise within the European context (EC and EUREKA programmes).

Reducing risk and uncertainty

The need to speed up these developments and make environmental issues a priority area of public R&D funding is particularly evident in relation to the risks, but also the uncertainties associated with emerging environmental threats, notably the issues of ozone depletion and global warming. Most experts believe that, in spite of the numerous uncertainties (see Box 41), R&D and government policy should be directed towards reducing emissions of greenhouse gases (the "precautionary principle" for risk avoidance), and that this should be a central policy concern for the 1990s and beyond (NAVF, 1990). However, it is being urged by some experts (e.g. Schelling, 1988) that policies be designed now to deal with climate change. As a result, the prospects for global climate change pose three rather different types of challenge:

i) A better understanding of the longer term processes, trends and possibilities is needed, with advances in monitoring techniques and a strong effort in basic research and modelling;

ii) Assuming that the need to limit global climatic changes is recognised, innovation and diffusion of technology are necessary to reduce aggregate net emissions of the principal groups of polluting gases;

iii) Given the strong probability of some global climate changes on any assumptions about current policy responses, technological developments will be required to adapt to these changes.

Uncertainty, risk, and the threat of irreversible consequences carry in their wake different degrees of optimism and pessimism in public attitudes to technology and to society's ability to manage it. It will not be known with any reasonable degree of confidence for some years what is actually likely to happen. As for the scope for adapting to climate change, much of the technology required is simply not in place, and some of it will require continued further developments in (relatively) basic science and associated transfer sciences. To complicate the problem, the likely time lags in climatic and ecological changes may still give mankind some 20 years before any major effects are felt.

Related to issues of optimistic versus pessimistic attitudes are judgements about the choice of different policy strategies. This is ultimately a question of attitude to risk. On the whole, observers are advocating use of the precautionary principle: it is prudent to insure against the longer-term risks by making policy changes and considerable S&T investment now (Pearce *et al.*, 1989). Moreover, this view is strengthened by an assumption that such "insurance" investments will be socially desirable even if the global climate risks do not materialise.

For various reasons, however, the S&T community is ambivalent about its priorities in relation to these important, emerging long term problems with very uncertain complex outcomes. One problem concerns involvement in practical policy advice and in relation to environmental lobbies. Another concerns the interdisciplinarity implications of the research involved and the implications for peer review and career development (NAVF, 1990).

Industrial R&D and the "environment industry"

The supply of equipment and services for pollution abatement and environmental protection is a growing industrial activity. The "environment industry" is not easily defined. It includes firms which produce pollution abatement equipment and a range of goods and services for environmental protection and management. It spans a variety of industrial products and services which have not been statistically classified and or which data are limited. The OECD estimates that turnover in these product groups in the OECD countries is currently around $185 billion annually. The world-wide market is estimated at $200 billion, with a projected rise to $300 billion in the year 2000.

A very rough estimate places this "environment industry's" related R&D at some $10 billion annually. As much as 80 per cent of all environmental R&D within OECD countries thus appears funded by firms, a much higher share than for overall R&D, although some estimates provided in the national reports prepared for the OECD survey indicate a somewhat lower involvement of the private sector in the R&D efforts (e.g. 55 per cent for Germany). This points to the capacity of industry to respond to the opportunities of what will necessarily become a new growth sector as laws and regulations are applied more strictly and the consciousness of environmental issues by firms and the public grows.

Table 33 gives estimates of the product structure of the environment industry by segment and by region. Over three-quarters of industry output is equipment produced for environmental purposes, primarily end-of-pipe pollution abatement equipment. Environmental equipment can be subdivided into four main types of products according to end-use: water and effluents treatment, waste management, air quality control and other (primarily land reclamation and noise reduction). Environmental technologies which are incorporated in industrial processes – generally termed clean technologies – are not included in this classification. Environmental services, which now account for almost a quarter of industry output and often relate to the installation of clean technologies, are listed as a separate category.

OECD research, as yet unpublished, further suggests that the environment industry, and perhaps espe-

Table 33. **Main components of the environment industry, 1990[1]**

Percentages

	North America	Europe	Japan	Total OECD
Equipment	74	76	79	76
Water and effluent treatment	24	34	22	29
Waste management	25	15	22	21
Air quality control	12	17	25	15
Other (land remediation, noise)	13	10	10	11
Services	26	24	21	24
Total	100	100	100	100

1. Estimated share of value of output.
Source: OECD estimates.

cially its R&D, is heavily concentrated in the United States, Japan and Germany, which accounted for 43 per cent, 16 per cent and 15 per cent, respectively, of estimated 1990 OECD turnover. About half of the environment industry's turnover is accounted for by large firms, which allocate an estimated 8-10 per cent of turnover to R&D. There are however several routes of access to the industry and so different categories of firms. Three groups can be identified in particular:

i) The well-established firms producing for the industry's oldest and most mature segment, e.g. the production of equipment for water and effluents treatment within both public facilities and manufacturing industry. This group includes Alfa Laval, the German firms Bilfinger and Bayer and Steinmuller and the French firms Cie Générale des Eaux and Lyonnaise des Eaux.

ii) In many product groups, the industry provides fertile ground for start-up and entrepreneurial ventures with smaller environment firms ranging from high-technology suppliers of chemicals, instruments and consultancy services to low-technology producers of recycling bins and suppliers of waste transportation services. Most small and medium-sized environment enterprises are specialised, owner-managed and offer a limited range of equipment and services. In total, there are estimated to be some 30 000 such firms in North America, 20 000 in Europe and 9 000 in Japan.

iii) Diversification into production of environmental equipment and services is now seen as a potential growth area by large firms in the chemicals, engineering and electronics industries. In the United States, firms such as DuPont, Dow, Westinghouse and Hewlett Packard have established subsidiaries for, respectively, toxic waste management services, plastics recycling, waste processing services and environmental laboratory instrumentation. The German firms Deutsche-Babcock, Robert Bosch and Siemens are among the firms developing new lines of environmental equipment, while BASF, Bayer and Hoechst have formed a venture to develop methods for recycling plastic wastes. In France, major firms such as Alsthom, Fenwick, Saint-Gobain and Ciments Lafarge are estimated to have over 20 per cent of their turnover in environmental products. Many of the large chemical companies such as Ciba-Geigy of Switzerland and metals companies such as Metallgesellshaft of Germany are active in the market for water and waste treatment equipment. Japan's leading environment firms are the large diversified industrial groups Mitsubishi and Hitachi.

4. PROGRESS IN ENVIRONMENTAL PROTECTION THROUGH END-OF-PIPE POLLUTION CONTROL TECHNOLOGIES

The results of environmental policies in OECD countries since the early 1970s have been mixed. Major progress has been achieved against large point sources for air and water pollution, but the degree of improvement varies depending on national policy and the type of pollutant. This section reviews some of the results achieved through the widespread adoption of end-of-pipe pollution control technologies in the areas of air, water, and solid waste. Some of these improvements have, however, been offset by increases in the overall level of economic activity or consumer behaviour that have generated additional quantities of pollution. Micro-toxicity is an increasing problem. Progress against pollution resulting from diffuse sources such as driving and the use of disposable products and packaging has been variable.

In the area of *air* pollution abatement, end-of-pipe pollution control technologies have resulted in considerable reduction of aggregate sulphur dioxide (SO_2) from *stationary sources* in most OECD countries, despite the expanding use of the fossil fuels that are responsible for the levels of discharge (Table 34). Substantial reductions in aggregate emissions have also been achieved for carbon monoxide, reactive hydrocarbons and particulate matter in most OECD countries (OECD, 1989n). The quality of air, measured in terms of SO_2 and particulates, has improved significantly in most important specific OECD locations. In the United States, SO_2 emissions decreased by 17 per cent in the decade from 1979 to 1988, essentially through the use of desulphurisation technologies and the use of low sulphur coal. Within the European Community, many

Table 34. **Total SO₂ emissions**
Thousand tonnes

	1970	1975	1980	1985	Late 1980s	Change from 1980 (%)	Per unit of GDP (kg/100 US$)[1]	Per capita (kg/cap.)
						Late 1980s	Late 1980s	Late 1980s
Canada	6 677	5 319	4 643	3 704	3 800	−18	9.7	146.4
United States	29 400	25 900	23 400	21 100	20 700	−12	4.7	84.0
Japan	4 973	2 586	1 263	..	835	−34	0.6	6.8
Denmark	574	418	447	340	242	−46	4.1	47.2
Finland	515	535	584	371	305	−48	5.1	61.7
France	2 966	3 328	3 339	1 475	1 272	−62	2.3	22.8
Germany	3 739	3 331	3 191	2 431	1 306	−59	1.9	21.3
Italy	2 830	3 331	3 211	2 086	2 070	−36	4.4	36.0
Netherlands	772	385	461	271	256	−44	1.9	17.3
Norway	171	137	142	97	65	−54	1.0	15.4
Portugal	116	178	267	198	205	−23	8.7	19.9
Sweden	930	696	502	273	199	−60	1.8	23.6
Switzerland	125	109	126	95	63	−50	0.6	9.4
United Kingdom	6 327	5 310	4 847	3 718	3 664	−24	7.0	63.1
OECD	64 600	57 900	53 000	42 200	39 900	−25	4.1	48.3

1. 1988 GDP at 1985 prices and exchange rates.
Source: OECD (1991e).

countries have also significantly reduced SO_2 emissions. Germany has achieved a lowering of emissions through the adoption of advanced pollution control equipment; France obtained even larger reductions through the expansion of its nuclear power industry.

Slower progress has been obtained in lowering nitrogen oxide (NOx) emissions because of technological obstacles. Despite recent improvements in the development of denitrification processes for stationary sources, the trend for NOx emissions in the United States tends to be flat, while emissions are up in Germany, France, and the United Kingdom (Table 35).

Perhaps the most significant feature in Tables 34 and 35 concerns the remaining disparities in emissions between OECD countries, rather than the trends. There are large problems about the comparability of national data. However, the tables suggest that, in the late 1980s, emissions per head of SOx and NOx in Japan were around 14 per cent and 22 per cent, respectively, of the OECD average: corresponding US emissions were 74 per cent and 81 per cent higher. Canada's SOx emissions were three times the OECD average.

In some areas of pollution control technology, Japan leads the world. This is especially true for flue gas desulphurisation equipment, denitrification equipment, and for the process of desulphurising heavy oil Japan has converted its investments in pollution control into a growing export business. The value of

exports for pollution control equipment and related services was estimated at Y 58 billion in 1986 (Weidner, 1989). As noted above, it has also begun a vast effort to develop cleaner substitutes for CFCs.

Regarding *mobile energy sources,* a recent OECD/International Energy Agency expert panel meeting highlighted the environmental problems posed by the continuing reliance on the automobile:"From the environmental viewpoint, motor vehicles and their fuels are now generally recognised as responsible for more conventional (nitrogen oxides, carbon monoxide, hydrocarbons and particulate matter) and toxic (benzene, lead, polyaromatic hydrocarbons, etc.) air pollutants than any other single human activity. The rising emissions of greenhouse gases (CO_2 and others) are considered by many to be incompatible with national and international policies being envisaged in the context of the global warming problem." (OECD, 1990*f*).

Carbon monoxide (CO) emissions have been lowered in many OECD countries in spite of large increases in vehicle miles travelled. From 1979 to 1988, US emissions dropped by 25 per cent despite 33 per cent growth in vehicle miles travelled. In Europe, the results were mixed, depending on the stringency of standards for automobiles. Germany, for example, achieved reductions similar to those of the United States. Japan has adopted some of the most stringent standards for CO, but has also experienced a rapid expansion of its passenger car fleet (Japan, Environmental Agency, 1988).

Table 35. **Total NOx emissions**

Thousand tonnes

	1970	1975	1980	1985	Late 1980s	Change from 1970 (%)	Per unit of GDP (kg/100 US$) [1]	Per capita (kg/cap.)
						1970-Late 1980s	Late 1980s	Late 1980s
Canada	1 364	1 756	1 959	1 959	1 943	42	4.9	74.9
United States	18 300	19 200	20 400	19 800	19 800	7	4.5	80.4
Japan	1 651	1 781	1 400	..	1 176	−29	0.8	9.6
Denmark	..	178	241	259	249	..	4.2	48.5
Finland	..	..	264	251	270	..	4.6	56.6
France	1 322	1 608	1 834	1 579	1 605	21	3.1	31.6
Germany	2 381	2 571	2 980	2 959	2 831	19	4.3	46.7
Italy	1 410	1 507	1 585	1 555	1 641	17	3.4	27.3
Netherlands	456	464	558	547	560	23	4.2	37.9
Norway	159	176	192	203	223	40	3.6	53.7
Portugal	72	104	166	96	116	61	5.2	11.8
Sweden	302	308	394	398	388	28	2.9	37.4
Switzerland	149	162	196	214	189	27	1.8	27.6
United Kingdom	2 510	2 427	2 442	2 402	2 636	5	4.9	44.0
OECD	32 300	34 700	37 700	36 200	36 600	13	3.8	44.3

1. 1988 GDP at 1985 prices and exchange rates.
Source: OECD (1991*e* and 1991*i*).

On the other hand, reduction of lead emissions from mobile sources represents one of the greatest environmental success stories. In the United States, for example, total lead emissions to the air decreased by 93 per cent from 1979-88. These reductions stem in large part from the transition to unleaded gasoline in automobiles, a straightforward substitution technology.

Most relevant indicators of *water* quality for important river basins show considerable improvement over the 1970-85 period (OECD, 1991*d*). Some progress has been achieved in the area of water pollution control by using pollution abatement technologies for industrial effluents and by constructing waste water treatment facilities. Many water treatment facilities have moved from simple to relatively advanced forms of treatment technologies. The percentage of the population served by waste water treatment facilities has increased markedly within most OECD countries (Table 36). However, the percentages remain much lower for some countries.

Less success has been achieved with *marine environments:* the data are more difficult to interpret and trends are mixed. In particular, the overall environmental situation of some regional maritime environments (e.g. the Mediterranean Basin, the Baltic Sea and the North Sea) is giving rise to increasing concern.

Control of both *domestic and industrial wastes* has lagged behind the regulation of air and water pollution in OECD countries. The quantities of both continue to increase, and the problem has proved particularly resistant to solution through end-of-pipe technologies.

Table 36. **Population served by waste water treatment plants, 1970 to late-1980s**

Percentage served

	1970	1975	1980	1985	Late 1980s
Canada	..	49	56	57	66
United States	42	57	70	74	..
Japan	16	23	30	36	39
Austria	17	27	33	65	72
Denmark	54	71	80	91	98
Finland	27	50	65	72	75
France	19	31	43	50	52
Germany	62	75	82	88	90
Italy	14	22	30	..	60
Netherlands	..	45	72	85	92
Norway	21	27	34	43	43
Spain	..	14	18	29	48
Sweden	63	81	82	94	95
Switzerland	35	55	70	83	90
United Kingdom	..	..	82	83	84
OECD	34	46	51	57	60

Source: OECD (1991*e*).

Table 37. **Amounts of municipal waste[1] 1975-89**

kg/cap.

	1975	1980	1985	1989
Canada	..	524	635	625
United States	648	703	744	864
Japan	341	355	344	394
Austria	186	222	228	355
Denmark	..	399	469	..
Finland	..	..	449	504
France	228	260	272	303
Germany	330	348	318	318
Greece	..	259	304	314
Ireland	175	188	311	..
Italy	257	252	263	301
Netherlands	..	464	449	465
Norway	424	416	474	473
Portugal	..	213	231	..
Spain	..	215	275	322
Sweden	293	302	317	..
Switzerland	297	351	383	424
United Kingdom[2]	323	312	341	354
OECD	407	438	462	518

1. The definition of municipal waste may vary from country to country.
2. England and Wales only.
Source: OECD (1991*e*).

Hazardous wastes have presented a particularly thorny problem, from both a technological and a political point of view. For example, the production of hazardous wastes in the United States quadrupled over the decade from 1975-85. In the United States, awareness of this problem provided the impetus for the establishment of the Superfund programme. There has also been a lack of sufficient attention to cleaning up the hazardous waste sites that potentially place local communities at increased health risk.

In the area of domestic wastes, quantities expanded in the decade from 1975 to 1985 in the United States, while remaining relatively flat in France, Germany and Japan (Table 37). Recycling programmes have only been moderately successful. In the United States, a number of states have instituted bottle bills to encourage the re-use of glass and plastic containers. However, because of the absence of industrial technology, many of the returned plastic bottles are not recycled but placed in landfills. Similarly, the recycling of newsprint has been constrained by the inability of many paper processing facilities to use recycled paper as an input. In general, consumers are difficult to regulate, but changes are necessary in consumer behaviour, and in institutional organisation and technology.

In order to address the problems resulting from excessive reliance on an end-of-pipe strategy for pollution control, many OECD countries have recently begun to develop more broad-based and longer-term policies for pollution prevention or source reduction. As an example, in the Resource Conservation and Recovery Act, the United States Congress declared it national policy that "wherever feasible, the generation of hazardous wastes is to be reduced or eliminated as expeditiously as possible" (US, OTA 1986, p. 5). Similar policy statements on *pollution prevention* have been enacted in the EC and Japan. However, attempts to shift policy priorities from pollution control to prevention have often been hindered by the structure and commitments of existing regulatory programmes (Olderburg and Hirschhorn, 1987). One commentator has described the potential obstacles to pollution prevention in the European Community as follows: As long as Community policy for waste control pursued the traditional objective of regulating pollution it could count on a widespread consensus. The more it ventures into waste prevention, the more it will encounter difficulties stemming from the fact that implementation of preventive strategies often requires administrative reorganisation and large appropriations, threatens vested economic interests, and must cope with unresolved scientific issues. (Rehbinder, p. 91)

5. CRITICAL ISSUES IN POLICYMAKING

Discussions in the TEP programme have clarified notions of how technology is developed and diffused and how the links between technology and macro-economic developments are changing. This can help bring into better focus the environmental dimensions of technological growth. Almost two decades of experience with environmental policies and institutions and with efforts at control allow a measure of stock-taking. Considerable success has been achieved through diffusion of end-of-pipe environmental control technologies, but to date, regulatory policy has not provided a strong enough incentive for fundamental re-design of products and processes. As the preceding discussion has shown, there is now an important broadening of concerns, as emphasis has shifted from visible current problems towards longer-term and more problematic areas.

The increasing complexity of the challenges

Present environmental policies designed to control pollution by particularly visible polluters may still yield massive results, especially outside the OECD countries. However, policymakers should now turn attention both to other actors who participate in the development, diffusion and use of technology and who therefore contribute to environmental pollution and to problems which may require increasingly innovative solutions.

Problems for government agencies

Past commitments of government agencies to command-and-control style regulation and promotion of end-of-pipe technologies have been considered by many to have proved an obstacle to implementing a consistent policy towards the development and use of processes and products which do not produce wastes, discharges, emissions or pollutants. In the United States, the Congressional Office of Technology Assessment (OTA) estimated in 1987 that less than one per cent of annual environmental spending by federal and state governments was devoted to pollution prevention. In the future, means for controlling pollution arising from the consumption of products, discovering more cost-effective methods of abatement and dealing with the possible ramifications of global climate change must be found. Regulatory bodies in OECD countries are becoming aware that more comprehensive policies are needed to cope with today's more complex problems.

Concerns about rising costs have been reflected in the growing interest in using economic incentives either in place of, or in combination with, command and control style regulations. Economic incentives are believed to be a means of obtaining given levels of emissions reductions in a more cost-effective manner, because they give firms greater flexibility of response and provide more permanent incentives for technological change. In practice, economic incentive systems have usually been implemented along with regulatory instruments and have shown only modest cost savings (Hahn, 1989).

Governments would face enormous difficulties if they were to attempt to mandate the details of technical change in industry, because they lack the necessary technical and economic knowledge to make such decisions. Nor can industrial societies rely on the autonomous force of technologies to create their own solutions. Governmental policy can influence some of the incentives facing the firm; however, a wide variety of internal and market incentives also affect firms' choices about new product and process technology. Past industry experience suggests that the continued improvement of environmentally superior technologies will prove far more difficult than the mere adoption of end-of-pipe controls.

Industry and pollution

Just as the institutional structures of government agencies may have impeded some attempts at pollution prevention, a whole bureaucracy has grown up within large firms to comply with government mandated environmental requirements. This bureaucracy's expertise currently lies in devising compliance solutions, not in engineering improvements in new product and process technologies. It will become necessary for firms to recognise that to respond to the kinds of pollution prevention policies that are beginning to be developed requires changes in their own internal structures and procedures.

Many firms have demonstrated considerable reluctance to change the character of their core production processes for environmental purposes. It has been argued that this was a critical constraint for US automobile producers, who resisted air pollution control standards in the early 1970s and allied attempts to have them modify their engine technologies. According to some experts, "There is a formidable difficulty with the alternative technology option: A large number of people will have to change their hearts and minds about what type of equipment belongs under the hood of a car ... The automobile manufacturers, who are doing well with the internal combustion engine, are not destined to respond with deep enthusiasm to a programme which forces them into a major re-orientation of plant capacity on behalf of a project whose central effect is likely to be to localise more of the cost of emission control at the manufacturing stage." (Jacoby and Steinbrunner, 1973). Rather than modify engine technologies, American manufacturers added catalytic converters to the tailpipes of their vehicles. End-of-pipe solutions did not require changes in core production processes nor present risks of undesired changes in manufacturing processes.

Other firms have resisted product or process changes because they claimed that there was scientific uncertainty about the nature of the hazard or technical possibilities for devising substitutes. DuPont, the world's largest producer of CFCs, responded sceptically to Molina's findings linking CFCs to the destruction of the ozone layer in the mid-1970s. DuPont scientists claimed that scientific uncertainty precluded the immediate development of substitutes. It required more than a decade for the scientific controversy to be resolved. At that point, DuPont scientists faced the challenge of creating substitutes for an extremely useful and versatile product.

One expert has summarised some of the *internal constraints* that face firms seeking to develop and use environmentally superior technologies. He argues that instituting industrial pollution prevention may require changes in many aspects of corporate practice, including company and facility organisation, personal attitudes, corporate policies, rewards for workers, procurement practices, worker training, information collection and accounting systems, methods of economic analysis for capital investment, and research and development goals and criteria for success. Insufficient attention to these institutional issues may mean that feasible methods of pollution prevention will not be identified and used (Hirschhorn, 1990).

As a result, some firms with highly publicised waste reduction programmes only initiated them in response to strong public pressures. In the early 1980s, Dow Chemical came under intense public scrutiny because of their production of agent orange and the release of dioxin into waterways. Only after a highly publicised lawsuit by Vietnam veterans and confrontations with the United States Environmental Protection Agency did the company overcome internal resistance to an ambitious environmental programme. The institution of a pro-active approach to the environment required changes in senior management, incentive systems for managers, and procedures for allocating capital.

Union Carbide similarly drew public attention after the Bhopal tragedy, and their experience demonstrates the complexity of implementing a more preventive approach to environmental management. The company has established two very ambitious goals, the elimination of emissions of known or suspected carcinogens and the reduction of exposure outside its plants to levels of at least 1 000 times less than workplace standards (C. Smith, 1990). To achieve these goals, Union Carbide has been conducting audits, to ensure that the company's 1 200 plants world-wide comply with government standards, and risk reviews, to determine whether high-risk processes rely on "state-of-the-art technology". Union Carbide has also changed its internal incentives so that health, safety, and environmental objectives are part of the performance review of every Union Carbide manager. Attempts are now made to integrate environmental concerns into all aspects of the company's business.

Preliminary studies at the Massachusetts Institute of Technology's Hazardous Substances Programme show that some firms have overcome internal obstacles to innovation by linking their environmental activities to ongoing safety or quality programmes. DuPont, for example, has organisationally tied its environmental activities to the company's strong safety tradition. Following the Bhopal disaster, Union Carbide sought to integrate environmental actions with its safety and quality programmes. In the area of accident prevention, the Japanese chemicals industry has been extremely successful by combining safety with ongoing programmes on quality and productivity. In the period from 1973 to 1986, the Japanese chemical industry reduced the frequency of injury by nearly 80 per cent; it has attempted to reach the goal of zero accidents by innovations in quality and production processes (unpublished remarks to OECD Workshop on Prevention of Accidents Involving Hazardous Substances, January 1990).

"Diffuse polluters"

Everyday practices such as driving, farming and the use of disposable products and packaging are proving to be more intractable problems than industrial point-of-source pollution. Within present environmental policies, consumers and other groups (notably farmers) appear as diffuse sources of pollution which are difficult to control, although there has been a surge in "green" issues and "green" consumerism and politics. However, the limits of consumers' acceptance of environmentally friendly technology remain largely unexplored.

In some areas, such as personal transportation, the consumer's responsibility is mitigated by the mediation of other actors. Automobile manufacturers are producing and selling vehicles which are progressively less polluting: average energy use per kilometre has fallen substantially as have emissions of most airborne pollutants, and they continue to decline. In most OECD countries, individual vehicle emission targets, accepted by the auto manufacturers, seem also to be accepted by the final consumers. And, although the focus of environmental attention has largely been on emissions, the disposal of the automobile at the end of its life is also beginning to be considered; "whole life cycle" concepts address issues of waste and resource conservation. Within the EC, legislation making automobile manufacturers responsible for the disposal of their products is being considered and would lessen consumer responsibility for disposal. The automobile designer would be required to think about how the car would be retired and how the component pieces might be salvaged most effectively (Clark and Field, 1990). On the other hand, if a major shift in "modes" of transport, from private automobiles to "public" transport, and, perhaps, from air to terrestrial systems, should prove necessary, consumer attitudes will obviously be extremely difficult to change.

Farming presents a different problem. Agricultural pesticides which run off to water and contamination of ground water from landfills represent major sources of environmental contamination (OECD, 1989p). In most OECD countries, systems of permits and controls do not address diffuse sources of pollution like agriculture. Agriculture within the OECD countries has shown a remarkable capacity to respond to market signals, but this has generally been in the context of a relatively expansionist market situation. The current outlook for profitability and land values is now pessimistic. How easily farmers might respond to "negative" market signals is unclear: supply price responses may be asymmetric. Nonetheless, most OECD countries now feel strong concern about the effects of some four decades of agricultural "intensification" on environmental quality (soil and water) in vulnerable areas.

There is evidence of market failure: the present price structures of inputs to agricultural production are not forcing farmers are failing to take account of the longer term profitability implications of their present input-intensive agricultural practices. It has been argued that, in the United States, profitability has been as high in some normal arable situations with Low Input Sustainable Agriculture (LISA) as with input-intensive farm practices. However, most observers judge that LISA can only succeed if farmers come to believe in the environmental "philosophy" of its objectives (Benbrook, 1990). Outside a change in prices (energy, chemical inputs, etc.), there is little that will oblige them to do so.

The cost of effective environmental policies

The relationship between environmental objectives and economic policy priorities appears to be both central and complex. Expenditures for "pollution abatement and control" (PAC), essentially operating expenses and investment expenditure by government and the goods producing business sector for control, but excluding conservation and renewal, have been rising in most OECD countries, although not clearly as a share of GDP (Nicolaisen et al., 1991). Per capita pollution control expenditures have risen in the last decade in the United States and especially in Germany; they have only slightly increased in France and decreased in the United Kingdom. As Table 38 shows, a best estimate of the magnitude of PAC places expenditures at a national average of between 0.8 and 1.5 per cent of GDP for most OECD countries in the mid-1980s. With the exception of the United States, PAC expenditure has tended to be greater in the public sector than in private enterprises. The share of PAC investment in total national investment tends to be slightly higher than for total PAC expenditures. Further, expenditures by "households" might represent an additional 20-25 per cent of the total (OECD, 1991d). More than 40 per cent of total PAC expenditures involved water treatment in major OECD countries.

As a means of appreciating the magnitude of the impacts of these expenditures on macro-economic growth and to the gravity and urgency of environmental threats, modelling analysis suggests that "the negative effects on GDP growth have ranged from negligible to a moderate 0.2 per cent per annum in the case of the United States over the period 1973-82" (Nicolaisen et al., 1991). It should be emphasised that these

Table 38. **Total pollution control expenditures in the mid-1980s**

Per cent of GDP

	Public	Private	Total
Canada	0.89	0.36[1]	1.25[2]
United States	0.60	0.86	1.47
Japan	1.17	0.08[1]	1.25[2]
Finland	0.52	0.64	1.16
France	0.56	0.33	0.89
Germany	0.78	0.74	1.52
Italy	0.13	..	..
Netherlands	0.95	0.30	1.26
Norway	0.54	0.27	0.82
Sweden	0.66	0.27	0.93
United Kingdom	0.62	0.62	1.25

Note: Although a standard questionnaire was used to collect the data, these data are not fully comparable.
1. Excludes running costs.
2. Partial total.
Source: OECD (1991*d*).

estimates are at market prices, and do not involve any environmental shadow pricing.

The Netherlands National Environmental Policy Plan (NEPP) represents a significant attempt to project the macro-economic consequences of policy acceptance of changed perceptions of environmental threats[3]. Essentially, it outlines, for a particular (small) country, the possible macro-economic and economic structural consequences of adopting policies which would conform to our current perceptions of global and local environmental threats (over the period to 2000 and beyond). While doubts may be voiced about the assessment of risks and some aspects of the plausibility of implementing the implied policies, the presentation of overall results gives an informative basis for policy discussion. It should be noted that the underlying analysis makes relatively static assumptions about technology development.

The results suggest that the economic costs of adopting "sustainable development" policies are fairly modest, even for a small country and assuming no international co-operation. Table 39 introduces three alternative scenarios, ranging from a continuation of present environment policies to a package of measures in which a mixture of emission-oriented and structural source-oriented measures is applied. Scenario III would meet all the major currently perceived environmental targets. Table 40 presents some of the results in terms of macro-economic indicators. The negative results over the longer term are very modest; it is important that these are evaluated at expected market prices and do not involve the natural resource pricing that would come from a better international assessment of environmental threats. With Scenario III,

aggregate GDP would be lower in 2010 than currently projected by a modest 4.2 per cent if the Netherlands acted alone, but might actually be slightly higher (0.5 per cent) if comparable measures were adopted internationally.

Nevertheless, it is widely argued, especially by the production sector, that pollution abatement expenditures have hitherto only dealt with the low cost problems, and that marginal costs for abatement will rise sharply. There is only anecdotal evidence for this,

Table 39. **Dutch emission scenarios**

Percentage changes

	Results in 2010		
	Scenario I	Scenario II	Scenario III
CO_2	+35	+35	− 20 to −30
SO_2[1]	−50	−75	− 80 to −90
NOx[1]	−10	−60	− 70 to −80
NH_3[1]	−33	−70	− 80
Hydrocarbons	−20	−50	− 70 to −80
CFC	−100	−100	−100
Discharges to Rhine and North Sea	−50	−75	− 75
Waste dumping	0	−50	− 70 to −80
Noise (leading to serious nuisance)[2]	+50	0	− 15
Odour[2]	+10	−50	− 60

1. Relative to 1980.
2. The changes for noise and odour refer to percent changes in numbers of people experiencing nuisance.
Source: NEPP (1989).

Table 40. **Costs of the Dutch emission scenarios**

Costs of the environment scenarios I, II and III[1]
in billions of 1985 guilders

	1988	2010		
		I	II	III
Gross annual costs	8.3	16.0	26.3	55.8
Annual savings	..	..	..	20[2]
Net annual costs	8.3	16.0	26.3	36.8
Net annual costs as % of GNP	1.9	2	3	4
Total investments in the period 1990-2010	..	100	200	350
Cumulative effect on real GNP	..	-1.3	-3.5	-4.2

1. See Table 39.
2. Savings in energy and raw materials; these are dependent on the development of energy prices. If the sudden 1985 price drop of 40 per cent were to be set aside, savings could amount to about Gld30 billion.
Source: NEPP (1989).

and it mainly concerns the established large existing polluters. They argue that the most cost-effective measures for controlling large point sources have already been undertaken and that future clean-up efforts will require far larger outlays. These firms also tend to feel that they can earn higher returns from other kinds of investments. They are particularly concerned about the costs of meeting global climate change targets.

In some areas, changing market conditions have created economic incentives for firms to modify their production processes in an environmentally sound manner, first in Europe (Piasecki and Davis, 1987) and, somewhat later, in the United States. The oil shocks of 1973 and 1979 induced investment decisions in energy saving, first within the bounds of existing technology, by "retro-fitting" energy devices to existing plant and heating systems, and only after several years' delay did firms begin to install new technology with a greater energy saving potential (Freeman, 1989). There is continuing evidence of "market failure", with firms not responding (or only slowly) to market signals (especially about energy saving).

Effective "carbon taxes", corresponding to most present OECD initial global climate objectives, have been advocated by one school of experts and are receiving serious attention in some OECD countries[4]. Such taxes might need to be of the order of $100 per ton of carbon, lifting the price of petroleum (from an assumed $20 a barrel) by some 60-70 per cent (Pearce, 1991). The tax would need to rise thereafter in real terms to meet longer term objectives. This would lead to important shifts in fiscal policies and to some structural adjustments. A major shift in fiscal policy towards environmentally based taxes need not affect economic growth, and could yield some economic efficiency gains. However, there could be potentially difficult problems of international policy co-operation: there will be "losers" as well as "winners" and there is plenty of scope for "free-riders". Technology considerations, outlined above, suggest that this central problem can be mitigated to the extent that "clean" technologies can be identified and promoted in the OECD countries to the point where new trajectories will shift production processes and competitivity, through falling costs of production and shifts in consumer tastes.

Existing evidence suggests that in the long term, investments in new products and processes can yield significant economic as well as environmental benefits[5]. Unlike compliance activities under traditional regulatory programmes, the goal of many waste reduction programmes in Europe is to improve industrial production as well as environmental performance. One study found that many European officials viewed waste reduction to be only a part of government efforts towards improvement of industrial capacity (Geisner et al., 1986).

In the United States, where the costs of waste management (end-of-pipe technologies) have nearly doubled over the last decade, waste reduction (pollution prevention) has become a highly attractive investment for some firms. Through a programme called WRAP (Waste Reduction Always Pays), Dow Chemical encourages employees to discover ways to reduce waste on the job. Employee suggestions have included process re-design, improved utilisation of by-products, operating procedures changes and optimisation of process control. From 1982 to 1988, Dow has reported returns of more than $50 million on investment in a variety of waste reduction programmes (Nelson, 1989).

In some industrial sectors, however, serious corporate obstacles to the implementation of more environmentally sound technologies remain. In spite of Southern California's draconian standards for air quality, the major oil companies have resisted public pressures to develop and employ alternative fuels or re-formulated gasolines on a major scale. Although the technology for producing alternative fuels is well-defined, these major oil companies are reluctant to make the billions of dollars in capital investments necessary to create a new refining system.

Targeting environmentally friendly technologies

There is increasing interest in the degree to which policies can usefully target and promote environmentally friendly technologies – whether processes or products. "Technology assessment" (TA) for environmental objectives[6] is attractive, because the analysis can explicitly include externalities and public goods issues and because the results are disclosed and placed in the public domain. On the other hand, environmental TA is difficult to perform. A relatively formal process needs relatively straightforward terms of reference, focused and limited in scope, while emerging environmental concerns are relatively ill-defined. As a result, there is a risk that formal TA procedures may tend to promote relatively conservative technological solutions and to favour existing polluting technologies over technological innovations.

"Targeting" green technologies and products need not – at least initially – be very constraining or problematic. Market signals are so distorted that "levelling the playing field" may not initially demand great precision. Thus, investment in almost any energy saving or alternative energy development, for example, seems environmentally friendly in relation to present levels of investment in petroleum and coal technology.

Seen in this light, decisions by policymakers in Europe and the United States to take into account the external costs associated with the use of coal and nuclear power and to factor it into utility decision-making can be seen as environmentally friendly. A recent study for the European Community on the external, or environmental, costs of Germany's coal and nuclear power found that costs for coal were between 2.4 and 5.5 US cents per kilowatt-hour (kwh), whereas for nuclear power they were 6.1 to 13.1 US cents per kwh. A 1989 study by the California energy commission concluded that "fossil fuel electricity generation technologies have significantly higher total societal costs than renewable energy resources" (Williams *et al.*, 1990, p. 67).

Concerns about the external costs associated with the use of fossil fuels in electricity generation have also led to a renewed interest in energy conservation. However, efforts by US utilities have often been hindered by regulatory policy that discourages conservation. In the current system, utility companies lose valuable resources when conservation lowers electricity demand. To correct this situation, public utility commissions in the states of Massachusetts and New York have reformed their rate-setting formulas to ensure that utility companies can earn money from programmes designed to save energy.

Because vested interests are now so embedded in the policy process, there may be considerable pay-off in targeting and promoting very specific technology innovations, even if they cannot be demonstrated to be environmentally the "best". Thus, for example, it would very plausibly be desirable to promote small electric vehicles for urban transport as a substitute for present internal combustion automobiles. Moreover, considerations of trajectories and the "embodiment" of technology in capital goods could also point to relatively specific targeting in terms of processes and specific goods.

A more complex example is provided by the nuclear power industry, where a series of incremental advances in technology have yielded improvements in both economic productivity and environmental health and safety performance. Especially in France and Japan, where investment in Pressurised Water Reactors (PWR) has continued, the nuclear industry has taken advantage of a number of technical advances that were not available when the original Westinghouse PWR design was developed. These include computer assisted design, advanced computers for controlling plant processes, fibre optic communication channels, increased fabrication of large parts in the shop rather than the field, and new construction materials. They have reportedly achieved much higher levels of safety and reliability than American facilities using a similar design.

This is an interesting case of trajectories and learning curves. It can be argued that some other nuclear power technologies could have been more environmentally friendly or that electric power development, including nuclear power, need not have been so dominated by economies of scale and integrated utility system preoccupations, the results might have been more environmentally friendly. The two best-known nuclear power "incidents", at Chernobyl and Three Mile Island, involved variants of PWR technology close to the (state-of-the-art) maximum scale. However, the actual course of development over the past two decades has more or less locked the industry into maximum size PWR reactors, thus rendering difficult the promotion by governments of competitive development of alternative nuclear technologies on environmental grounds.

6. SOME POLICY CONCLUSIONS

The environmental challenges that industrial societies confront today is a huge, multi-faceted phenomenon that cuts across economic sectors and regions. Everywhere it is integrally related to economic production, modern technology, life-styles, the sizes of human and animal populations, the surrounding environment and a host of other factors. If industrial societies are to overcome the limitations of existing environmental policy, they will need to develop and use more environmentally sound technologies, despite the difficulties inherent in the development of preventive products and processes.

The discussion above has emphasised the increasing internationalisation of environmental threats and the need for international policy responses. It is impor-tant to stress that notions of "sustainable development" do not imply community of interest in the world arena even though the threats are increasingly global. Moreover, as the "green" discussion in OECD countries shows, there is only partial community of interest within countries, while the social and political mechanisms for articulating the conflicts are fragile and only slowly becoming institutionalised. Outside the OECD area these mechanisms are even more fragile, partly, perhaps, because the conflicts are greater, but, more importantly, because institutions are weaker. The experience of OECD countries, for example in relation to "acid rain", suggests caution about the effectiveness of inter-governmental negotiation and agreements, although it emphasises the importance of international co-operation in research and the dissemination of its results. It also emphasises the importance of the enterprise sector in technology innovation and diffusion, but does not imply an important international role for inter-governmental agreements on technology transfer.

Within the OECD "community of interest", however insufficient is considerable and should be exploited fully, both through a greater use of economic instruments and through the rapid establishment of common R&D programmes among Member countries.

In the long run, the only affordable and effective way to diminish pollution of the scale and type societies confront today is to work "upstream" to change the products, processes, policies, and pressures that generate waste and give rise to pollution. This will require a stronger emphasis on basic research in universities and scientific institutions, in order to gain better understanding of the processes at work. In addition, social science research may have a crucial role to play in understanding capabilities for adaptation and in designing appropriate policy actions.

There is increasing emphasis on the integration of environment and economic policies. OECD governments have judged both that environment issues will need to influence macro-economic policies and that economic instruments should play a much greater part in environmental regulation. However, the adjustment of market price signals to environmental objectives (especially the emerging global concerns about fossil fuel use) may require very large adjustments to individual prices and the pricing of hitherto "free" resources. In particular, using the concepts outlined in Section 1, the shadow discount rate for "environmental investments" should, perhaps, fall massively in relation to market prices for "human-made investments". This would run counter to present policy priorities of integrating capital markets, and, in any case, poses difficult problems of defining "environmental investments". The problem is particularly acute to the extent that it is judged that maintaining rapid economic growth is essential to generate the technology innovations necessary for sustainable development.

It is important to emphasise that "concordance of objectives" and "community of interest" are limited in relation to perceived notions of "sustainable development". Global and inter-temporal conflicts of interest are dramatic. This limited "no-regrets" policy area is ill-defined. It seems potentially to be very large (see e.g. Nordhaus, 1991, and Hoeller *et al.*, 1991), especially if one is optimistic about technology innovation and diffusion in the OECD countries. Most interest and most government policies are directed towards this area, in order to "buy time" to deal with the more intractable and longer term issues. Governments recognise that it is prudent to invest in environment R&D now in relation to the more intractable problems. Nevertheless, it is widely judged that the discussion will quickly spill into the area of competing objectives.

Currently, concerns are voiced about how far implementing environmentally friendly policies would damage competitiveness, and even about the need for protectionist measures against "polluted" imports (with or without international agreement). This should not be a serious trade policy issue. Experience with the differential impacts of energy price changes since 1973 suggests both that there have been few negative effects on balance for individual countries and that technology innovations in particular sectors have largely neutralised them.

In the immediate future for OECD countries, the problems to be addressed are not only economic and technological. They also involve changes in the priorities and standard routines of governmental agencies and the firms they traditionally regulate. Among the salient technological issues are choices to be made between developing new technologies and improving existing ones, the attitude to be taken on the use of natural resources, the strategies to be adopted in the face of uncertainties, and the balance between economic and environmental demands. Policy efforts are needed, including the setting of fiscal and pricing incentives. The appropriate mix of regulatory and market incentives that spur firms in environmentally desirable directions must be identified. This will require both shifts in political priorities and more research on how to target the incentives to obtain the desired results.

Existing policies are not providing strong enough inducements for widespread development and use of environmentally sound technologies. A new policy mix of regulatory instruments, economic incentives, information dissemination, education and technology polices that are applicable to both the producers and consumers of new technology is required. Regulatory policies should be designed to give consistent signals with emphasis on pollution prevention, particularly where market forces are imperfect. A more open dialogue among regulators and firms, with greater flexibility in the implementation of environmental standards and removal of systematic obstacles to innovation, should be undertaken. At present, the weighty approval systems for new products and standards based on best available technology often discourage innovation. Also, firms should not be penalised for actions that move them beyond compliance with existing standards.

Economic incentives, such as pollution taxes or tradeable emissions permits, can be a valuable adjunct to regulatory policy and are probably most useful in situations where relatively long-term environmental goals are clearly established. Hand in hand with economic instruments, which provide incentives for pollution prevention, goes the removal of perverse signals like tax credits and subsidies which encourage environmentally unsound agricultural or resource depletion practices. Such incentives as taxes on waste disposal and fossil fuel consumption could also be used to ensure that firms maintain financial incentives to conserve energy and resources over the long term.

Technology information dissemination should be practised more consistently with environmental objectives in mind. There are endless opportunities: product labelling systems, requirements that firms disclose the quantities of hazardous substances they release into the environment, government provision to firms – especially SMEs – of information necessary for them to initiate pollution prevention activities. Research findings on company-run programmes for improving environmental technology and practice need to be evaluated and disseminated. Publication of information on the economic benefits of pollution prevention technologies would in itself spur industry initiatives. Such strategies need to be expanded nationally and international interfaces established and their results systematically evaluated.

Efforts should be made to integrate concern for environmental quality in curricula for engineers and corporate managers (as well as in the education of the public-at-large), so that it would become a natural part of the process of commercial technology development and application. Governments might explore the feasibility of encouraging firms to combine environmental concerns and ongoing quality programmes. In this way, the environment would become a continuous responsibility of workers at all levels of the organisation; practices in industry could be designed so that environmental concerns become as indigenous and natural as the best of the quality programmes initiated over the last decade. Multinational companies need to be encouraged to regulate their environmental conduct by applying the same set of environmental standards world-wide.

Technology policy can become an increasingly important instrument in attaining environmental objectives. More cross-sectoral, multi-disciplinary approaches are called for. Consistent strategies for funding research on technologies to improve industrial performance and environmental quality need to be expanded, and co-operative research programmes among countries should be promoted. Funding research and development is, however, not sufficient. Governments could also perform a valuable service by evaluating company-run programmes in areas such as waste reduction and energy conservation that have the potential to provide economic as well as environmental benefits and to publish information on those economic benefits. International organisations can serve a vital role as information clearing houses, by diffusing information about the effects of different national policies on the development and use of more environmentally sound technology.

NOTES

1. For a recent review of the issues, see the report of an international conference in held in Bergen, 8-12 May 1990, in Norwegian Research Council for Science and the Humanities (NAVF), 1990.

2. This strand of analysis has a history. Attempts were made at different times in the development of economic theory to integrate ecological issues into the main corpus of economic analysis. For a history of these attempts, see, in particular, Martinez-Alier (1987).

3. An English translation of the NEPP is available as *To Choose or To Lose,* Second Chamber of the States General, Session 1988-89, The Hague, 1989.

4. For a discussion of economic issues about carbon taxes, see e.g. Pearce (1991) and Nicolaisen *et al.* (1991).

5. For a discussion of long-term benefits of investments in environmental technology, see Jaffe and Stavins (1990).

6. This was a major theme of the Symposium on Technology Assessment, held in Vienna on 28-30 June 1989 under the joint auspices of the OECD and the Austrian Government.

Chapter 10

TECHNOLOGY AND GLOBALISATION

ABSTRACT

Globalisation represents a new phase in the process of internationalisation and the spread of international production. It refers to a set of emerging conditions in which value and wealth are increasingly being produced and distributed within world-wide corporate networks. Large multinational firms operating in concentrated supply structures are at the hub of these conditions. Globalisation is marked by a new ranking of the factors creating interdependencies. Today, foreign direct investment (FDI) in manufacturing and services rather than trade is leading internationalisation and is influencing strongly international location patterns for the production and exchange of goods and services. During the 1980s, two main factors accelerated the changes in earlier patterns of internationalisation. The first was financial deregulation and the increasingly marked character of financial globalisation. The second was the role played by the new technologies which act both as an enabling factor and a pressure towards further globalisation.

The transition from internationalisation to globalisation has been accompanied by a process of global concentration. As a result, in a growing number of industries in manufacturing and services, the prevailing form of supply structure is world oligopoly. This raises novel and previously unaddressed problems relating to the measurement of international concentration and possibly its control in the face of the possible dangers of international cartelisation.

Systematic studies of the impacts of globalisation on the location of innovation are not yet available. The chapter discusses the available evidence. As a result of the convergence of computer, communication and control technology, it has now become feasible for MNEs to install captive intra-corporate world-wide information networks, through which management can link production, marketing and R&D facilities around the world. Although many MNEs still locate their strategic R&D near their corporate headquarters, telematic networks offer such firms new opportunities for the internationalisation of R&D, as well as for the international sourcing of technological resources. The globalisation of R&D also takes the form of international inter-firm technology agreements. These alliances now often involve the most important firms in high-technology industries.

The implications of globalisation for the international system are not yet fully understood. However, four issues can be identified as already requiring special attention from government. The first concerns the marginalisation of developing countries within the globalisation process. The second concerns concentration as a world process and whether a global competition policy is needed and how it might be implemented. The third relates to the problems posed by support given by governments to firms engaged in global competition and the need for new rules and codes of behaviour. The fourth relates to standards and norms in a context of rapid and pervasive technological change, corporate concentration and globalisation.

INTRODUCTION

When OECD Ministers for Science and Technology met in 1987, their discussions were based on a definition of internationalisation as the wide set of processes and relationships as a result of which previously fairly separate national economies have become increasingly interrelated and are now economically interdependent to an unprecedented degree. These processes include the export and import of goods and services; outward and inward flows of direct investment and financial capital; outward and inward flows of embodied and disembodied technology; international movement of skilled personnel and transborder information flows.

Today the term "globalisation" is used widely. This report accepts that over the last few years both economic reality and political thinking have evolved significantly and that this new term is the semantic expression of this evolution.

The notions behind the term globalisation are still considered by many economists as being fuzzy, subjective and ill-defined, and the term itself as a "catchword". If this is so, then economists themselves are at least partly to blame. As was suggested during the TEP Tokyo Conference, jargon generally flourishes at times when concepts move more slowly than reality and leave large areas of important but "intellectually uncharted lands" (Lanvin, 1990a, p. 1). In the present case, some responsibility for this state of affairs can be ascribed to traditional trade theory. On the basis of the TEP Conferences and work carried out in parallel at the OECD Development Centre (see Chapter 12), this report adopts the position that while globalisation may have begun as a catchword, it nonetheless reflects the emergence of a new set of processes and relationships.

The term is used by authors in different ways: it is seen by some as a belated recognition of the internationalisation process by those who had failed to acknowledge its existence before the recent dramatic increase in the level and rate of growth of FDI in the late 1980s and the recent wave of international mergers and acquisitions; it is seen by others as simply representing an intensification of the earlier process. Other analyses again take internationalisation as their starting point but argue that globalisation possesses a number of qualitative features and ushers in a new phase of this process, differing in significant ways from earlier ones.

The data and analyses developed during the TEP tend to support the latter view. Thus, this chapter argues that the term globalisation should indeed be considered as reflecting, however inadequately at present, a number of qualitatively new features and relationships in the world economy. The evidence for this is set out and discussed in Section 1. Section 2 examines some of the effects of globalisation on the location of R&D before discussing the sourcing of scientific and technological resources, by large firms and the prevailing pattern of inter-firm agreements in high-technology industries among the world's largest corporations. Section 3 sets out a few current policy issues stemming from the advent of globalisation.

1. TOWARDS A CLARIFICATION OF THE TERM "GLOBALISATION"

Globalisation refers to the stage now reached and the forms taken today by what is known as "international production" (Michalet, 1985; Dunning, 1988a and 1988b; and Cantwell, 1989), namely value-adding activities owned or controlled and organised by a firm (or group of firms) outside its (or their) national boundaries. It pertains to a set of conditions in which an increasing fraction of value and wealth is produced and distributed world-wide through a system of interlinking private networks (Bressand, 1990a). Large multinational firms (MNEs) operating within concentrated world supply structures and capable of taking full advantage of financial globalisation are at the centre of this process.

According to the OECD summary of the Tokyo TEP Conference, as a result of globalisation: "Borderlines between previously fairly distinct channels and processes of international relationships have become increasingly blurred. Collaboration among firms combines technology, inputs and production across countries. Investment in foreign countries and companies spreads the location, assets and ownership of firms among nations. Trade has always created economic ties between countries, but today extensive trading within the same industries, notably as a result of intra-firm trade, integrates these industries at the international level in new ways. The linkage of the world's financial centres enables private savings and non-reinvestment corporate earnings to be pooled internationally and borrowers and lenders to be connected across borders" (Brainard, 1990, p. 2).

Contemporary technology lies at the root of the process, acting as an enabling factor and exerting pressure towards further globalisation. As a result of telematics, substantial changes have taken place in the economics of location, both within countries and, increasingly, across national boundaries. The convergence of computer, communication and control technology has opened up vast new opportunities for MNEs to deploy resources and operations on a truly world-wide scale. It has now become technically feasible for MNEs, banks, industrial and service firms to install intra-corporate world-wide information networks, through which headquarters' management can link together production and marketing facilities around the world (see Chapter 4). Firms can adopt network forms, building a wide variety of inter-firm alliances and linkages and developing corporate strategies involving new kinds of interactions with suppliers, customers and competitors.

Alongside its close association with the development of new forms of corporate organisation and governance and consequently "new style multinationals" (see Chapter 4), globalisation marks the advent of a new phase of internationalisation on account of, among other things:

- the very rapid growth of FDI;
- the predominance of FDI over trade in the key area of services;
- the role played by MNEs in world trade and the present volume of intra-firm trade;
- the emergence of highly concentrated international supply structures and so of global oligopoly as a result of restructuring and cross-border mergers and acquisitions.

The analysis begins by recapitulating some key concepts and earlier findings regarding internationalisation, before examining these points in turn.

Factors accounting for internationalisation

Until very recently (see Krugman, 1990), the orthodox trade theory premise of the immobility of productive resources across countries was never seriously challenged from within the paradigm (see also Chapter 11). As a result, international trade and foreign direct investment theory have developed as separate disciplines. Mainstream trade theory has considered FDI, if at all, as a source of disturbance explaining inconsistencies in the results of trade models. Attempts to bridge the gap between trade and investment have been made almost exclusively by industrial economists [e.g. Vernon's product life cycle theory (1966) which proposed a unified account of trade and FDI based on the observed behaviour of US firms in the 1960s] or by specialists of multinational enterprises [e.g. Dunning's "eclectic paradigm" (1981) and Mucchielli's "synthetic approach" (1985) which seek to reconcile the location advantages of countries and the ownership advantages of firms]. Likewise, international investment theory has, in the main, taken the initiative in the study of "intra-firm trade", e.g. the part of international trade which takes place in the internationalised corporate structures of multinational enterprises (see in particular Helleiner, 1979).

The theory of foreign direct investment and international production offers three main explanations of the factors accounting for the process. This report considers that they all shed light on the process and will not discuss how they might be ranked or unified:

i) FDI and the organisation of production in foreign locations first represent responses by firms to market imperfections in the form of obstacles to trade and disparities in economic conditions between countries. Such obstacles and disparities include tariff and non-tariff barriers, wage and interest rate differentials, unequal access to technology, and the fact that national economies are at varying stages of development. These obstacles and disparities are one major reason why national enterprises set up delocated production and become MNEs. The underlying rationale for the change is the opportunity for MNEs to play on inter-country differences while eliminating them inside their corporate "territory" through intra-group flows of productive resources (management expertise, capital and technology), internal pricing procedures, intra-group standards, etc. (Michalet, 1985).

ii) FDI and the organisation of international production are an outcome of domestic capital accumulation and the building of fairly large domestic multi-plant firms. The surge of internationalisation from the mid-1950s onwards was the making of large US firms possessing large corporate savings which were faced with insufficient investment opportunities at home (Steindl, 1952). Oligopolistic reaction first by other large US firms (Knickerbocker, 1973) and subsequently by non-US MNEs then kept the process of FDI and delocated production in momentum. The rise of multinational banks and the establishment of totally internationalised financial markets (e.g. financial globalisation, see Aglietta *et al.*, 1990) accelerated the trend towards internationalisation. These institutions now play a central role in globalisation, but corporations must nonetheless have the necessary internal finance to attain MNE status and so have gone through prior processes of growth and consolidation.

iii) FDI and the organisation of international production can be accounted for in terms of transaction cost theory (see Chapter 3). MNEs emerge because of market imperfections, but also as a consequence of the particular advantages offered by the "internalisation" of international transactions and the saving of a wide range of transaction costs through multinationality and the organisation of international production (including the international sourcing of productive inputs). As argued by Dunning (1981, 1988a and 1988b), the benefits accruing to firms from the organisation of international production through MNE group structures, which represent a response but also a

major cause of "market failure", stem both from the creation and usage of specific assets through technological innovation and of organisation of economic activity in other ways than through market co-ordination (e.g. hierarchies and inter-firm relationships, see Chapter 3). The capacity to internalise successfully on an international level offers MNEs a particular source of ownership advantage which can be enjoyed by no other category of firm (see Box 42).

Phases in the process of internationalisation

The literature on the growth of FDI and the spread of production has identified a succession of phases in the process of internationalisation (see, for example, Dunning, 1988b, Chapters 3 and 13; and Michalet, 1989). MNEs emerged in the last quarter of the nineteenth century following technological breakthroughs leading to the discovery and commercialisation of new products and production methods, improved methods of transportation, communication and storage, and the need for new materials and sources of energy to meet the rapidly increasing requirements of industry in the leading capitalist economies. Uncertainties regarding the capability of suppliers (mainly located in developing countries) to meet the requirements of rapidly growing demand made it profitable for firms to invest in the production of the intermediate products themselves. In mining, oil and agricultural raw materials, the new MNEs initiated backward vertical integration at an international level. It has been estimated that, by 1914, at least $14 billion had been invested in enterprises or branch plants in which either a single or a group of non-resident investors owned a majority of substantial minority equity interest, or which were owned or controlled by first generation expatriates (Dunning, 1988b, p. 73). This amount represented about 35 per cent of the estimated total long-term international debt at that time.

A number of economies, particularly developing countries, and some sectors, particularly capital-intensive primary resource producing sectors, were dominated either by affiliates of MNEs or by foreign entrepreneurs who both financed these activities and organised the supply of technology and management for them. This form of FDI had a strong influence in shaping international trade patterns, both by its trade-creating effects and by the specialisation patterns of countries which it helped to crystallise. In many cases, the "vocation" of colonial or economically dependent countries as exporters of one or two primary products was created, or at least significantly strengthened, as a result of foreign investment.

During the first decades of the twentieth century, the organisation of international production began to

---

Box 42. The ownership advantages stemming from successful internalisation

"This advantage reflects the ability of hierarchies to organise related productive activities more efficiently than the market. It stems not so much from the possession of a particular asset, but from the economising of transaction costs when such transactions are under the same governance. Such an advantage includes better access to, and ability to assimilate, information flows; the economies of synergy and the spreading of overhead costs; the capacity to practise price discrimination; the diversification of risks; the ability to protect property rights; and the efficient scheduling of production consequent upon internalising intermediate goods or services markets. For example, a branch plant of an existing firm can benefit from the research and development or marketing facilities of its parent company, which a *de novo* competitor would have to reproduce for itself. Because of their high transaction costs, the rights to these assets cannot be bought and sold, that is, externalised (...), for their value depends on their being centrally co-ordinated and controlled. Nevertheless, they often confer an important economic leverage on the firms possessing them" (Dunning, 1988a, p. 11).

take place in response to somewhat different opportunities and constraints in the new science-based manufacturing industries (heavy electrical equipment and chemicals in particular). As R&D became a more important link in the value-added chain, manufacturing firms began to integrate forwards through FDI in order to protect against the theft, dissipation or inappropriate use of proprietary technology. International oligopology and indeed international cartel-building took place on the basis of international cross-investment, international cross-licensing and patent pools as well as of identifiable market-sharing agreements by countries and continents (Newfarmer, 1985).

The surge of tariff protection after the First World War also stimulated international production through FDI in other sectors of manufacturing, causing firms to replace exports by local production even if expanding markets in host countries later justified local production on economic grounds. The rapid growth of US multinational investment in Europe, Latin America and parts of Asia in the 1950s and 1960s also took place on this basis, in some cases in the context of conscious import substituting policies. In addition, US foreign investment was also the result of a novel variety of non-price oligopolistic competition between large American rival corporations based on the building of productive capacity in foreign markets (Hymer, 1960; Knickerbocker, 1973; Caves, 1974).

The 1960s and early 1970s saw widespread dismantling of trade barriers (notably tariffs) in major industrial nations, particularly in the EC. The developing countries similarly helped create free trade areas, e.g. in Latin America and East Africa, and towards the end of the 1970s, in Asia. However, no fall of FDI ensued because of many sources of market imperfection other than tariffs, including the level of concentration and oligopoly reached in domestic economies and the advantages and constraints on firms competing in technology-intensive industries to internalise their activities internationally. FDI and trade both grew. During this period there was a regular increase in intra-firm flows of goods and services among affiliates operating in different countries or between subsidiaries and parent companies [see e) below]. International transfer of technology through MNEs grew markedly (Vickery, 1986). For a few years, the organisation of international production was approached along the lines of totally standardised "world production" (e.g. the short-lived "world car" concept). International sourcing became the norm in a growing number of industries, and there was a marked increase in the propensity of MNEs to establish central control over sourcing, marketing and innovatory activities.

With the demise of the Bretton Woods system, the second oil shock (1979), the onset of marked economic instability and the emergence of radically new technologies, changes in the organisation and governance of MNEs begin, along with changes in patterns of FDI,

notably the increase in the "new forms of FDI (Oman, 1984 and Oman et al., 1989). The period has the characteristics of a transition. Dunning has shown that it is at this time that the ownership-specific advantages of MNEs begin 'to reside *more* in their ability to create, acquire and *effectively coordinate the use of resources across national boundaries,* than in the privileged possession of *individual assets*" (Dunning, 1988b, p. 333; emphasis added). But effective control was still thought by most MNEs to require generally full equity ownership of subsidiaries. In sectors where this was not thought to be necessary (e.g. petrochemicals, see Chesnais, 1989a), or in some newly industrialised economies seeking new types of partnerships, joint ventures or contractual agreements became the dominant form of investment. In others, especially those utilising the most advanced technology and capable of exploiting the economies of common governance, full ownership was the preferred modality. This was the case in particular for firms which were quick to understand the advantages of telematics (see Chapter 4, Section 2).

FDI, mergers and acquisitions

An examination of the factors currently shaping globalisation must begin with the role played by the rapid growth of FDI, which serves to highlight the part played by firms. Firms are active agents of internationalisation and, in the case of MNEs, the architects of many dimensions of the resulting world system. For firms, internationalisation does not mean loss of autonomy and power as it may for governments and other segments of national communities; it simply means changes in some of the characteristics of the markets in which they operate.

Direct investment flows of major OECD countries, which had stagnated in the first half of the 1980s, has grown rapidly in recent years, both absolutely and relative to trade flows. The dynamism of FDI can best be illustrated by comparisons with trade, gross fixed capital formulation (GFCF) and gross domestic product (GDP) (see Figure 29). From 1983 to 1989, FDI by OECD countries grew at an annual average growth rate of 31.4 per cent, nearly three times faster than trade (11.0 per cent) and GFCF (11.9 per cent) within OECD and over three times as fast as GDP (10.4 per cent). Since 1985, the gap between the rate of growth of FDI and that of trade has widened drastically, leading one author to suggest that "as a means of international economic integration, FDI is in its take-off phase: perhaps in a position comparable to world trade at the end of the 1940s" (Julius, 1990, p. 36).

FDI by OECD countries has tended to be increasingly concentrated within the OECD, in particular between the three poles of the "Triad", namely the United States, the EC and Japan (Ohmae, 1985).

Figure 29. Trends in foreign direct investment[1,3], GDP, total trade[2,3] and GFCF in the OECD area, current prices

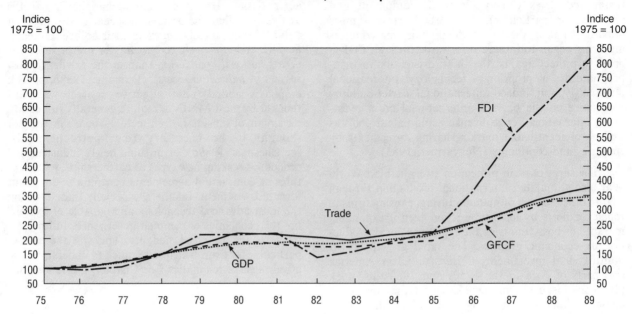

Indice
1975 = 100

Indice
1975 = 100

1. Average value inward and outward investment.
2. Average value of exports and imports.
3. Includes intra-OECD.

Source: OECD, STIID Database, September 1991.

Recent calculations made by the United Nations Centre on Transnational Corporations show that "between 1980 and 1988, intra-Triad foreign-direct-investment stock nearly tripled, from $142 billion to $410 billion. In 1980, the stock within the Triad accounted for 30 per cent of the world-wide stock of inward investment; by 1988, intra-Triad stock had increased to an estimated 39 per cent of world-wide inward stock. Intra-Triad trade also grew more rapidly than world trade, increasing from 13 per cent to 17 per cent of world trade over the period. Thus, in terms of both foreign direct investment and in terms of trade, interactions within the Triad have outpaced interactions in the rest of the world, and between the Triad and the rest of the world, indicating a faster rate of integration among the Triad than between the Triad and the rest of the world." (UNCTC, 1991, p. 36).

Table 41 shows that the growth of FDI within the OECD has been marked by a number of structural developments. The increase in Japan's share in OECD outward investment is striking, but France and Sweden also doubled their share in comparison with the decade of the 1970s. The United States' share in inward

investment jumped dramatically in the first part of the 1980s, but in the most recent phase the EC has again become an almost equally important destination of investment. The share of EC countries in OECD inward investment climbed to above 45 per cent in 1988-89, with the largest increases recorded by the United Kingdom, Belgium-Luxembourg and the Netherlands. The importance of EC countries as a destination of foreign investment flows is not a new phenomenon, however, and their share in OECD inward flows at the end of the 1980s is still below those recorded in the 1970s for most EC countries (shown in Table 41), with the exception of Italy and Spain. The late 1980s witnessed an increasing importance in the level of intra-EC flows of FDI (OECD, 1990*i*) in preparation for the Single Market.

Figure 30, prepared by the UNCTC, shows that, in terms of *stock*, the most important FDI relationship within the Triad is between the EC and the United States, which together accounted for 79 per cent of intra-Triad FDI stock in 1988. However, Japan's share of intra-Triad stock more than tripled in the 1980s, from 5 per cent in 1980 to 16 per cent in 1988.

Table 41. **Share of major OECD countries in outward and inward direct investment, 1971 to 1989**

Percentage of total OECD flows

	1971 to 1980	1981 to 1984	1985 to 1987	1988 to 1989
	Outward investment			
United States	46.4	20.1	25.3	16.9
Canada [1]	3.9	8.0	5.4	3.5
Japan [1]	6.2	13.2	16.2	27.6
EC [2]	41.8	56.0	50.1	47.6
Belgium-Luxembourg [1]	1.1	0.4	1.8	−1.0
France [1]	4.8	8.1	6.5	10.8
Germany	8.6	9.7	9.4	8.7
Italy [1]	1.2	4.6	2.7	2.6
Netherlands [1]	6.4	7.8	5.6	4.9
Spain [1]	0.4	0.9	0.6	0.9
United Kingdom	19.2	24.6	23.4	20.6
Sweden [1]	1.6	2.7	3.0	4.4
	Inward investment			
United States	33.8	62.8	59.3	51.4
Canada [1]	3.3	−2.4	1.3	2.7
Japan [1]	0.9	0.9	1.2	−0.6
EC [2]	61.5	38.3	37.3	45.8
Belgium [1]	5.5	3.6	2.3	4.6
France [1]	10.1	6.4	5.7	6.6
Germany	8.4	2.9	2.2	2.9
Italy [1]	3.4	3.5	3.0	3.7
Netherlands [1]	5.1	3.0	3.2	3.9
Spain [1]	4.2	5.7	5.9	6.1
United Kingdom	24.7	13.2	15.0	18.0
Sweden [1]	0.5	0.5	0.9	0.7
Memorandum items:				
Total of above countries				
Outward Investment				
level [3]	28.9	36.0	83.0	142.2
as a share of exports of goods and services	3.6	2.3	4.4	5.3
Inward Investment				
level [3]	16.7	30.4	56.2	127.0
as a share of imports of goods and services	2.1	2.0	3.0	4.7

1. Exclude reinvested earnings.
2. Data for the European Community exclude flows of Denmark, Greece, Ireland and Portugal. Data include intra-Community flows.
3. US$ billion, annual average.
Source: OECD (1990*i*).

The rapid growth of FDI inside the OECD included an acceleration in *cross-border acquisitions and mergers* in the 1980s. Figure 31 shows recent trends in acquisitions of US firms by foreign corporations in terms of numbers of transactions; a more qualitative analysis would probably reveal more clearly still the scale of foreign-led merger activity in the United States. Table 42 sets out data on industrial deals (mergers and acquisitions of majority holdings; acquisitions of minority holdings; and establishment of jointly-owned subsidiaries) conducted by firms from EC countries) which point to a rapid pace both of domestic corporate restructuring and of internationalisation. The number of mergers and acquisitions of majority holdings in 1988/89 was more than four times higher than the level recorded by 1982/83. Mergers and acquisitions of majority holdings increased significantly after 1987, suggesting that the prospect of the Single Market provided additional impetus to the process.

Figure 30. The flow of intra-triad investment, 1985-89 and the 1988 level of stock

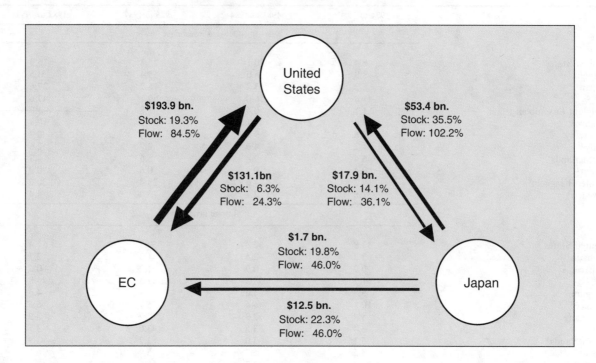

Note: Dollar figures show 1988 outward stock; percentages show average annual growth rates, stocks and flows. Stock growth rates are for 1980 to 1988. Flow growth rates are for 1985 to 1989.
Source: UNCTC (1991).

Table 42. Industrial deals of the 1 000 largest firms in the European Community: number of operations and geographical breakdown[1]

	1982/83	1983/84	1984/85	1985/86	1986/87	1987/88	1988/89
	Mergers and acquisitions of majority holdings						
Number of operations	117	155	208	227	303	383	492
National share[2]	50.4	65.2	70.2	63.9	69.6	55.9	47.4
Community share[3]	32.5	18.7	21.2	22.9	24.8	29.2	40.0
Non-Community share[4]	17.1	16.1	8.7	13.2	5.6	14.9	12.6
	Acquisitions of minority holdings						
Number of operations	33	54	67	130	117	181	159
National share[2]	60.6	68.5	67.2	67.7	71.8	63.5	64.2
Community share[3]	27.3	14.8	14.9	15.4	17.9	20.4	23.3
Non-Community share[4]	12.1	16.7	17.9	16.9	10.3	16.0	12.6
	Establishment of jointly-owned subsidiary						
Number of operations	46	69	82	81	90	111	129
National share[2]	50.0	46.4	48.8	42.0	32.2	40.5	43.4
Community share[3]	17.4	15.9	18.3	24.7	17.8	27.9	27.9
Non-Community share[4]	32.6	37.7	32.9	33.3	50.0	31.5	28.7

1. Data collected from the specialist press regarding operations involving at least one of the 1 000 largest firms of the Community, ranked according to their financial data.
2. Operations of firms from the same Community member state.
3. Operations of firms from different Community member states.
4. Operations of firms from Community member state and third countries which affect the Community market.
Source: OECD (1990*i*).

Figure 31. **Growth of acquisitions of US companies by foreign firms**

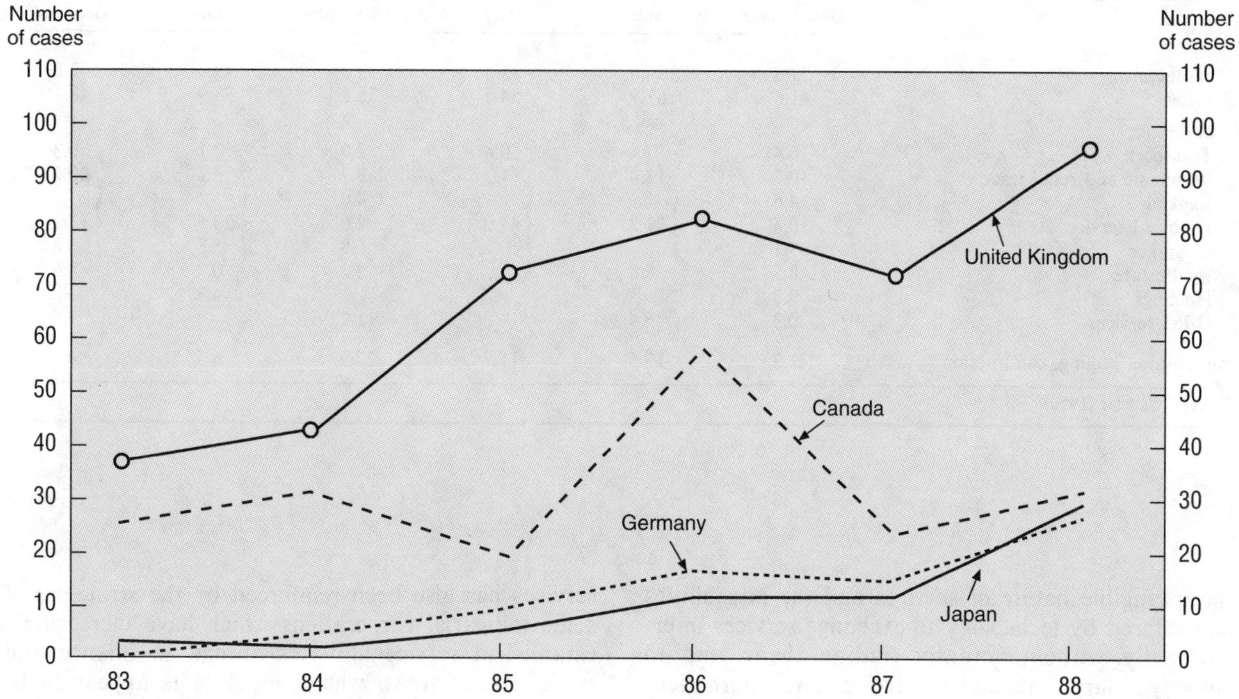

Source: *Mergestat Review*, W.T. Grimm and Co.

Data on the geographical distribution of these operations, while showing considerable fluctuations in individual years, confirm their dominant intra-Community nature. The EC share of mergers and acquisitions of majority holdings and of the establishment of jointly-owned subsidiaries reached record levels in 1988/89. The EC share of acquisitions of minority holdings has increased steadily since 1983/84, although to levels that are still below those recorded in 1982/83. The increasing weight of cross-border operations with other Community firms is primarily reflected in a contraction in the share of national deals.

The process of international mergers and acquisition was facilitated considerably during the 1980s by financial deregulation and the large-scale role played by globalised financial and monetary markets. It remains to be seen whether the process will continue at the same level in future years[1].

FDI and internationalisation in services

The central role played by foreign direct investment and the strategies of large firms in the globalisa-

tion process is clearer still in the service industries. One of the paradoxes revealed by studies on the internationalisation of the service sector[2] is the striking stability, as captured by traditional balance of trade data, of its contribution to world trade. Over a fifteen year period, from 1970 to 1985, the share of services in world trade only increased from 29 per cent to 32 per cent. This apparent stability stands in stark contrast to the important structural shift experienced by most industrialised economies, where services have become the major source of domestic output and employment.

The relatively small contribution of services to trade also stands in sharp contrast to their contribution to FDI. Services represented about 25 per cent of the total world stock of FDI at the beginning of the 1970s. At the end of the 1980s, this share was close to 50 per cent, while services accounted for 55 to 60 per cent of annual foreign investment flows (UNCTC, 1991, p. 15).

The fact that the unquestioned "tertiarisation" of advanced economies has not altered significantly the aggregate pattern of international trade reflects to a certain extent the inadequacy of the indicators used to measure international trade flows. It is due in part to

Table 43. **Share of manufacturing, services and other sectors in total book value of FDI**
End-1987, in percentage

	United States	Japan	Italy	United Kingdom	France	Germany
Manufacturing	37.3	26.6	33.8	33.3	49.9	43.8
Services	41.4	61.9	54.5	38.8	38.9	48.1
of which:						
Transport	0.4	7.4	0.4	2.0	2.1	0.3
Wholesale and retail trade	10.7	12.4	7.7	8.7	9.1	20.2
Banking	4.6	} 21.2	} 45.0	4.0	} 23.5	} 13.1
Financial services	10.4			2.0		
Insurance	3.1			6.9	5.8	3.8
Real estate	0.5	8.8	..	3.2	0.2	8.6
Holdings	9.3	..	..	..	4.3	..
Other services	0.3	5.4	..	11.2	..	1.5
Agriculture, mining, construction	21.3	19.5	11.7	27.9	6.5	4.3

Source: Sauviat (1990).

the intangible nature of services and the possibilities now offered by technology to exchange services internationally without actually trading them, and to embody a larger amount of services into internationally traded goods.

The relatively low share of services in international trade in fact reflects at least three phenomena. First, because they imply a direct relationship with customers, services by definition are less tradeable than goods and certainly less traded as final products; second, a significant share of the services produced in advanced economies are "intermediate services" embodied in goods before they are exported; and third, a relatively large proportion of "internationally transacted" services are not actually "traded" in the GATT sense (in some service industries intra-corporate exchange of services has increased sharply). It has been estimated, for instance, that 70 per cent of international data transactions ("transborder data flows") were intra-corporate: for a country like Canada, the proportion even reaches 90 per cent (Sauvant, 1986).

Radical technological change in computing and telecommunications has hastened the blurring of the traditional frontiers between manufacturing and services. It has been argued that the true "service revolution" is the convergence of manufacturing and marketing through integration of product development and customer relations (Bressand, 1990). The growth of FDI in services outside the financial sphere is due simultaneously to firms originating in the service sector and to industrial corporations which have increasingly diversified into this area. Indeed the typical global corporation has considerably increased the focus and technology-based proximity of its "core" manufacturing activities but has also diversified into high growth lucrative service activities. The trend towards FDI in

services has also been reinforced by the strategies of some industrial corporations which have increasingly established service affiliates abroad in finance and trade-related areas, where investments appear to be designed to strengthen and internalise corporate functions, rather than diversify into services *per se*.

Consequently, internationalisation in services has increasingly taken place through FDI rather than through exports (Dunning, 1988b, Chapters 9 and 10; Sauviat, 1990). Such investment includes both wholly owned or majority-controlled FDI and a range of the "new forms" for which the service industries have in fact often been a laboratory (e.g. international franchising in hotels and catering). As shown in Table 43, the size of foreign investment in services is striking for the large OECD countries, notably for Japan which is generally thought of mainly as a home country of industrial corporations. The table also reflects the ambiguous character of the notion of services and the unsatisfactory state of the available statistics. Banking, financial services and insurance account for a large fraction of the recorded FDI, but the differences in national figures obviously reflect strong discrepancies in the definition of branches.

The single most important destination for FDI in services during the second half of the 1980s was the European Community. Growth in this region alone accounts for much of the rapid expansion of world services' foreign direct investment during that period. From 1984 to 1988, cumulated investment flows (excluding reinvested earnings) in services in the EC by both third countries and firms from the EC exceeded by far those in all other sectors and amounted, respectively, to ECU 28 billion and 16 billion for third countries and ECU 34 billion and 21 billion for intra-Community foreign direct investment.

The creation of the Single Market has provided a powerful incentive for MNEs active in the service industries to invest in the Community. While the EC started dismantling its internal barriers to trade and foreign direct investment in goods some thirty years ago, obstacles to foreign direct investment and trade in services remained largely untouched until the late 1980s. The announcement of the Single Market and the process now underway to remove those obstacles has led to a massive reorganisation of service industries in the Community, thereby creating substantial interest in the large and dynamic community market on the part of service industry MNEs from all major OECD countries.

The role of MNEs in world trade and intra-firm trade

MNEs now play a very important role in world trade, both in their capacity as large firms exporting either from their home economy or from one or several foreign countries and because of a large and growing system of intra-group flows of goods across national borders. Two forms of trade are characteristic of MNEs: sales of final products by foreign affiliates and intra-firm trade of international goods or final products within internationalised group structures. Students of MNEs have been interested for many years in the peculiar nature of intra-firm flows (Helleiner, 1979, 1981; Lall, 1979; Little, 1987; Julius, 1990). Conservative calculations suggest that MNEs are involved in at least 40 per cent of total OECD manufacturing trade. The figure for the United States is 80 per cent (1988),

for the United Kingdom also 80 per cent (1981) and for Japan 40 per cent (1983). Nearly a third of these countries' exports are made up of intra-firm flows, that is, trade in goods and services among the different components of the same multinational group, or again, products which circulate within the "internal" confines of MNEs. It will be noted that the figures are not recent; in the case of Japan they almost certainly understate current reality.

Ongoing work by the OECD underscores that there is still insufficient knowledge about the way the global operations of MNEs affect the structure of world trade. Traditional methods of collecting and presenting foreign trade data have great difficulty in measuring the role of MNEs in trade flows. This fact cannot, of course, be dissociated from the tenet of orthodox trade theory and its assumption regarding the immobility of productive factors across national borders: nowhere does globalisation have more the appearance of an "uncharted land" than in the field of the relationships between FDI, the role of MNEs and the structure of international trade.

Only a few countries publish detailed and reasonably reliable data concerning intra-firm trade. In the case of the United States, data is available up to 1988, broken down by trade by US based parent corporations, US affiliates abroad, foreign affiliates based in the United States, and intra-firm trade affecting US imports and exports whether by US firms within their group structures or by foreign MNEs within theirs. As noted, the most recent year for which detailed figures are available for Japan is 1983, and some UK information is available for 1981. Table 44 assembles the available statistics, covering these three countries.

Table 44. **Data on intra-firm trade for selected OECD countries**

		United States		United Kingdom		Japan	
Total home-based MNEs and foreign subsidiaries (% share)	Export	1982[2]	31.8	1981[1]	30	1983[1]	31.8
		1985[1]	31			1986[3]	38
	Import	1982[2]	38.3			1983[1]	30.3
		1985[1]	40.1				
Home-based MNEs (% share)	Export	1982[2]	22.4	1981[1]	16	1983[1]	30.6
		1985[1]	29			1983[3]	38
		1988[3]	32				
	Import	1982[2]	17.6			1983[1]	18.3
		1985[1]	16.1			1983[3]	40
		1986[3]	18				

1. UNCTC (1991, Table VI.I, p. 92).
2. Hippel (1990, Table 1, p. 497; Table 2, p. 499; Table 3, p. 500).
3. Julius (1990, Table 4.1, p. 74).

Analysis undertaken by the OECD reveals the following patterns:

i) The combined totals of exports by home-based MNEs and the subsidiaries of foreign MNEs are set against each country's total exports and imports. In the case of the United States, nearly a third of exports and 40 per cent of imports are made up of flows between units controlled by the same group. Thus, they have a fairly significant influence on the balance of trade. In the case of the United Kingdom and Japan, the share in total exports of intra-firm flows is roughly the same as for the United States.

ii) Intra-firm flows between a home-based MNE and its foreign subsidiaries are more export-oriented than import-oriented in both the United States and Japan (British MNE figures are available for exports only).

iii) In the case of foreign subsidiaries located in the United States or Japan, however, the dissymmetry between exports and imports is reversed, foreign subsidiaries being larger importers than exporters. The implication is that the presence of foreign subsidiaries has a measurable influence on the balance of trade. Where United Kingdom exports are concerned, the share of foreign subsidiaries and that of home-based MNEs is practically equal. This is partly explained by the large extent to which the economy of the United Kingdom is multinationalised; the country is both a major investment source and host to foreign investment.

As a result of corporate strategies, the size of intra-firm flows varies from one country or region to another, which means that their impact on the foreign trade of their host country varies accordingly. In the case of US MNEs, for example, 90 per cent of the exports of their Canadian-located subsidiaries go to the United States. The situation in South East Asia is somewhat similar. Eighty-five per cent of the exports of US subsidiaries in the region are towards the United States (UNCTC, 1988, p. 96). This is part of a global strategy of foreign sourcing of intermediary products and consumer goods by US MNEs. US subsidiaries in Europe follow a different pattern: they export 85 per cent of their output to other European countries. Latin America lies between the two: 50 per cent of the subsidiaries' exports go back to the home country, while 45 per cent are sold in Latin American markets.

The size of intra-firm flows also varies strongly from industry to industry (see Table 45). The variations are particularly striking in the case of Japan where the figure for transport equipment stands well above the average, due to the development of sophisticated group policies by Japanese automobile manufacturers. While in OECD countries, foreign affiliates of these firms were geared to domestic markets during the

Table 45. **Size of intra-firm trade by industry**

Percentages

	United States (1982)		Japan (1983)		United Kingdom (1981)
	Exports	Imports	Exports	Imports	Exports
All industries	32.9	40.8	23.7	27.7	27.0
Petroleum	23.4	24.1	42.4[1]	28.8[1]	..
Mining	..	..	4.5	11.5	..
Manufacturing	38.6	63.0	29.9	20.9	..
Food and beverages	19.3	22.0	18.3	2.3	27.4
Textiles	..	..	2.7	5.2	..
Wood, paper and pulp	..	..	0.4	22.0	..
Chemicals	36.4	..	19.5	8.0	48.1
Metals	16.7	..	2.2[2]	1.8[2]	21.5
Non-electrical machinery	51.4	74.2	12.8	20.0	..
Electrical machinery	32.5	55.7	24.8	41.9	..
Transport equipment	43.6	..	45.3	34.3	49.8[3]
Precision instruments	..	..	38.7	32.1	..
Other manufacturing	36.2	42.8	..	..	18.3
Wholesale trade	12.2	12.8	18.2	30.6	..
Other industries	24.6	49.4	12.3	22.1	21.2

1. Petroleum and coal products.
2. Iron and steel.
3. Motor vehicles.
Source: United Nations Centre on Transnational Corporations (UNCTC), based on official national sources.

Figure 32. Specialisation and intra-firm trade by South East Asian subsidiaries of Toyota

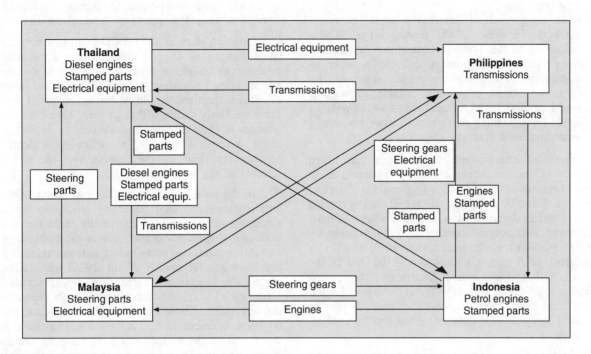

Source: Far Eastern Economic Review (1989, p. 73).

1980s, in South East Asia these affiliates were part of closely integrated regional networks (see Figure 32) aimed at obtaining plant specialisation, economies of scale on a regional basis and high quality/low cost production directed simultaneously towards Asian and world markets.

The new forms of international supply structure

The transnational merger movement of the 1980s represents the climax of a process, initiated in the 1960s, which has led to world or international oligopoly[3]. This concept may be a usefull analytical tool for discussing "global competition" between MNEs, or more precisely, cross-border rivalry between large firms which have built domestic and international network structures (see Chapter 4).

World oligopoly is not, of course, an entirely new form of supply structure. In petroleum and in several non-ferrous metal mining and processing industries (e.g. aluminium), world oligopoly has long been a key feature of supply. What is new is the current rapid extension of global oligopoly and the fact that it now

constitutes the dominant form of supply structure in most R&D intensive or "high-technology" industries, in many scale-intensive manufacturing industries, and in an increasing number of service industries. In R&D-intensive industries, the only exceptions are in fact those industries where supply structures are even more highly concentrated (e.g. space launchers, long-range civil aircraft, or highly specialised military products) and where the disappearance of one of the few remaining competitors would increase the possibility of outright monopoly.

World oligopoly and the "global competition" or "cross-border network rivalry" which accompany it, are the outcome of two related but nonetheless distinct processes: internationalisation and the extension of international or delocated production; and the process of industrial concentration resulting from domestic and cross-border acquisitions and mergers. It occurs:

i) when, in a given industry, industrial and technological development places extremely strong constraints on firms (notably the need to recoup large R&D costs) to produce for world markets as distinct from even the largest domestic markets, and when there are significant opportunities for world-wide sourcing of key inputs to production, notably scientific and technological advances made in foreign countries;

ii) when, as a result of "mutual invasion" of home markets, (Erdilek, 1985) among large MNEs belonging to the Triad, concentration becomes a truly international process and the number of oligopolistic rivals (in the proper sense of the term, e.g. those effectively capable of waging world or global competition) drops world-wide to levels corresponding *pari passu* to those previously associated with domestic oligopoly.

Work on international concentration has fallen seriously behind the pace of change in the world economy. Measures of concentration (e.g. market shares of the top 4, 8 and 20 firms) are still largely being carried out on a purely domestic basis, at a time when the most significant indicator of concentration has probably become global or world market shares[4]. A number of industrial sector studies, undertaken by the OECD, the EC, the OECD Development Centre and the United Nations (UNCTC and UNCTAD) have collected some data. But these scattered efforts should be replaced by systematic data collection and by analytical work.

Available case study material (see Tables 46, 47 and 48) suggests that the share of the largest 5, 10 or 20 firms in total world assets or sales would furnish a relevant measure of global concentration in an industry[5]. This would show the product groups and industries where studies should be carried out assessing the *degree to which market contestability effectively exists* (Baumol and Lee, 1991). It would also provide one meaningful index of the power of a small group of firms to influence business decisions, build collective entry barriers (e.g. through the protection offered by their technological advance), and, when conditions permit, to limit price competition in regional (e.g. European) or in world markets.

Ongoing work by the OECD point to a very diversified picture of the competition and consolidation processes at the global level. In industries as diverse as automobiles, computers, pharmaceuticals and international construction, the global market shares of the top firms are fairly stable, although there have been many changes in their identity and nationality. However, in some narrower industries, in particular in electronic components, there is now concern over the declining number of key suppliers.

In the past, oligopoly was seen to be synonymous with cartelisation and generally studied principally through price theory. Of course the main manifestations of global interdependence were the collusive price and international market-sharing agreements set up by the international cartels of the 1920s and 1930s in chemicals, electrical engineering, pharmaceuticals, tyres, etc. Today, this focus should only represent a starting point. There can be other manifestations of oligopolistic interaction. It is the *mutual market dependence* and so the *mutual recognition* among the large firms belonging to an industry or product-specific concentrated supply structure which carry the "hallmark of oligopoly" (Caves, 1974; Buckley and Casson, 1985; and Newfarmer, 1985). This is why price collusion may not be the only expression of oligopoly and why inter-firm co-operation, which has developed extensively alongside oligopolistic competition, in areas as important as technology and the setting of industrial standards, must be scrutinised with care.

Table 46. **Firm structure in electronics, 1987**

Per cent share of total world output (first 10 firms)

Computers		Telecommunications		Semiconductors	
Firm	Share (%)	Firm	Share (%)	Firm	Share (%)
IMB (US)	51	AT&T (US)	23	IBM (US)	10
Digital (US)	8	Alcatel (FR)	9	NEC (JP)	8
Burroughs (US)	5	Siemens (GER)	7	Texas Inst (US)	7
Control Data (US)	4	ITT (US)	7	Hitachi (JP)	7
NCR (US)	4	N. Telecom (CAN)	7	Motorola (US)	7
Fujitsu (JP)	4	Ericsson (SWE)	7	Philips (HOL)	6
Sperry (US)	4	IBM (US)	7	Toshiba (JP)	5
Hewlett Packard (US)	4	Motorola (US)	7	AT&T (US)	4
NEC (JP)	3	NEC (JP)	6	Fujitsu (JP)	4
Siemens (GER)	3	GTE (US)	5	INTEL (US)	3
Total	90	Total	85	Total	61

Source: OECD Estimates.

Table 47. **Other selected data on world concentration**

Automobile	1984	78 per cent of world output by 12 firms
Glass components for automobiles	1988	53 per cent of world output by 3 firms
		88 per cent of world output by 7 firms
Tyres	1988	85 per cent of world output by 6 firms
Dataprocessing/DRAM	1987	65 per cent of world output by 5 firms
		100 per cent of world output by 10 firms
Dataprocessing/ASIC	1988	54 per cent of world output by 4 firms
		100 per cent of world output by 12 firms
Medical equipment	1989	90 per cent of world output by 7 firms
Petrochemicals:		
Polypropylene	1980	34 per cent of world output by 4 firms
		50 per cent of world output by 8 firms
Polysterene	1980	51 per cent of world output by 4 firms
		69 per cent of world output by 8 firms
ABS	1980	55 per cent of world output by 4 firms

Source: OECD, Development Centre and EC/FAST industrial case studies.

Table 48. **Examples of world concentration in the service industries**

Number of firms	Share of world market (percentage)	Countries of origin (figures in brackets refer to number of firms from a given country)
Reinsurance Market (1986)		
4	30.3	Germany (1), Switzerland (1), US (2)
8	40.7	Germany (1), Switzerland (1), United States (4), United Kingdom (1), Sweden (1)
16	53.6	Germany (5), Switzerland (1), US (5), UK (1), Sweden (1), Japan (3)
32	70.6	Germany (6), Switzerland (2), US (11), UK (1), Sweden (1), Japan (7), France (3), Italy (1)
Computer Services (1988)		
4	33.3	US (4)
8	54.4	US (7), France (1)
Advertising (1989)		
4	25.7	US (2), Japan (1), UK (1)
8	43.9	US (4), UK (3), Japan (1)
16	60.7	US (10), UK (3), Japan (2), France (1)
Strategic Business Management Consulting Services (1989)		
4	53.7	US (4)
6	62.2	US (6)
15	80	US (14), Germany (1)

Source: Data on reinsurance market from UNCTAD (1989). Other data from Sauviat (1990), Institut de Recherches Économiques et Sociales (IRES), Paris, from professional journals and consulting firms.

2. SOME EFFECTS OF GLOBALISATION ON THE LOCATION AND ORGANISATION OF INNOVATION

Systematic studies of the impact of globalisation on the location and organisation of innovation are not yet available, but they are imperative for better understanding of the effects and implications of globalisation on national systems of innovation. This section discusses the limited and sometimes contradictory evidence at hand.

Internationalised R&D and the world-wide sourcing of scientific knowledge

The advent of telematics has increased the scope for internationalisation of R&D by MNEs. Up to the mid-1970s, US firms often gave, as a reason for not spreading more of their R&D activities abroad, the difficulty of ensuring adequate supervision and control (Mansfield, 1974; and Pearce, 1989). For obvious reasons, the telecommunications and computer and data-processing industries were leaders in using the new methods of control, and they have probably seen the greatest development in world-wide organisation of corporate R&D and in international sourcing of scientific and technical resources. As early as the mid-1970s, several large firms in these industries had a kind of international technical system with a foot in several national systems which allowed them to ensure the flow of technology within international group structures.

IBM is a prime example. To ensure that the R&D undertaken in geographically distant affiliate laboratories was consistent, IBM had set up, by the mid-1970s, a vast telecommunications network for pooling the group's technological resources. Laboratory computers were interconnected on a world-wide basis, and a single data bank was set up at the corporate R&D headquarters (Michalet and Delapierre, 1978). With the emergence of world-wide telecommunication networks, IBM has been followed by an ever greater number of large corporations. A survey of forty US and European MNEs indicates that among the seven main reasons given by firms for adopting the new international telecommunications equipment in their operations and undertaking the organisational changes required in corporate structures was the possibility of "international implementation of R&D capacities generated by increased interaction among affiliates and headquarters and greater division of labour, according to the technical requirements and scientific endowments of countries and affiliates" (Antonelli, 1984, p. 15).

There are strong differences among industries in the extent to which firms organise their R&D internationally and the type of R&D they carry out abroad. The pharmaceutical industry has one of the most highly internationalised patterns of corporate R&D. Most large drug companies have research centres of one kind or another in a number of countries and, increasingly, products arriving on the market have been developed as part of an integrated global, rather than a national or regional, innovation programme. Table 49 shows that by the early 1980s, firms from all major OECD pharmaceutical industries had set up R&D facilities in other OECD countries. Case study material on the pharmaceutical industry (Burstall

Table 49. **Estimated R&D expenditure by pharmaceutical companies in various countries in the mid-1980s**

Million $

	Nationality of company								
	Belgium	France	Germany	Italy	Netherlands	United Kingdom	Switzerland	United States	Japan
Belgium	30	10	..	..	..	..	..	30	..
France	..	400	50	..	..	20	20	80	..
Germany	15	..	700	..	..	..	..	50	..
Italy	..	10	20	120	..	20	30	50	..
Netherlands	..	..	..	..	80	..	..	10	..
United Kingdom	..	20	20	..	10	400	20	120	..
Switzerland	..	..	10	..	..	..	450	..	..
United States	..	10	100	..	10	50	250	2 000	..
Japan	..	..	20	..	..	..	30	100	700
Total	45	450	950	120	100	500	800	2 500	700

Note: Because of problems of allocation, columns do not add to total.
Source: Burstall and Dunning (1985).

Table 50. **International biotechnology agreements between US and non-US firms**
1981 – 1st Quarter of 1986

	Acquisition	Venture capital	Contract R&D	Joint R&D	License/ Production	Distrib./ Marketing	Joint Prod. Marketing	Total No. contracts[1]
Belgium	1	1	1	3	–	1	1	9
Denmark	–	1	2	2	1	1	1	7
Finland	–	1	–	1	3	–	3	–
France	4	1	6	3	4	6	5	21
Germany	1	1	7	6	14	18	3	31
Italy	–	3	1	3	2	1	1	10
Netherlands	2	3	–	2	–	2	1	9
Norway	–	1	–	–	–	–	–	1
Spain	–	1	–	1	–	–	1	2
Sweden	3	3	5	4	5	6	3	19
Switzerland	1	2	11	4	5	6	5	26
United Kingdom	4	11	4	5	4	5	5	35
Total Europe	16	28	38	33	36	49	27	173
Japan	2	24	30	24	40	59	24	141

1. Some contracts may involve more than one type of agreement.
Source: Yuan (1987).

et al., 1981) also indicates that a larger part of this internationalised R&D is directed towards basic research than in any other industry. In the case of food processing (OECD, 1979b), internationalised R&D is also quite large, but limited to development and to product and process adaptation to local manufacturing conditions.

MNEs can also organise the international sourcing of scientific knowledge in other ways (Chesnais, 1988b), in particular through R&D contracts with foreign universities and agreements with small firms in other countries. In some cases, transborder agreements have a high degree of visibility (e.g. the Hoechst-Massachussetts General Hospital Laboratory Agreement). In others, links will be organised domestically through local affiliate R&D facilities; these may be quite modest in size, and one of their major *raisons d'être* may be interface or observation post functions. A survey on the motivations for US MNEs to set up foreign laboratories confirms that the most frequently mentioned objective was the desire to "have a window on foreign science", first and foremost in Europe. Other objectives include access to special skills not easily available in the home country, developing new sources of technical concepts or simply establishing, on an international scale, corporate activities concerned with science and technology (Fusfeld, 1986).

The increased opportunities for organising the sourcing of technological resources on an international basis and, in particular, for establishing in their turn a window on the world's most advanced science base, that of the United States, explains why in the 1970s and 1980s large European firms, rapidly followed by

Japanese ones, also took this road. This is especially true for the chemical and chemicals-related industries; the need to gain direct access to the US science and technology base in the field of pharmaceuticals and biotechnology, in particular, motivated much of their investment [see Howells (1990) for recent case studies of UK pharmaceutical MNEs and Wortmann (1990) for German corporations]. Innovation-related investment has involved the form of take-overs of existing firms, whose R&D facilities the new parent firms have often expanded; the creation of new laboratories and a wide range of inter-firm technological co-operation agreements, notably with the small genetic engineering firms (see Table 50).

The locations chosen by MNEs for their foreign R&D activities are shaped by the merits of national scientific capacity. In the case of the United Kingdom there has been strong foreign investment in pharmaceutical R&D owing to the high quality of British science in the life sciences and in chemistry. In the case of Germany, the R&D activities of foreign MNEs are concentrated in the electrical engineering and electronics industries; this again is a reflection of German excellence in these areas (Wortmann, 1990).

Patenting in the United States by MNEs and the Pavitt-Patel arguments

The extent to which firms internationalise their technology-related activities has recently been studied by Patel and Pavitt (1990). They argue that the extent to which the internationalisation of R&D has occurred

225

may be measured, as a first approximation, as the proportion of technological activity controlled by a foreign firm, as distinct from a domestic one, which has led to patenting in the United States. This indicator would reflect the capacity of foreign firms to monitor and absorb indigenous technological capacity, in particular in basic research. Similarly, the importance of the foreign technological activities of nationally owned firms as measured by patents would reflect a country's capacity to benefit, through the foreign subsidiaries of its MNEs, from the results of research undertaken in other countries.

The first column in Table 51 compares the proportion of each country's US patenting originating from the subsidiaries of foreign MNEs. The authors claim that, on the basis of this particular indicator, large foreign firms still appear to play a relatively small role in national technological activities; only in Canada and the United Kingdom do they account through their affiliates for more than 20 per cent of total national patenting in the United States; Belgium stands out as a special case. The second column shows the percentage of US patenting by the foreign subsidiaries of nationally controlled firms as a proportion of total patenting by their country in the US. As might be expected, it shows that large firms based in small countries undertake a higher proportion of their technological activities, as measured by patenting, outside their home countries. Patents obtained by firms through foreign affiliate laboratories amount to more than 70 per cent of the national total for the Netherlands, more than 25 per cent for Switzerland and more than 16 per cent for Sweden.

Taken together, the two measures indicate a wide range of situations. In Belgium and Canada, foreign-controlled domestic technological activities are much greater than domestically controlled foreign technological activities, whereas for the Netherlands and Sweden, the situation is reversed. Contrary to what the authors seem to suggest, given the size of the United Kingdom's technology base and the long tradition of UK science, the two percentages shown for the United Kingdom indicate a high degree of internationalisation in R&D. When Europe is considered as a whole, the degree of internationalisation of R&D is lower than for most of the European countries taken individually, but still higher than for either Japan or the United States.

The Pavitt and Patel data is *output* data. It is subject to the limitations inherent in the patent indicator as well as in the procedure used for consolidating the patents granted to corporate groups[6]. The study confirms that many corporations, with a growing number of exceptions – significant ones in the case of Dutch, Swiss and UK firms – continue to locate their strategic R&D in laboratories in the home country. However, use of the patent indicator has to be completed by the information available about the ways MNEs are sourcing knowledge inputs from basic or university science globally through a variety of international industry-university relationships and transnational inter-firm agreements.

Inter-firm technological alliances in the setting of globalisation

The overall growth of inter-firm agreements bearing on technology, whether domestic or international, was discussed in Chapter 3 (see in particular Table 7 and Figure 5), along with the factors that account for this growth. Here the focus is on international agreements which involve large firms from the three poles of the Triad: the United States, Japan and Europe. Yet another term, "techno-globalism", has been coined in reference to this triadic technology networking (Ostry, 1991; OECD, 1991*h*).

To some degree, of course, international inter-firm technological agreements existed in the past. International cross-licensing between large corporations, which remains a fairly basic form of technical co-operation agreement in some industries, was already a significant feature of the chemical and heavy electrical equipment industries in the 1920s and 1930s. The 1930s also witnessed the establishment of at least one large and very effective international research consortium among a number of major oil companies (Freeman, 1982). Similarly, joint ventures have long existed and became increasingly important from the late 1960s onwards. However, the recent wave of inter-firm agreements has involved a much wider and more flexible range of arrangements (in particular non-equity

Table 51. **Foreign-controlled domestic technology compared to nationally-controlled foreign technology**

Based on US patenting, 1981-86

Home country	US patenting from domestic based foreign subsidiaries (as % of country's total US patenting)	US patenting by foreign subsidiaries of domestic firms (as % of country's total US patenting)
Belgium	45.7	16.5
France	11.8	3.8
Germany	11.5	8.5
Italy	11.2	3.0
Netherlands	9.5	73.4
Sweden	5.4	16.7
Switzerland	12.5	27.8
United Kingdom	22.3	24.5
Europe (average)	7.4	9.3
Canada	28.1	12.5
Japan	1.2	0.5
United States	4.2	4.4

Source: Patel and Pavitt (1990).

agreements, agreements formed for a single project, and corporate venture capital agreements between large corporations and small high-tech firms). Even authors like Porter and Fuller, who do not see a significant recent increase in actual numbers, stress the novel or "strategic" features of contemporary coalitions or alliances and their qualitative departure from earlier models of inter-firm co-operation (Porter and Fuller, 1986). In the face of recurrent financial and economic turbulence and of rapid and radical technological change, the new forms of agreements offer ways of ensuring a high degree of operational flexibility, which is particularly valuable for international operations.

Studies by business management specialists have identified some other distinctive features of alliances formed by the modern MNE. As often as not, the partners are of equal size and stature; they are usually firms with complementary assets, preferably operating in different market segments but drawing on common generic technologies. The paramount motives are generally the need to protect or increase the core assets of the enterprises – be they technology, organisation, management or entrepreneurship (or a mixture of these) – and to gain or improve access to global markets. Coalitions may also be prompted by the need to pool resources to meet rising R&D costs, to reduce risk or to strengthen the competitive position of a group of firms against their rivals. In some contexts, efforts by leading firms to maintain or regain international competitiveness within the Triad, or possibly to establish oligopolistic technological entry barriers, can now only take place through collective effort. Thus, the largest firms in the electronics industry – the VLSI project in Japan, the Micro-electronics and Computer Technology Corporation (MCC) in the United States and ESPRIT in Europe – have all been involved in large nationally or regionally based joint agreements for creating and sharing R&D and technology. These agreements have been aimed at improving international competitiveness but may have simply increased the oligopolistic character of supply (Mytelka, 1991a).

Inter-firm agreements involving the world's largest firms in important industries or product groups, and their implications for national systems of innovation, must be set in the context of the present trend towards high levels of world concentration. These agreements between otherwise competing firms must be examined from the standpoint of their effects on entry barriers and on access to technology by firms which are not parties to them. Control over best practice technology by leading firms in an industry creates an industrial barrier to entry by newcomers, as do the classic factors identified by authors such as Bain (1968), Blair (1972), Scherer (1970) and others, e.g. economies of scale, level of marketing expenditures and other absolute cost advantages, etc.

Contemporary innovation theory stresses the importance of considering "appropriability regimes", i.e. the degree to which the core competences of firms can be protected (ranging from "tight" regimes, where technology can be successfully shielded against imitations, to very "weak" regimes, where it is almost impossible to protect, see Chapters 1 and 2). Periods of rapid and paradigmatic technological change will often involve, simultaneously, a destruction of the previously dominant technological design, a strengthening of the appropriability regime in favour of the most innovative firms and the creation of transitory opportunities for market entry. Other components of entry barriers, such as the level of sunk costs (Gaffard, 1990a and 1990b; Flamm, 1990; Baumol and Lee, 1991) or a strong hold over distribution networks (as in pharmaceuticals), may offer oligopolists a considerable degree of protection against would-be entrants, in particular smaller domestic firms, but they are still confronted by a strong challenge.

The tendency of new generic technologies to modify, blur and eventually break down previously well-established boundaries between industries creates particularly strong pressure on firms to co-operate with oligopolistic rivals. The impact of generic technologies means that in key instances, if barriers to entry are to be effective, they will no longer be erected around closely defined and possibly quite narrow industries or product groups. They will have to be built around *clusters* of technology-related industrial and/or service activities and product groups, organised around vital basic core technologies (notably within the area of micro-electronics and biotechnology) (LARA/GEST, 1985). A step in this direction is the shift in the structure and composition of corporate assets within large firms and the emergence of a more technology-based corporate strategy, i.e. what Mytelka (1984) calls the move from "product-based" to "knowledge-based oligopolies".

This hypothesis has been considerably strengthened by the work undertaken at MERIT in Maastricht (Hagedoorn and Shakenraad, 1990a and 1990b) on the structure of strategic alliances in core technology industries[7]. Applying fairly sophisticated analytical techniques (including non-metric multidimensional scaling and cluster analysis) to the 2 700 agreements in the MERIT data base, scholars have been able to identify clusters of alliances between oligopolistic rivals and to study their evolution over time, with respect to density and to the degree of stability of partnerships. The overall results of their analysis of the IT industries, both at an overall level and on the basis of a more detailed five sub-sector breakdown, are given in Tables 52 and 53.

In order to identify elements of structural centrality in cluster-type networks, the MERIT group also computed a network density index. This density index is defined as the ratio of the actual number of links between companies (k) to the possible number of links n(n-1)/2 where n denotes the number of points in

Table 52. **Result of cluster analysis for information technologies in 1980-84 and 1985-89**

	1980-84	1985-89
Information technologies	Three clusters: (Europe, US, Japan)	Basically the same pattern as in 1980 and 1984
Computers	Five clusters: – TRW, AMP, etc., group including Sperry, Toshiba, Mitsubishi, IBM – Cluster around CDC Smaller groups with japanese firms in the centre – Fujitsu – NEC – Matsushita, Oki, Kanematsu	Seven clusters: – Honeywell Bull, NEC, Hitachi – Fujitsu and STC – Olivetti (with Stratus, etc.) – Group CDC (with Elbit, Cray) – Philips, IRI, GEC, Nixdorf – AT&T, Unisys, Toshiba, Mitsubishi – Cogent, Sequent, Nsube, Float PS
Industrial automation	Several small groups: – GM, Westinghouse, etc. – Dainichi Kiko, etc. – Siemens, Fanuc, etc. – Bosh and Shoun – Olivetti and Allen Bradley, etc. – IBM group – Cincinnatti Milacron cluster – Yascava, CGE, Toshiba	Relative large groups: – Rockwell (with DEC, Koreans, Olivetti, Fiat, Honeywell) – Group GM (with Fanuc, Fujitsu, Tandem) – Shoun Toyota, Nippon Autom.
Micro-electronics	Two main clusters: – INTEL, Fujitsu, Matra-Harris, NEC, Siemens, AMD, IBM – Motorola, Thomson, Philips, Matsushita, Schlumberger, Hitachi	Three main clusters: – An combination of the INTEL/Motorola groups from the previous plus Toshiba – A group with AT&T (and SUN, Samsung, Lucky, CTNE, Olivetti) – The HP, RCA, Kodak, TI, GE, IBM, etc. consortium
Software	Some rather separate groups: – US Consortium with Sperry, NC, DEC, etc. – European firms (Philips, Siemens, STC, Bull) – Microsoft, IBM, Olivetti – Hitachi, Telex, etc. – Digital Research, Exxon, Gould, NEC	Four main clusters: – A US aerospace group (with Boeing, Northrop, etc.) – A US group with Unisys, DEC, NCR, etc. (same group as a previous period) – A Japanese cluster – US/Europe group with several subgroups: Volmac, CAP-Gemini, Sema – Microsoft, IBM, Olivetti – AT&T, SUN, – Philips, BSO, Nixdorf, Bull, STC
Telecom	Some largely national groups: – Japanese firms (plus IRI, Rolm, Mitel, IBM, GTE) – German/Scandinavian cluster, plus ITT – AT&T, plus Philips and Olivetti – UK firms (Racal, Plessey, GEC, British Telecom) – DEC, XEROX, INTEL	More international groups: – IRI, AT&T, Toshiba, Olivetti – Siemens, IBM, Philips, Fiat, Bosh, CGE, Daimler, ITT – Racal, Matra Ericsson, GEC, Plessey – NT, DEC, BT, STC – Some "baby" Bells – Japanese firms and Bellcore, GTE and C&W

Source: Hagedoorn and Schakenraad (1990*a*).

Table 53. A comparison of the top ten firms in information technology by number of strategic links in 1980-84 and 1985-89

| | Information Technologies | | | | | Computers | | | |
	1980-84		1985-89			1980-84		1985-89	
1.	Motorola	53	Siemens	134	1.	Sperry	27	Olivetti	22
2.	Siemens	51	Philips	127	2.	IBM	19	CDC	19
3.	IBM	48	Olivetti	110	3.	CDC	18	Unisys	17
4.	Sperry	47	IBM	108	4.	Olivetti	17	Bull	14
5.	Fujitsu	46	HP	96	5.	Fujitsu	15	Philips	13
6.	Olivetti	42	DEC	95	6.	NEC	12	Fujitsu	12
7.	CDC	41	AT&T	90	7.	Burroughs	11	NEC	12
8.	INTEL	41	Thomson	83	8.	Toshiba	10	SUN-Micr.	11
9.	Philips	40	Fujitsu	78	9.	Du Pont	10	DEC	10
10.	NEC	39	Motorola	68	10.	3M	10	Hitachi	10

| | Industrial Automation | | | | | Micro-electronics | | | |
	1980-84		1985-89			1980-84		1985-89	
1.	GM	8	GM	20	1.	INTEL	34	Thomson	51
2.	Mitsubishi	8	IBM	20	2.	Motorola	23	INTEL	46
3.	Dainichi	6	ABB	13	3.	Philips	20	AMD	42
4.	Siemens	6	Dainichi	11	4.	Thomson	19	Motorola	40
5.	Westinghouse	6	Tandem	11	5.	Toshiba	18	Philips	39
6.	ACME-C	5	FANUC	10	6.	Siemens	17	TI	37
7.	Asea	5	Rockwell	10	7.	Fujitsu	16	Siemens	36
8.	Daimler	5	Siemens	10	8.	NEC	16	IBM	30
9.	FANUC	5	Westinghouse	10	9.	Exxon	15	Toshiba	27
10.	IBM	4	C. Milacron	9	10.	AMD	14	AT&T	26

| | Software | | | | | Telecommunications | | | |
	1980-84		1985-89			1980-84		1985-89	
1.	CDC	18	HP	47	1.	Siemens	17	Siemens	45
2.	NCR	16	DEC	45	2.	AT&T	15	CGE	32
3.	Honeywell	14	Siemens	36	3.	ITT	14	Sumitomo	29
4.	Motorola	14	Bull	34	4.	Fujitsu	10	Mitsubishi	28
5.	HP	13	AT&T	33	5.	IBM	10	Fujitsu	27
6.	Speery	13	Philips	32	6.	Plessey	10	AT&T	26
7.	Allied	12	SUN-Micr.	31	7.	Hitachi	9	Philips	26
8.	AMD	12	NCR	29	8.	ANT	8	IBM	24
9.	DEC	12	Volmac	29	9.	NEC	8	NEC	23
10.	Harris	12	Olivetti	28	10.	Olivetti	8	Ericsson	20

Source: MERIT, CATI database.

the network. The network density in information technologies moved from 23 per cent in the first half of the 1980s to 40 per cent in the second half. This means that in the years since 1985, 40 per cent of all theoretically possible links between the 45 firms are empirically observed. In other words, as Hagedoorn stresses, one can speak of a very dense network in information technologies, which, of course, is also partly due to the broad scope and interrelatedness of fields of information technologies.

Figure 33 reveals a set of fairly distinct clusters. To improve readability, lines have been drawn between firms whose proximity exceeds a given threshold value: thick solid lines indicate very strong co-operation (seven co-operative agreements or more). In the 1980-1985 period, some largely national clusters can be identified (Figure 33a). The first cluster consists of US semiconductor and computer manufacturers, the second contains some European firms, and the third one is a Japanese cluster. The figure also shows that during the first half of the 1980s there were a number of corporations with rather strong interrelationships. For example: the networks around CDC, Sperry, Intel, Motorola, Siemens, NEC, Fujitsu and Philips. All these firms ranked among the top 10 corporations engaged in alliances in IT in that period (see Table 53). In the second half of the 1980s (Figure 33b) a significant increase in the density of networks as well as in the number of agreement between major firms can be observed. This is true both of national and

Figure 33a. **The structure of strategic partnering in information technologies, 1980-84**

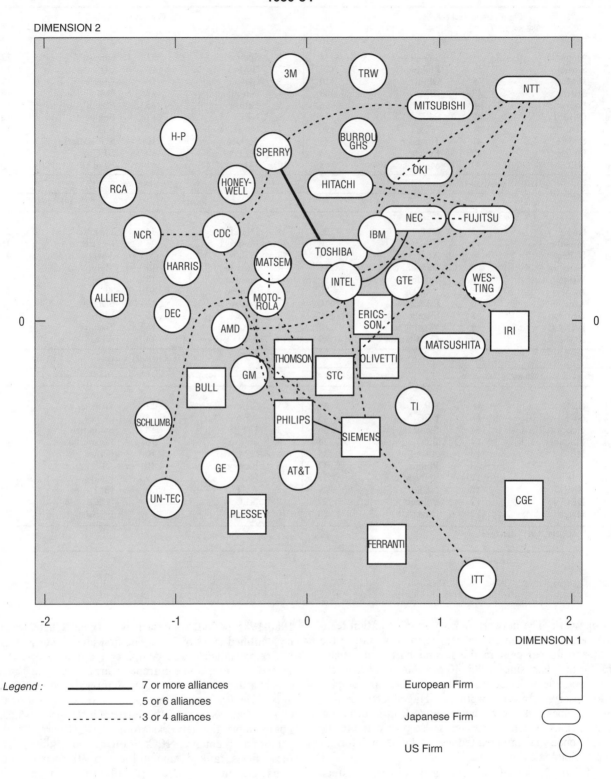

Source: Hagedoorn and Schakenraad (1990a).

Figure 33*b*. **The structure of strategic partnering in information technologies, 1985-89**

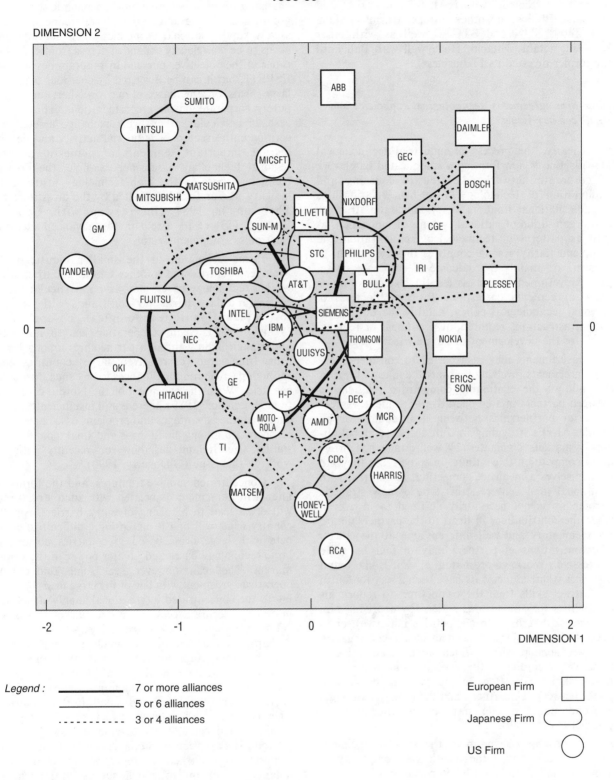

DIMENSION 2

0

-2   -1   0   1   2

DIMENSION 1

Legend : ──────── 7 or more alliances
          ───────── 5 or 6 alliances
          ········· 3 or 4 alliances

European Firm ☐

Japanese Firm ⬭

US Firm ◯

Source : Hagedoorn and Schakenraad (1990*a*).

regional (e.g. EC) groupings, but also to a lesser extent of alliances between firms belonging to different poles of the Triad (Siemens with IBM, Intel and DEC, and of course Philips and other European firms; Fujitsu with Intel, SUN and STC as well as with other Keiretsu, notably Hitachi; Honeywell with Bull until the former divested its IT business).

Inter-firm agreements, technological capacities and small country issues

Clearly, the process of industrial technological globalisation is now forming a complicated patchwork of corporate networks directed to sourcing and exploitation of technology. Inter-firm relationships straddle national frontiers. They affect the structure and organisation of national systems of innovation, as well as patterns of technological specialisation and complementarity among countries in ways which, at present, are mainly the outcome of corporate strategies. Governments must assess the extent to which these agreements correspond to the objectives of national technological policy, notably the building of the infrastructures required, at a national or regional level, for the development of the new technologies.

Governments may also be led to consider inter-firm agreements involving innovative domestic firms in the light of the continuing competitive relationship between partners in a co-operative agreement. Agreements or partnerships between near equals, or by smaller MNEs struggling for full access to global oligopoly markets dominated by well-established firms, allow competition to continue. In some cases they are in themselves a means of competing. In contrast with traditional joint ventures, this type of new strategic partnership is not necessarily destined to last, even when the contributions of firms to the partnership are complementary and well differentiated at the outset. The motivations of partners may in fact be quite aggressive. Work co-ordinated at INSEAD found instances where agreements have been a way for a firm to "extract skills from the competitor; to reduce his capabilities for autonomous action (by depriving him of some skills critical to competitive initiative); or to make him increasingly dependent on the continuation of a partnership within which he is steadily losing ground". Accordingly, the central issue in strategic partnerships may often be the evolving balance of power between the partners, and the threat of strategic encroachment of one partner on the other(s) (Hamel et al., 1989).

Everything known about the structural dimensions of technology – the length and complexity of accumulation processes, the importance of use-producer relationships and feed-backs, the numerous synergies resulting from proximity (e.g. Route 128, Silicon Valley, etc.) – suggests that the development of regional technological development poles (De Bresson,

1986) is one of the main paths to successfully restructuring economies under the pressure of technological change. In Europe, where regional economic structures are increasingly marked by the interrelatedness of (relatively small) national economies, such development needs to be consciously expanded across frontiers. This is one of the objectives pursued by programmes such as ESPRIT, but it can be thwarted by decisions taken by large firms on the basis of narrow short-term and purely corporate business considerations. This is the case, in particular, when firms exchange assets concerning entire sectors of industrial activity and deeply disrupt important segments of domestic and/or regional technology bases. For example, the French technology base in the high technology instruments industry was profoundly affected by the divestiture by Thomson in 1987 (in a swap with General Electric, US) of its assets in the advanced medical electronics equipment sector.

Firms belonging to the smaller industrialised Member countries and to other OECD countries with weaker technology bases, may have a particular interest in, and face particular constraints with respect to, international technological co-operation (Walsh, 1987). Such co-operation has so far been undertaken mainly through partnerships among large firms from large (or relatively large) countries and through participation in alliances led by such firms and countries. With the exception of the Nordic countries (e.g. Hakansson, 1989), there is still insufficient experience with the advantages and problems associated with co-operation among small firms and small countries of similar size. Ireland has, however, recently initiated work on this issue (O'Doherty, 1990).

Concentrated supply structures, and the forms of inter-firm agreement associated with such structures, will often lead to a raising of entry barriers. Small-country firms will find it increasingly difficult to compete with large firms based in countries possessing large technology bases and greater power in the world market. They can, however, seek some kind of co-operative agreement with one of the large firms belonging to the concentrated international supply structure, in order to secure access to a wider market and/or to catch up in some areas of technology – in exchange, perhaps, for the sharing of technological expertise in another area (e.g. manufacturing technology in exchange for design expertise). One of the objectives of the EUREKA project is to facilitate this type of co-operation between large and small European firms. The chance a small-country firm has of achieving this is likely to depend on its bargaining power; this is related to its experience and the degree of dominance it exercises in its market, the appropriability of its technology, the extent to which successful innovation depends on complementary assets and whether the technology is characterised by a dominant design or by technological flux.

Globalisation is neither a mirage nor an over-stated idea. It is a portentous reality which has begun to give birth to a number of new policy issues. While some of these existed previously, they have been singularly sharpened by the developments which occurred during the 1980s.

The exclusion of developing countries from the globalisation process

One of the most disturbing phenomena of the 1980s was the trend shown by all the indicators most likely to reflect North-South transfer of technology – FDI, capital goods exports, receipts for royalties and fees as recorded by technological balance of payments – either to fall or to stagnate (see Chapter 12, Table 59). The indicator for which the most reliable and recent data exists is FDI. Figures recently published by the UNCTC show that while FDI has grown more rapidly than trade as noted above, this growth has been concentrated to a degree unprecedented in the history of Western economies among the advanced industrialised countries. Globalisation has taken place, but FDI has by-passed entire continents or large parts of them. The group of least developed countries *taken together* amounted only to 0.1 per cent of total world investment inflows and 0.7 per cent of inflows to all developing countries reaching barely $200 million in 1989. The marginalisation of a number of African countries has been particularly significant.

The evidence presented throughout this report makes it impossible to view this trend as a *cyclical* phenomenon, even one in which the downswing might be of long duration (Chesnais, 1990a). As argued by Lanvin: "Because globalisation is technology-driven and technology-focused, it tends quasi-naturally to perpetuate and reinforce the phenomena of integration and exclusion of most previous technology-based changes in the world economy" (1990a). Over the past decade, technology may have spread internationally at a more rapid pace than ever before. But the way in which it has spread has contributed to broaden even further the existing gap between countries and groups of countries with different levels of access to technology. If OECD countries and some Asian NIEs may be, despite a number of problems (see Chapter 12), on their way to increasing technological homogenisation, the rest of the world is increasingly excluded.

The structure of information networks (here, telematics infrastructures and conditions of access) and the issue of selective by-passing must be considered in this setting. Contrary to frequent misconceptions about the "information society", ITs which form the main technological basis of the globalisation pro-cess are not inherently user-friendly. In this instance, accessibiliy is dependent on organisation and so on the economic and political decisions which shape it. Expert opinion is that, in the current context of deregulation and privatisation, the most advanced telecommunications networks could generate "islands" of activity and growth and "semi-arid" or "desert zones" (Solomon, 1990).

As international competition tends to depend increasingly on information infrastructures, access to the network becomes of vital importance. Lanvin has argued that one of the clearest examples is furnished by INTELSAT, which currently provides two-thirds of international telecommunications traffic and almost all international television traffic. As stated by its charter, the purpose of INTELSAT is "to continue the development of the telecommunications satellite system with the view of achieving a single global commercial telecommunications satellite system as part of an improved global telecommunications network". INTELSAT averages the rates it charges over routes with heavy, medium and light traffic in order to afford access to all potential users of its global infrastructure (see Article V of the 1973 Agreement). It thus consti-tutes an important foothold into the global information economy for peripheral countries (Lanvin, 1990a).

However, such a structure can only survive if it is shielded from the type of pricing policy which has emerged as a result of deregulation and privatisation. At the world level, the currently dominant approach would logically drive the providers of international telecommunications towards the heavy routes (where economies of scale can easily be reaped), leaving INTELSAT with the less profitable (or even costly) routes. The emergence of new private providers of international telecommunications systems might thus lower average prices for international telecommunica-tions, and even lower them in a spectacular manner on the most densely used routes, while making access to satellite-based networks more expensive for developing countries. Issues dealing specifically with NIEs and developing countries will be treated at greater length in Chapter 12, but this example serves as a reminder that the overall setting for that discussion is the analysis of globalisation presented in this chapter.

Globalisation and national systems of innovation

Some of the implications of globalisation for national systems of innovation have been discussed in relation to small countries and small-country firms in a world where inter-firm agreements among large corpo-rations have become increasingly frequent. Other dimensions require careful attention.

The scale of the international merger and acquisition process was noted above. Innovation systems are rooted in production systems (Andersen and Lundvall, 1988). Vital components are embedded in firms (see Boxes 20 and 21, Chapter 3). During the previous phase of internationalisation it was recognised that through foreign direct investment MNEs could create but also destroy industrial R&D facilities and other innovation-related national capacities. The stress laid in this report on learning processes and complex user-producer relationships involving economic and geographical proximity suggests that careful attention must now be paid to the effects of merger and acquisition operations on the innovation system in its widest sense, e.g. not only on formal R&D, but also on the vital tacit components of design, engineering and production located knowledge, which are embodied in firms and can easily be destroyed through "savage" finance-led merger operations.

Globalisation and concentrated international supply structures

Investments and disinvestments of formal business enterprise R&D facilities and shifts in the international location of corporate R&D must continue to be the subject of careful analysis, but these shifts must now be analysed in the context of the new international supply structures.

The data set out earlier in this chapter demonstrate the need to begin assessing the extent and possible implications of the highly concentrated international supply structures which are tantamount to world oligopoly and which may carry risks of leading to cartelisation (see the conclusion to Chapter 8). A first step is to collect appropriate data. A starting point requires assuming that in certain industries or product groups there exists, basically, a single globalised industry or product group with a single globalised supply structure. Concentration ratios should be calculated for these, however approximate the initial calculations may be.

It has been suggested (Soete, 1991; Ostry, 1991) that highly concentrated supply structures could have serious consequences for international competition in situations where: i) the products concerned represent critical intermediate inputs to other production (e.g. semiconductors and other micro-electronic products); and ii) concentrated supply structures coincide with the nationality of firms. In such a hypothesis, which has ceased to be simply academic, a situation could emerge where an exploitable advantage for a foreign firm or another country could have serious, widespread and long-term consequences. The risk would be especially high if the foreign supplier were a monopoly or a cartel, and if high sunk costs reduced the credible threat of entry (Flamm, 1990,

pp. 225-292). Consequently, a major policy issue facing the OECD is whether competition policies at domestic or EC levels are now enough and whether the need is not for global competition policies. The existence of a supra-national form of competition policy aimed at counteracting the emergence of world-wide cartels among global firms would directly undermine the reason just examined for strategic government support, namely dependence on "foreign" monopoly pricing. But the establishment and implementation of a global competition policy is obviously a formidable task.

Government support to domestic firms engaged in global competition

Global competition is not waged exclusively among firms; it often involves some kinds of participation by governments. As observed in a notable contribution to the Tokyo TEP Seminar, governments are concerned about the competitiveness of their economies and of their corporations and, to a large degree, "they still equate the two" (Ostry, 1990b). Consequently, they may be led to support their firms through a range of measures and instruments which have been characterised as representing a novel brand of "offensive high-technology protectionism" (Ostry, 1990b; and Ernst and O'Connor, 1989). As Ostry argues, "the threat to multilateralism stems not from a replay of the 1930s, but from the absence of rules to mitigate and contain these new pressures. In the absence of *rules* a system based on *power* is the only alternative. Such a system, which could quite likely be a form of negotiated managed trade within the Triad of the United States, Japan and the European Community, would result in continuing instability and uncertainty" (1990b, p. 5).

These are not new issues for the OECD. The use of a variety of government instruments in order to gain advantages in trade in high-technology products was examined in a special study requested by the Council in 1982 and finished in 1985. During the discussions, a fairly long list of issues were identified (see Box 43).

While some of the issues may have been slightly resolved and others have lessened in intensity, on the whole this list probably remains pertinent. Overall, only slow progress was made on these issues in the 1980s. One reason for this, insufficiently recognised then but now much clearer, relates to the blurring of the earlier "clear" distinction between science and technology and so of the "public good" area where government support and funding are considered legitimate, as opposed to the "close-to-market" technology where they are not (see Chapter 1). When the frontiers between science and technology are blurred and innovation is seen as an interactive process, it becomes increasingly artificial to set boundaries between "pre-competitive" and competitive R&D. This makes rules yet more necessary but even harder to establish.

Papers prepared for the Tokyo Seminar point to a list of trade, investment and technology-related policy instruments used to support domestic firms which is almost identical to the list established at the end of the OECD study on trade-related issues in high technology products. New proposals proposed by Ostry for work in this area would benefit from the earlier experience. If work is undertaken again, the OECD appears to be the most adequate forum for a new round of studies which would build on the experience of the 1982-85 work and on the current assessment of globalisation and its implications (Hufbauer, 1989).

Standards in the context of globalisation

Given the role played in the process of globalisation by telematics and the truly global nature of the telecommunications system, this area has been chosen to discuss some issues regarding standardisation in the two-fold context of pervasive rapid technological change and globalisation. The strategic implications of IT standardisation are now broadly recognised to be enormous because they will determine the future of individual firms, affect the competitive advantages of countries, and even influence the development of whole technologies and their diffusion.

In terms of organisation and procedures, standardisation in information and communications technologies has become a highly complex process reflecting the extent and diversity of its strategic implications. Here, the standardisation process has become infinitely more complicated than in earlier periods as a result of three major developments: i) the cross-fertilisation between IT and other technologies (in particular, but not exclusively, telecommunications), which has brought an ever-widening circle of standardisation agencies to be concerned about this area; ii) the growth of international and regional IT standardisation bodies, which has added another institutional layer to the international system; and iii) the recognition of the central importance of the testing and certification stage, which has led to the creation of new standardisation bodies.

Since formal procedures are slow, both inherently and because of increased complexity, standardisation is increasingly prepared at national and international levels by informal deliberations. Very often, consensus between the major actors (such as manufacturers, service providers, large industrial and service users) is reached at the informal stage. Increasingly often, the resulting standard is actually implemented before a formal decision is made. This makes it very difficult for participants in the final phases to influence the outcome if they have not been involved from the start either because they are not invited to participate in, or because they are not aware of, the informal stages. In particular, some governments may not be aware of the significance or even of the existence of these informal discussions before they are over.

The scale of R&D coupled with the irreversibility of subsequent large scale investments and the increasing returns reaped by the systems adopted early on (see Chapters 1 and 2) are now also favouring the establishment of "anticipatory standards", developed during the R&D phase, thus largely pre-empting competition between alternative technologies. The blurring of the frontiers between R&D and standardisation activities now calls for a new generation of committees and organisations at national and international levels, at the interface of R&D and standardisation. Anticipatory standards will also contribute to removing users' representatives from the standardisation process, since by its very nature an anticipatory standard will anticipate the needs, the demands and even the existence of users who do not yet exist.

The ways in which this contradiction between the required participation of users and their *de facto* absence will be solved will no doubt be decisive for the future of standardisation in the IT area. It points to a new two-fold role for governments in standard setting, as potential clients for new technologies and as representatives of the public good and the interests of future users.

A recent OECD report concludes that "governments may legitimately be expected to become more directly involved in the IT standardisation process, for a number of reasons ranging from the representation of the interests of users who are not able to be involved themselves, to the promotion of a standard which might be of public interest, for example because it may ensure greater compatibility and inter-operability between equipments" (OECD, 1991*b*, p. 7). The capacity of the larger governments to enter in the standardisation process seems reasonably well ensured. For smaller countries and developing countries this is not the case, and the issue of how their interests could be represented in the process remains unresolved.

NOTES

1. Some useful cautions concerning the dangers of extrapolating the M&A and FDI trends of the 1980s into the future are made in du Granrut (1990).

2. For a discussion of these methodological problems in TEP Conferences, see Lanvin (1990*b*); and Sauviat (1990).

3. Early uses of the notion of "world oligopoly" or "international oligopoly" can be found in Vernon (1974); Cotta (1978); Chesnais (1982) and Dosi (1984*a*).

4. For an interesting discussion of this point, see Newfarmer (1985).

5. A measurement of the share in production and sales of the first 4, 8 and 20 leading firms has since the 1930s been considered as providing an initial approximation as to the dominant form of competition on industry or product group. Most authors have taken the view that where the share of the four biggest firms in production, sales or turnover of an industry group or product class is less that 25 per cent, competition still exists in a relative, if imperfect, state, in the sense that one firm or group of firms is not an established power on the market for any length of time. When the four largest firms hold more than this 25 per cent share, oligopoly begins. Between 25 and 50 per cent, this oligopoly is loose and unstable, but beyond that point, it becomes firm and clearly established. See in particular, Blair (1972) who analyses the main conclusions of the United States Senate Subcommittee on Anti-trust and Monopoly, of which he was secretary for a long period. Kaysen and Turner (1971) define the three types of situation as follows: "unconcentrated oligopolies, type II oligopolies, type I oligopolies".

Bain (1968) proposes a classification of *very highly concentrated oligopolies* where the top 8 firms control over 90 per cent of the market and the top four 65-75 per cent; *highly concentrated oligopolies* where they control 85-90 per cent and 60-65 per cent respectively; *high to moderate concentrated oligopolies,* where corresponding control is 70-85 per cent and 50-65 per cent.

6. The procedure used for consolidating the patents granted to corporate groups seriously weakens the findings on patent data regarding the extent and pace of internationalisation. Regular changes occur in ownership of firms through mergers and acquisitions. To make the analysis manageable, authors identify corporate groups at a particular point in time. The patent data for the entire period under survey are then consolidated on this basis. No changes in ownership links during the period are then allowed for. This means that when the internationalisation of R&D is achieved through acquisition, this is not recorded as a change in the geographical composition of the firm's technological development since the affiliate has been considered as part of the corporate group at both the beginning and the end of the period. The true extent of any trend over time towards the internationalisation of technological activity is likely to be understated more or less seriously. For instance, a study of the foreign location research of Swedish firms conducted by Hakanson (1981) suggested that around 60 per cent of the personnel they employed in foreign R&D facilities worked for affiliates which had been acquired by the parent firm.

7. See also the important study by Caicarna *et al.*(1989) with particular focus on Italian corporations.

Chapter 11

TECHNOLOGY AND COMPETITIVENESS

ABSTRACT

In micro-economics, competitiveness refers to the capacity of firms to compete, to increase their profits and to grow. It is based on costs and prices, but more vitally on corporate organisation, the capacity of firms to use technology, and the quality and performance of products. At the macro-economic level the concept of competitiveness is more elusive. It may be defined as the degree to which, under open market conditions, a country can produce goods and services that meet the test of foreign competition while simultaneously maintaining and expanding domestic real income. Approaches to macro-economic competitiveness using cost and price indicators are now complemented by analyses of the factors underlying long-term shifts in market shares and changes in international trade patterns.

In this perspective the analysis of competitiveness calls for holistic approaches focusing on the relationships between firms and national environments. Numerous studies have shown that business strategies – both successful and unsuccessful – follow country and industry-specific patterns. The structural characteristics of a national economy or a particular domestic industry appear to exert a strong influence on firms' performance. This hypothesis is considerably reinforced by TEP analyses. The notions of "structural competitiveness" and of the "diamond of national competitive advantage" stress the part played, alongside the role of firms, by organisational and institutional factors relating to the interactive nature of the innovation process, the systemic features of technology, the learning processes associated with innovation and the proper use of human capital. The effects of military R&D on competitiveness can usefully be examined in this context. The competitiveness of firms in manufacturing is likewise seen to depend crucially on the quality of their interaction with the business service industries as well as on the performance of key infrastructure services.

The role played by technology in determining competitiveness at the micro-economic level and by national environments and their structural attributes at the macro-economic level open up totally new vistas on the foundations and the motor of international trade today. Absolute trade advantages in a range of technologically related industries can now occur as a consequence of the processes discussed in this report and may explain the existence of large quasi-permanent imbalances in manufacturing trade among industrialised countries.

Competitiveness can no longer be discussed independently of the trends examined in Chapter 10. Globalisation tends to reinforce the cumulative character of innovation-based competitive advantages for large firms, but it may be weakening the resource base and organisational cohesion of domestic systems of innovation. The competitiveness of large network firms now stems increasingly from their ability to make the best use of R&D and human capital resources located in several countries. This may allow virtuous circles of interaction and feedback between the accumulation of physical capital, the development and improvement of human capital, technological accumulation and the competitive performance of firms to occur at the level of specific industries or industry clusters and at specific business locations or "sites". But globalisation will tend to dislocate economy-wide networks and interactive mechanisms, thus creating a further important challenge for nations and governments.

INTRODUCTION

From the standpoint of analysis and policy, approaches to competitiveness are changing as a result of phenomena and processes examined during the TEP exercise and discussed throughout this report. Technology and other innovation-related phenomena, along with corporate organisation and the proper use of human capital in all the phases of the production process, now represent one of the main pillars of competitiveness. The other pillar is, of course, formed by price and cost determinants. An overall assessment suggests that "over the medium to long-term, factors related to technology and to productive capacity for delivery are empirically found to have played more important roles than cost or price competitiveness" (Yoshitomi, 1991, p. 22).

These features are not simply the attributes of individual firms, but also, to a large extent, those of national or local environments where organisational and institutional developments have produced conditions conducive to the growth of the interactive mechanisms on which innovation and the diffusion of technology are based. National environments conducive to competitiveness will also be shaped by the level and pattern of tangible and intangible investment within national economies. Competitiveness thus provides an important yardstick for assessing government policies with respect to the level and structure of public investment and the relative priority given in particular to investment in defence, education and infrastructure.

The structural dimensions of competitiveness and the role now played by technology have important implications for the interpretation of trade patterns in the 1980s. Trade among highly industrialised countries with strong technology bases is strongly supply-driven. Technology-based economies of scale and scope are open only to firms capable of gaining and preserving strong footholds in foreign markets. This provides, of course, an ever stronger rationale for preserving an open trading system. But it also explains why trade flows in manufactured products among industrialised countries can quite easily experience large and increasingly permanent imbalances over long periods of time and why the theorem of comparative advantage (in the *strict* sense given to it by trade theory) has come under heavy strain. Today absolute trade advantages, observable across a range of technologically-related industries which benefit from cumulative interactive processes and experience dynamic scale efficiencies, represent an identifiable outcome of the processes analysed in this report.

Historically, the notion of competitiveness has been associated with a relatively specific set of ideas and policies, in which technological and organisational factors are not given a central role. These earlier approaches are still the ones most familiar to many macro-economists. This chapter therefore begins by examining the earlier ideas and policies (Section 1), before going on to discuss structural competitiveness (Sections 2 and 3). In Section 3 the relationships between technology, competitiveness and trade are examined. Section 4 raises issues related to globalisation and the competitiveness of sites. Finally some analytical and policy-related conclusions are drawn.

1. THE NOTION OF COMPETITIVENESS: ANALYTICAL PROBLEMS AND APPROACHES

International competitiveness and the relationship between technology and competitiveness have been the object of considerable analysis and discussion since the 1960s. Some economists may judge that the area has been quite thoroughly studied and that everything useful has already been said. Others challenge the very foundations on which most earlier work was based.

A major difficulty arises from the fact that the notion of competitiveness originated in micro-economics and was later transferred, with some awkwardness, to the level of national economies. A further difficulty arises from the fact that macro-economic competitiveness has traditionally been viewed as being principally, if not exclusively, a question of prices, costs and exchange rates and so of appropriate macro-economic policy. Today, however, emphasis is placed increasingly on the macro-institutional and structural dimensions of national competitiveness.

Micro-economic or firm-level competitiveness

In micro-economics, the term "competitiveness" refers to the capacity of firms to compete and, on the basis of their success or "competitiveness", to gain market shares, increase their profits and grow. The factors which contribute to micro-economic competitiveness have long been a special concern of managerial and industrial economics. These disciplines use a wide range of indicators (market shares, profits, dividends, investment, etc.) to assess the competitiveness of firms. Corporate surveys and industrial case studies carried out over the last 20 years have found that:

i) in most industrial branches and sectors competitiveness cannot simply be viewed as centred on prices and the cost of inputs, notably labour inputs (e.g. wages and indirect labour costs)[1]; and

ii) a variety of non-price factors lead to differences in the productivity of labour and capital (scale economies, process systems, size of inventories, management, labour relations, etc.) and in the quality and performance of products[2].

These micro-economic components of competitiveness emerge reinforced from the analyses in earlier chapters of this study. Successful competition requires quality of products (as much as any inherent novelty, which competitors can often quickly imitate); superior process technology and a superior organisation of production; speed of delivery and quality of after-sales services. The competitiveness of firms today is largely shaped by the various aspects of corporate organisation (see Chapter 4) that command the effectiveness of industrial R&D and other innovation-related investments (see Chapters 1 and 5). At firm level, factors contributing to competitiveness thus include: the successful management of production flows and raw material and component stocks; the successful organisation of effective interactive integrating mechanisms between market planning, formal R&D, design, engineering and industrial manufacture; the capacity to blend in-house R&D and innovation-related activities with R&D co-operation with universities and other firms; the capacity to incorporate closer definitions of demand characteristics and the evolution of markets into design and production strategies; the capacity to organise successful inter-firm relationships with component and materials supplier firms upstream and with retailers downstream; and finally the steps taken by firms to enhance workers' and employees' skills through investments in vocational training as well as to establish greater degrees of worker responsibility in production.

Some remarks may be made regarding the strategies adopted by firms in relation to final demand and marketing and the implications that appropriate producer-consumer relationships may have for competitiveness. A recent study based on survey data and interviews with firms has focused on corporate performance with respect to commercialisation in the United States, Japan and Europe (Nevens, et al., 1990, p. 3). The study finds that critical differences between high-performing and low-performing companies were sustained over periods of several years and were as great in Japan as in America or Europe. It identifies a strong correlation between a corporation's competitiveness and its ability to commercialise technology. In many markets – such as copiers, facsimile machines, computers, automobiles, semiconductor production equipment and pharmaceuticals – industry leadership does not rest only on technology per se but also on superior commercialisation skill. In such R&D-intensive industries "companies that are first to market with products based on advanced technologies command higher margins and gain market shares. Companies that spin out variants more rapidly and leverage their core technologies across more markets earn higher returns" (Nevens, et al., 1990, p. 155). Other work suggests that the type of corporate competitive strategy now required for the full exploitation of technological potential also requires making a distinction between strategies based on products (often single products) and strategies based on a proper recognition of the generic features of key contemporary technologies and the exploitation of a firm's core technological competence. The latter allow higher degrees of flexibility in decisions concerning the successful marketing of a range of distinct but technically interconnected products (Giget, 1990).

Flexible production technology and the part now played by information processing in product design require a shift away from the traditional product-centered conception of manufacturing and marketing based principally on economics of scale (see Chapter 4). In the context of an "information-intensive production system" (Willinger and Zuscovitch, 1988), the competitiveness of firms rests increasingly on the capacity to exploit economies of scope and of variety. Once customisation ceases to be synonymous with handicraft, earlier product differentiation strategies will prove to be inadequate. As Bressand argues (1990b, p. 3), "Customising products to individual buyers' needs now plays an increasing role in the search for competitiveness. (...) The shift from traditional economies of scale to mass-customization implies that the management of commercial relationships, notably relationships with customers, becomes an essential aspect of the search for competitiveness" (see Box 44).

This is not an easy change for firms to make. Modern industrial societies have assigned a passive and reactive role to consumers in the innovation process (Gervasi, 1991). Initiative in the exchange of information regarding the design of products and the assessment of consumer preferences is in the hands of manufacturers and has been dealt with principally by consumer polls and advertising. In contrast with many inter-firm buyer-seller relationships in capital equipment and intermediary goods, the final consumer market has up to now been organised unilaterally by the producer side. Once the rate of consumer learning is recognised as being a long-run variable of structural growth, new relationships with consumers, along with a modified conception of what a "product" is, may be a requisite not only for micro-economic competitiveness but also for the overall performance of economies engaged in a process of structural change (Lundvall, 1991, p. 15).

Competitiveness in a macro-economic context

Transposing the notion of competitiveness from the sphere of micro-economics (i.e. that of the firm) to the macro-economic level (that of nations) is an enterprise marked by considerable difficulties and intellectual dangers. Many economists still hold to the position that "the analogy between the competitiveness of a firm in a domestic economy and the competitiveness of a country in the world economy, which is often the basis of popular theorising, is tenuous indeed" (Reidel, 1980). Agreement on appropriate indicators of macro-economic competitiveness is much harder to reach (Mathis *et al.*, 1988). Are they simply balance of trade

and balance of payments indicators, or do they include other key indicators of performance such as growth of GNP, the rate of inflation, the level of employment and social indicators? Are cost and price indexes meaningful indicators of competitiveness?

A widely held approach to the competitiveness of national economies views macro-economic competitiveness as being principally, if not exclusively, a question of prices, costs (in particular wages) and exchange rates. Relative unit labour cost (RULC) indices were for a long time considered as representing trustworthy indicators of competitiveness. Well into the 1980s, in fact, many economists took the position that international competitiveness was determined mainly, if not solely, by export prices and that these prices were mainly determined by unit industrial input costs, notably wages. They formed their policy recommendations on this basis. In terms of economic policy, this approach to international competitiveness led to:

i) measures concentrated essentially on wage costs and on labour productivity (and sometimes solely on wage costs); and

ii) the view that devaluation was a way to achieve gains in competitiveness. (The view was relatively widely held through the late 1970s; even after 1985, the policy directed to depreciating the dollar in real terms was still thought by some observers and policy makers to be capable *per se* of restoring US competitiveness).

In the late 1970s, as incongruous or "perverse" relations were repeatedly detected in empirical studies comparing the movement of cost and price indicators with the export performance of countries, the price and cost approach to macro-economic competitiveness began to be subjected to closer scrutiny. A particularly influential 1978 study by Kaldor was later followed by others (e.g. Kaldor, 1981; Kellman, 1983); for the United States and the United Kingdom they showed that, contrary to the prediction of accepted theory, a drop in relative unit wage costs and in export prices (as affected by exchange rates) had gone hand in hand with significant and totally "perverse" losses in world export market share for manufacturing. By contrast, in Japan and Germany, a rise in relative unit wage costs and export prices had occurred in parallel with an equally "perverse" increase in export shares.

Subsequently, a study of the US situation by the Brookings Institution which confirmed the Kaldor "paradox" triggered off an interesting debate among US experts. Some authors attempted to defend the validity of the traditional position by arguing that the German and Japanese performances could be due to i) cutting of profit margins by firms in export markets; ii) government export subsidies; iii) dual pricing for domestic and foreign markets; and iv) "dualism" in the drive for productivity by firms, with priority being given to export goods. Other authors pointed to the role of "unmeasured, non-price aspects of competitiveness", in particular "technological change, economies of scale and new markets, but also quality, service, financing and a better adaptation of products to foreign users" on the part of non-US firms[3].

The price/RULC approach has been the object of further empirical investigation. Using historical data for essentially the same period as Kaldor (1961-73), Fagerberg has developed a model of international competitiveness which relates growth in market shares to three sets of factors: technology, price and the ability to compete in delivery on the basis of adequate prior investment. Empirical testing of the model shows that Japan's large gains in market share during this period can be explained by a combination of a rapid increase in technological competitiveness, a further catching-up effect associated with the international diffusion of technology and a high level of investment.

More generally, the Fagerberg results show that "the main factors influencing differences in international competitiveness and growth across countries are technological competitiveness and the ability to compete on delivery. Regarding the latter, the findings point out the crucial role played by investment in creating new production capacity and exploiting the potential offered by diffusion processes and growth in national technological competitiveness. Cost-competitiveness does also affect competitiveness and growth to some extent, but less so than many seem to believe" (Fagerberg, 1988a, pp. 370-371).

Despite the findings of this and other research (Owen, 1989) the cost and price approach still forms the basis of much government policy for competitiveness, to the detriment of policies based on a better understanding of the role of technology, investment and organisational change. This is so, despite the many methodological problems entailed in constructing adequate cost and price indicators[4] and despite the fact that in many manufacturing industries wages now represent only a relatively small proportion of total costs (often below 25 per cent and in some cases below 20 per cent).

Transformations in the world monetary and financial system (the disappearance of fixed exchange rates, the establishment of the European monetary system, etc.) mean that changes in exchange rates as a policy instrument to improve competitiveness are much harder to use than in earlier days. It is also unlikely to have more than a short, passing effect on competitiveness. On the contrary, long-run trends in relative exchange rates will ultimately be explained by the competitive position of countries. This form of adjustment will, of course, only be gradual. In many trade models, imbalances are rapidly eliminated through exchange rate adjustments (or, if exchange rates are fixed, by domestic income or price adjustments).

This approach is also losing ground. Today the emerging view is that "exchange rate movements or changes in the terms of trade will follow in the wake of trade imbalances, but with a fairly long lag which may often leave a country room for a sustained improvement in competitiveness so long as domestic productivity growth is pursued. Countries which have a large number of industries in which they have a relatively high rate of innovation by international standards tend to experience a systematic appreciation of their currencies over long periods, while less innovative countries witness persistent trade deficits and long-term currency depreciation" (Cantwell, 1989, p. 181).

Many research groups have grappled with these issues. They have sought to establish a definition of competitiveness which would reflect the findings of the Kaldor "paradox" and express, however imperfectly, the fact that competitiveness for a country must be reflected *simultaneously* in growth in income, levels of employment as high as its direct competitors, and an acceptable balance-of-payments situation. Box 45 sets out a working definition of competitiveness compatible with this view.

National environments and structural competitiveness

Dissatisfaction with earlier approaches to the macro-economic dimensions of competitiveness does not imply that the organisation of national economies is not important[5]. On the contrary, as numerous recent references to "national competitive environments" attest, national economies and their institutional set-up do matter. Some authors even suggest that since "the competitiveness of the firm depends not only on its *own* competitive strength but also on the *interaction* of its capabilities with the capabilities of the external environment in which it operates, competitiveness among

Table 54. **Factors accounting for changes in market shares, 1961-73**

	Japan	United Kingdom	United States
Growth in market share (value predicted)(1)+(2)+(3)	103.3	−16.2	−29.8
(1) Technology (TG)	66.9	6.9	−0.6
(2) Costs (RULC)	−0.9	0.8	1.6
of which :			
• through volume (ME)	−7.8	7.9	12.4
• effects on terms of trade (TERMS)	6.9	−7.1	−10.8
(3) Delivery (total)	37.4	−23.9	−30.8
of which due to:			
• diffusion (TI)	20.9	15.9	7.3
• investment and demand (INV, W)	16.5	−39.8	−38.2

Source: Fagerberg (1988a).

firms is also competition among systems" (Ostry, 1990*a*, p. 4). This can be interpreted in several ways. As far as TEP is concerned, findings point to the competitive trade advantages (see Section 3 below) which accrue to countries as a result of the factors discussed in earlier chapters. These include the quality of inter-firm technical co-operation, user-producer relationships and sub-contracting linkages and relationships; the level of tangible and intangible investment; and the nature and quality of the interfaces and support that firms receive from public bodies and institutions at regional, local and national levels, in areas such as infrastructures, the supply of trained personnel and R&D.

Recognition of the systemic or structural characteristics of competitiveness occurred in several steps. One of the earliest comprehensive statements of the new approach was given by a 1981-84 OECD study on Science, Technology and Competitiveness[6] which identified the competitiveness of national economies as being something more than the collective or "average" competitiveness of its firms. Industrial case studies often showed – as indeed did the one launched by the OECD on the machine-tool industry (Sciberras and Payne, 1985) – that successful and unsuccessful business strategies followed country-specific patterns. On the other hand, work by French economists belonging to the "*école de la régulation*"[7] reached the conclusion that "competitiveness is the expression of a global property (both micro and macro-economic) specific to each national economy – the efficiency with which each country mobilises its factorial resources and, in so doing, modifies the technical and social characteristics of industrial activity" (Mistral, 1983, p. 2). This suggests that the structural characteristics of a domestic economy influence the performance of firms (see Box 46). Consequently, the OECD study put forward the notion of "structural competitiveness" as a way of reflecting the fact that, while the competitiveness of firms will obviously reflect successful management practice by entrepreneurs or corporate executives, their competitiveness will also stem from country-specific long-term trends in the strength and efficiency of a national economy's productive structure, its technical infrastructure and other externalities on which firms build.

The study subsequently carried out by the MIT Commission on Industrial Productivity, presented at two TEP-related conferences[8], also adopts, broadly speaking, a holistic approach. It relates the problems of US industrial productivity and so of US competitiveness to "a set of organisational patterns and attitudes", which it designates as "systemic rigidities". The implication is that "change, if it is to occur, will have to take place on a broad front, involving firms, government, educational institutions and organised labour". The MIT study pays particular attention to six sorts of deficiencies: outdated corporate strategies; short corporate horizons; weaknesses in technological development and production within corporations; neglect of human resources within US society as a whole and by firms in particular; failure to co-operate within firms and between firms; a cross-purpose relationship between government and industry.

The MIT study differs from the OECD study in that it focuses closely on corporate behaviour and examines the wider systemic features through their impact on the attitudes and strategies of corporate management. For this reason, the MIT study is less clearly holistic than the OECD work, and it has been criticised for only presenting a "shopping list" (Reich, 1990). If by this is meant that it does not establish an overall ranking and still less one on an industry by industry basis, the reproach is founded. Otherwise, it is an unfair assessment, all the more unfortunate because it could deter potential readers of a study which has a number of merits. It is one of the first reports with high intellectual and political visibility to show the need to shift the focus of analysis from the cost of capital to a study of the ways in which corporations are financed and the *nature of the institutions and arrangements that influence capital supply,* i.e. what is sometimes called the "micro-structure of capital markets", a point to be returned to below.

Porter's study of the competitive advantage of nations, carried out over the same period as the TEP, explicitly takes a holistic approach and a position on competitiveness that in many ways converges with the view taken in the TEP (1990b, p. xiii). The author's starting point recalls that of the OECD study, namely, the fact that despite the importance of management and appropriate corporate strategies, "national environment seems to play a central role in the competitive success of firms. With striking regularity, firms from one or two nations achieve disproportionate world-wide success in particular industries. Some national environments seem more stimulating to advancement and progress than others" (p. xii).

The Porter study focuses on the analysis of competitive advantage and successful environments at the level of industries and industry clusters: "nations succeed not in isolated industries, however, but in clusters of industries connected through vertical and horizontal relationships. A nation's economy contains a mix of clusters, whose make-up and sources of competitive advantage (or disadvantage) reflect the state of the economy's development" (1990b, p. 73). In turn, a nation apparently achieves international success in a particular industry by reason of four characteristics that shape the national environment in which local firms compete and that promote or impede the creation of competitive advantage: factor conditions; demand conditions; related and supporting industries; and firm strategy, structure and rivalry (see Box 47).

2. FURTHER DIMENSIONS OF STRUCTURAL COMPETITIVENESS

The work carried out in the TEP has helped to highlight two related characteristics of national environment and their competitive attributes. First, although "national environments" are shaped by historical developments and depend to some extent on government policies, this does not mean that firms do not have an important responsibility in building them. While firms may be tempted to view the national environment as an "external" given, this environment is in fact strongly affected by the collective views and decisions of the business community and the management of firms (Hollingsworth, 1990). Second, national environments will furnish a climate conducive to competitiveness to the extent that co-operation, in various forms, is recognised and valued by a nation's people in general and by firms in particular as representing a specific source of increasing returns (Ferguson, 1991).

Three specific dimensions of structural competitiveness were identified in the course of the TEP as being important and requiring greater attention and research. They are: the micro-structures of capital markets and the role of banks in relation to the long-term funding of innovation; the relationship between military expenditures, especially for R&D, and competitiveness; and finally, the place of services in relation to manufacturing.

The micro-structures of capital markets

The TEP has witnessed renewed interest in a number of analytical and policy-relevant issues concerning the funding of industrial R&D and other costly innovation-related investments. However, the focus of analysis has shifted away from country-specific arrangements for the public provision of long-term finance on favourable terms for corporate R&D to the organisation of capital markets and the way this organisation shapes corporate governance. The OECD study on industrial investment in a financial economy carried out in 1988 reflected this new focus. It noted that while "[i]n the new financial context, credit suppliers' increasing tendency to lend to the most solvent, reliable enterprises and investors' preference for buying shares in firms offering high returns combined with the least possible risk are improving the quality of capital allocation to industry. (...) There, nevertheless, is a feeling in certain government, industrial and even financial circles that investors' thirst for the highest possible returns may sometimes be unfavourable to the long-term expansion of industrial enterprises. When investors demand high returns in the short term, they pay more attention to rises in the price of their shares on the market (i.e. to rises in their value in the eyes of other investors, which can bring them capital gains) than to actual longer-term value of their capital to the firms using it (i.e. to industrial returns)" (OECD, 1989s, p. 37).

The TEP has confirmed the importance of these issues for the competitiveness and growth of national economies. Presentations made at TEP symposia by members of the MIT Commission on Industrial Productivity argued that "the cost of capital may be less important than the nature of the institutions and arrangements that influence its supply" (Dertouzos *et al.*, 1989, p. 61). On the basis of the evidence collected in the OECD study, the situation can be described as follows. Japanese firms, and to a lesser extent European ones, tend to raise more capital from financial institutions such as banks and less through open-market sales of securities. Equity capital accounts for 60 per cent of total company assets in the United States and Canada, 50 per cent in the United Kingdom, 40 per cent in Germany, 30 per cent in France and 20 per cent in Japan. Banks and insurance companies provide long-term finance for Japanese industrial firms in ways that they do not in the United States. Typically, they own equity and are represented on the board of directors of the companies to which they lend. It is also common for industrial firms with long-standing commercial ties to own each other's stock (OECD, 1989s).

A recent assessment by a Japanese expert confirms that "financial institutions' stock holdings coupled with the main bank system appear to have contributed in part to the development of long-sighted strategy of business investment and active innovations by productive firms. Cross-stock holdings have even sharpened the separation of power between management and ownership through sheltering long-sighted management of member firms from short-sighted interests of individual stock holders. The main bank system has substantiated the promotion of such long-term strategy of innovative business investment and R&D expenditure. (...) The main bank monitors the borrower's corporate performance on a basis of long-term relationship, takes an active strategic role in financing the borrower's innovation through internal information exchange, and performs as last resort lender in the case of the borrower's extreme financial difficulties." (Yoshimoto, 1990, p. 8).

In contrast, a large and growing share of the capital of US firms is owned through the public stock market by mutual funds and pension funds, whose assets are in the form of a market basket of securities. "The actual equity holders, the clients of the funds, are far removed from managerial decisions. Their agents, the fund managers, have no long-term loyalty to the companies in which they invest, and they have no representation on their boards. Although some fund managers invest for the long term, most turn over their stock holdings rapidly in an effort to maximise the current value of their investment portfolio, since this is the main criterion against which their own performance is judged" (Dertouzos *et al.*, 1989, p. 62). North American experts consider that US firms respond to this financial environment by maximising their short-term profits in the belief that the market will penalise them for taking the long view. The wave of take-overs and leveraged buy-outs, many of them hostile, which marked the second half of the 1980s, contributed to the creation of a very strong focus on short-term returns by corporations working in unsheltered financial environments.

Within the EC, at least three types of organisation of capital markets and corporate ownership and control coexist: the UK model, with a situation very close to that of the United States, the German model, which resembles in some respects that of Japan, and an intermediate model (France, Italy and Spain). Recent studies prepared for the European Commission (Muldur, 1990; and Booz-Allen Acquisition Services, 1989) document the marked differences in take-over activity in the EC over the 1985-1988 period and examine the reasons for these differences, i.e. the obstacles to mergers. These include a long list of *structural* differences (especially size and sophistication of equity markets and the role of banks in corporate ownership and control) and a variety of *regulatory* differences (e.g. anti-

Table 55. **Foreign direct investment as a percentage of gross fixed capital formation, excluding housing**

Annual averages per period

	1961-65	1966-70	1971-75	1976-80	1981-85	1986-88
United States						
Outward FDI	4.2	4.4	5.3	5.8	2.5	4.7
Inward FDI	0.4	0.7	1.1	2.7	4.1	8.3
Japan						
Outward FDI	0.5	0.5	1.4	1.1	1.8	3.9
Inward FDI	0.3	0.1	0.2	0.1	0.1	0.05
Germany						
Outward FDI	1.4	2.2	3.0	3.5	4.0	6.8
Inward FDI	3.1	3.0	3.0	1.4	1.1	1.0
United Kingdom						
Outward FDI	5.9	6.2	7.9	14.6	15.2	25.3
Inward FDI	3.9	3.9	6.3	10.4	6.7	11.3
France						
Outward FDI	..	..	2.0	2.6	3.5	7.0
Inward FDI	..	..	2.7	2.9	2.0	3.8

Source: du Granrut (1990).

trust, company law, labour law, stock market regulations governing take-overs, etc.).

The ease or difficulty of take-overs is certainly one of the factors which shapes the level of inward investment in an economy (see Table 55). This in turn contributes to the degree to which the structural cohesion of a national economy is affected by the process of globalisation. In the course of the 1980s, globalisation was driven, to a not insignificant extent, by the financial sphere. This is why the approach taken to the long-term governance of domestically owned firms by different national banking systems and thus the role they play in preserving the structural cohesion of the domestic economy have become so important. In an era of finance-led globalisation, privatisation and deregulation, much of what remains of the capacity of countries to shelter long-term innovation, ensure domestic investment (see Table 55) and preserve the structural cohesion of national economies may lie as much with the banking system as with governments.

Military R&D and competitiveness

The principal purpose of military R&D is the development of improved weapons. This implies that defence programmes should be judged principally, if not exclusively, on their degree of success in that task. However, broader economic considerations have also become part of the discussion. Today, two main issues underlie the strong public debate about military R&D. First, for some countries (see Table 56), outlays are large in absolute and even larger in relative terms; these countries have also dropped behind in competitiveness since the 1970s. Second, during the 1980s, arguments about the beneficial spin-off or spillover effects of military R&D re-emerged. There was also discussion of the use of military R&D as an instrument (supposedly beneficial) of government support to industrial innovation and of the use of security regulations for controlling foreign access to R&D results (including, naturally, the participation of foreign firms in military R&D co-operation) as a potentially useful way of preserving temporary monopoly positions[9].

The assessment of recent research dealing totally or partly with the issue is quite severe. Econometric studies carried out during the 1980s, in particular in the United States, conclude that the pay-off from military R&D in terms of productivity and competitiveness has at the least been very small and at the worst negative. Industrial case studies suggest that in the aircraft, electronics (semiconductors) and computer industries, which undoubtedly benefited from military expenditure in the 1950s and 1960s (although probably more through captive and well remunerated markets permitting protracted learning and cost reduction processes than through R&D support *stricto sensu*), the growth and sophistication of civil business and consumer demand has now outstripped the military market. Today, the direction of decisive technological flows has reversed. The defence sector depends strongly on the results of competitive industrial R&D, notably in micro-electronics and advanced chemicals. Conversely, significant military influence on the shaping of technological trajectories is now identified as representing a competitive handicap in civilian markets[10].

For a number of very good reasons (see Box 48), as soon as the innovation and development of generic technology find a strong base in normal market-oriented industrial R&D, military R&D will simply represent an economic burden involving quite high opportunity costs. Military technology has developed over the years a number of distinctive features that reflect the particularities of defence market conditions and of military demand (Mellman, 1983; Stowsky, 1986; Reppy, 1990). These features include: emphasis on performance over cost and reliability; a related emphasis on high and often "baroque" technology; long development cycles; inefficient production processes; and the secrecy associated with security classification.

All these features accentuate the non-generic, highly customised features of military technologies and their very low degree of transferability to civilian uses. Consequently, they aggravate the obstacles to the intersectoral diffusion of technology related to military R&D (Chesnais, 1990*b*), and they represent a heavy burden on all the countries with large military expenditures[11].

Competitiveness and services

To date, insufficient attention has been paid by economic research to the relationships between services and competitiveness. A major difficulty resides in the fact that services represent a very unsatisfactory grouping of activities: the tertiary sector was born as a

Table 56. **Military R&D, selected countries, 1987**

	Military R&D			
	£ million	as % of GDP	as % GERD	as % govt. R&D
United States	22 708	0.88	32.5	68.6
United Kingdom	2 232	0.54	23.5	48.4
France	1 953	0.47	20.8	43.2
Germany	660	0.14	5.0	11.6
Italy	216	0.05	4.2	7.0
Japan[1]	279	0.02	1.2	4.4

1. 1986 data supplied by the Japanese Government. UK Cabinet Office, ACOST, *Defence R&D: A National Resource*, pp. 13-14. Other countries: UK Cabinet Office, *Annual Review of Government Funded R&D*, 1989, Tables 6.8-6.10.
Source: Reppy (1990).

Box 48. Some distinctive features of military technology

The technologies developed in the military sector have several distinctive features, reflecting the peculiarities of defence market conditions and military demand.

Emphasis on performance over cost and reliability

Because the military are interested primarily in battlefield performance, the development efforts have concentrated on such characteristics as firepower, accuracy, speed and range, rather than robustness, production feasibility, maintainability and low cost. The relative neglect of cost considerations is reinforced by the widespread use of cost-plus contracts and the lack of competition once a weapon has been approved for production.

Long development cycles

As weapons systems have become more complex and government regulation more pervasive, development cycles have lengthened. In the absence of wartime urgency, there is a tendency to retain projects in development in order to add the latest technology.

Resistance to reappraisal

Once a military-industrial network has been set up to build a particular weapon system, the futures of the military user, the prime contractor and associated firms and institutions all depend on follow-on projects to provide increased capability without threatening the *raison d'être* for the weapon. Thus a technological trajectory will take the form of successive models of a given weapon, leading to a situation in which incremental improvements become ever more costly (Kaldor, 1981).

Inefficient production processes

The production technology for military systems does not match the level of technology embodied in them. Military R&D is concentrated on product development to the almost total neglect of manufacturing technology. As a consequence of the high cost and complexity of modern weapons, governments can afford to buy only a relatively small number of a given system. Small production runs rarely justify investment in the latest in manufacturing technology or permit much in the way of learning curve economies.

Obstacles to technology transfer associated with security objectives

Restricting knowledge of important technologies to those with a "need to know" is necessary from the point of view of national security, but it limits diffusion of new developments outside a very small group of defence contractors, complicates the management of international collaborations, and serves as a further deterrent to the entry of new producers into the market.

"residual", and extremely heterogeneous activities continue to be lumped together for statistical convenience (Sauviat, 1990, pp. 5-7). Macro-economic indicators of national competitiveness in services are particularly unsatisfactory[12]; they have become even more so with the acceleration of foreign direct investment and a variety of non-equity alliances which are a central component of growth in many service industries. The difficulties of measuring productivity in services offer another example of the problems that stem from extending the use of analytical tools developed for manufacturing to services.

The papers discussed during the TEP suggest that the advent of an information-intensive economy is now creating the basis for the homogeneity previously lacking for the service industries. These industries increasingly have in common the fact that the centralisation and treatment of information is as at the heart of their activity. One expert has argued that the micro-economic competitiveness of firms in service industries now lies in their capacity to collect, process and supply different specialised forms of information (Sauviat, 1990, p. 46). Another suggests that the important new dimension is less the development of the tertiary sector, already the largest one in many economies for several decades, than the growing role of "customised" relationships – as opposed to the arm's length sale of pre-determined products – as a key factor in success in creating and exchanging value (Bressand, 1990b).

With respect to structural competitiveness, both insist on the need to view manufacturing and services as now representing closely interconnected activities rather than two quite different "sectors". Manufacturing certainly matters, but the competitiveness of firms in manufacturing now depends crucially on the quality of their interactions with the business services industries as well as on the performance of key infrastructure services: transport and delivery systems and telecommunication systems. With regard to the first, the efficiency with which business services firms collect, treat and supply specialised information and meet the needs of user firms in manufacturing is a component of structural competitiveness. The role played by business services firms in helping to raise the standards of management practice in manufacturing firms (in areas such as finance, accounting, corporate organisation, work organisation, use of IT, etc.) is important in this respect. Here again, the US approach stresses arm's length market relationships and offers fairly standardised packages to user firms, while European and Japanese firms adopt a "customer-specific" approach (Sauviat, 1990, p. 42).

Some aspects of the long-term relationships between firms and banks have already been discussed above. The nature of downstream relationships with wholesale and retail trading firms likewise play a role in competitiveness. In Japan, the *sogo soshas* are a complement to the *keiretsu*. They offer manufacturing firms durable, trustworthy and firm-specific relationships based on reciprocity and exchange of information. This procedure contrasts with the standardised, anonymous and opportunistic behaviour which often characterises the dealings between manufacturing and wholesale and retail trading firms in export markets. Some German and Italian firms have found ways different from those of the *sogo soshas* to overcome factors affecting competitiveness negatively, in some cases, vertical integration of export wholesale affiliates, in other cases, a novel approach to franchising (De Laubier, 1986).

3. TECHNOLOGY, COMPETITIVENESS AND TRADE

Important new perspectives on the foundations and the motor of international trade emerge once the role of technology and national environment are recognised in determining productivity and competitiveness at the micro-economic level, and their structural importance for competitiveness. Comparative advantage (in the broad or familiar sense) is no longer the *ex post* result of a simplistic, almost unchangeable set of national factor endowments, but a situation open to change, slowly in some cases, quite rapidly in others. The examples of Japan (Yoshitomi, 1991), and to a lesser degree of Korea and Taiwan, demonstrate that the basis of national competitive advantage can be chosen and within given constraints, built through conscious action on the part of firms and countries. Such advantage can also be lost, at least in part, in times of radical technical change, if the foundations of technological accumulation are undermined by shortsightedness on the part of firms or inappropriate government policies.

The findings of the TEP accord with the analysis undertaken both by a part of the new trade theory (Krugman, 1990) and by the trade theorists now looking at international trade from the standpoint of the economics of technical change (Dosi *et al.*, 1990). The analysis in this report points to a situation where the cumulative process arising from technological advance, from the increasing returns stemming from appropriate corporate organisation and effective inter-firm co-operation, and from differences in the level and efficiency of private and public investment can produce absolute trade advantages ranging across a spectrum of technologically-related industries.

The theory of international trade is one of the most complex and possibly one of the most controversial areas of economics, all the more since it is now in a process of redefinition. Consequently, the aim of this section is simply to present a few pointers to the interface between the thrust of this report and different approaches to trade theory.

The organisation of production as the starting point for trade analysis

The basis of trade theory as developed and passed on to future generations by Ricardo and Mill is that countries trade in order to take mutual advantage of their differences: the capacity of a country to produce cloth better than wine, or wheat more cheaply than machine tools and *vice versa*. In classical economics, this capacity was defined as being based on labour productivity and the accumulation of capital rather than on natural resources and other "gifts from God"[13]. Subsequent theories eliminated these production capacity and productivity-enhancing foundations of comparative advantage; they reduced the differences generating trade to a set of very simplistic "factor endowments" – natural resources ("land"), capital and abundant and cheap labour. One consequence of the elimination of the reference to a theory of production is the type of statement made by neo-classical theory regarding constant returns to scale or perfect competition, as the dominant form of supply structure.

The new trade theory has begun to re-establish the link with the organisation of production by stressing the role of economies of scale as a cause of trade, by recognising imperfect or oligopolistic competition as the dominant form of market structure (implying firms of different size, with different competitive attributes and different market power), and by recognising the "arbitrariness" of capital and technological accumulation on sites, in locations or possibly countries whose "natural endowments" as identified by neo-classical trade theory do not necessarily point to such accumulation. The analysis recognises furthermore that as a result of technological learning and R&D investment in particular, economies of scale are not static but *dynamic,* so that cumulative past output and investment determines current productivity (Krugman,1990, p. 107). The new trade theory clearly states that "most of world trade is the product of industries that one can have no hesitation in classifying as oligopolies" (Krugman, 1990, p. 185). In combination, the existence of oligopoly and the role of dynamic learning curves stress the importance for firms of securing their home market through their own collective action and the action of public authorities in order to establish large-scale production, sell enough over time to move down the learning curve and earn enough to recover the costs of R&D. As Krugman points out, there is a natural alliance between the new trade theory, with its emphasis on increasing returns and imperfect competition, and the view that technological change is a key factor driving specialisation. The failure of neo-classical trade theory to integrate the role of technology in any meaningful way is, of course, a major cause of the current profound reappraisal.

Orthodox trade theory and technology

Initial recognition of the need to re-examine the traditional views of factor endowment date back to the so-called "Leontief paradox", i.e. the demonstration that, contrary to the expectations of a factor endowment theory based on labour and capital, US exports of manufactures were clearly more labour-intensive (and less capital-intensive) than their imports. Many studies attempted to fit the Heckscher-Ohlin-Samuelson trade model to economic reality by extending the number of "factors", in particular by introducing qualified labour ("the human capital variable") and technology (using domestic R&D expenditures and later patenting as proxies) into factor endowment and factor proportion theory (see Dosi and Soete, 1988). Empirical studies generally took the form of simple regressions which added independent variables associated with technology, skills and human capital to the capital and labour variables which had failed adequately to explain US trade patterns in earlier empirical tests of the simple two factor Heckscher-Ohlin-Samuelson model.

During the 1960s, these tests appeared to yield satisfactory results. Empirical studies incorporating human capital or human skills and technology variables into the traditional model produced evidence of what seemed to be a close relationship between technology and US international trade. In the 1970s, the tests increasingly told a different story[14]. Data on the growth of output showed that, in the United States, the intensity of R&D (or the level of technological sophistication) in an industry was positively correlated with its simple growth rate. The direct relationship was observed to be statistically significant in the 1960s but increasingly insignificant after 1968. Similarly, human capital remained a basis of comparative advantage for the United States in relation to developing countries but ceased to have this characteristic *vis-à-vis* OECD countries.

The collapse of an identifiable statistical relation between R&D expenditures and export performance within the framework of comparative advantage trade analysis can be attributed to the foundations on which these studies were built. The factor endowment approach essentially reduced technology to the status of a static production "factor" with which some countries, notably the United States, were better "endowed" than others. All the important features of technology (discussed in earlier chapters) were ignored by the approach, which also had little to propose at the level of government policy except continuing and possibly increased expenditure on R&D.

Furthermore, orthodox trade analysis leaves little or no room for corporate strategies. The simplified hypotheses on which comparative advantage trade theory is built do not allow for the existence of firms in their contemporary form. It assumes that the activity of firms and the competitive advantages they strive for cannot modify initial endowments and create productivity "gaps" between countries[15].

Technology gap trade theory

One of the earliest challenges to the traditional approach came from technology gap trade theory. As initially proposed, the model basically extended a simplified version of Schumpeter's innovation cycle model into the area of international trade (Posner, 1961). Its point of departure was a situation in which a firm developed a completely new product in response to a perceived market demand and the market responded favourably to the innovation. As demand grew, competing firms, at home and abroad, would react by attempting to imitate the new product. Imitation takes time, and during the years in which the innovating company maintained its monopoly control over the technology, it might hold a commanding position in both domestic and international markets. As a result, a one-way flow of exports might build up from the innovating country. Sooner or later, however, competitors' efforts at imitation would be rewarded, and the original innovator's market expansion would be checked. If an industry abroad became a successful imitator and had cheaper factors of production at its disposal, traditional "comparative advantage" based on factor endowments might come into its own again.

Subsequent research showed that this might take a long time to occur. Because production learning curves and the dynamic scale economies with which they are associated come into play, the original innovator's advantage may prove to be much more lasting and to form the basis for technological cumulativeness beyond this point. Industrial case studies – e.g. plastics by Freeman (1963), synthetic materials by Hufbauer (1966), electronic capital goods by Hirsch (1965), and chemical process plant by Freeman (1965) – supported the technology gap hypothesis. These studies established the existence of imitation lags, dynamic economies of scale, process innovation, and a range of scientific and technical activities in a variety of industries. The role attributed to such factors implied that the temporary monopoly position of firms – and so the resultant technology gap – would be rather more permanent than that initially hypothesised by Posner (see Dosi and Soete, 1988). The studies showed that imitation lags could be prolonged if R&D threshold costs were high. As long as the initial innovators were able to maintain a flow of process innovations related to scale economies and new generations of products, the competitive efforts of would-be imitators might be repulsed. Case studies of the electronics and semiconductor industries carried out by Golding (1972); Sciberras (1980); Dosi (1984) subsequently showed such process innovations to be particularly important. All these findings concur with those of the TEP and have been confirmed by further work.

Dynamic growth efficiencies and absolute competitive advantages

The factual and theoretical material discussed throughout this report suggests that competitiveness results from the successful establishment of systemic foundations which enable countries to capture the cumulative and dynamic features of technological advance. Once the importance of these features has been recognised, it becomes necessary to move from the rather unrealistic realm of static allocative efficiency to the much more plausible arena of dynamic growth efficiency, of which competitiveness is a major

result (Dosi *et al.*, 1990). Even within static models, however, the effects of technology-related factors are identifiable. One paper presented at the Paris TEP Conference set out a model combining a Schumpeter-Vernon type theory of temporary monopoly situations based on product innovation and the "new" trade theory approach. The model shows the conditions under which temporary product monopoly allows export price setting at high price levels with effects on the terms of trade and on income between trading countries (Guellec and Ralle, 1990).

The foundations of dynamic growth efficiency rest on the cumulative and increasing returns associated with technology and the interactive and organisational factors discussed throughout this report. Once the competitive effects of generic technology with strong inter-industry and feed-back relationships between complementary technologies has become significant, countries which have succeeded in reaping the results of these interactions may demonstrate high levels of international competitiveness in a fairly wide range of partially interconnected industries. The extent of intra-industry trade may in certain circumstances be an indicator of such a process. This will be the case if for historical reasons strong industrial-*cum*-technological complementarities and inter-industry linkages and exchanges have been built up domestically rather than through an international or regional (as in Europe) division of labour. Much that is known about Japan's industrial system and the *keiretsu* business group organisation suggests that this is a plausible explanation of the relatively low level of Japan's intra-branch trade (see Table 57).

When differences in technological and organisational proficiency between countries are industry-specific and limited to given sectors, the existence of technological gaps does not rule out comparative advantage completely. This point has been made again by Dosi *et al.* (1990, p. 147) who show "how in principle an evolutionary micro-structure, continuously yielding inter-firm and international asymmetries, can be directly linked to a classical (Ricardian) approach of comparative advantage: any country will find its comparative advantage in the sector where its technological gap is proportionally smaller (or the lead greater) and *vice versa*".

This may cease when trade advantages occur across a broad range of technologically interdependent industries. Here, absolute trade advantages are founded on the technology-based "structural competitiveness" discussed above. In this hypothesis, the break with traditional theory is much more radical (see Box 49).

The challenge to trade theory is also a challenge to policy, one of the strongest perhaps now facing OECD countries in their relationships with one another inside the OECD area. The mechanisms for returning to equilibrium postulated by traditional theory have

Table 57. **Intra-industry trade in selected countries**

Intra-industry trade index points[1]

	Index basis SITC[2] categories						
	3-digit						4-digit
	1959	1964	1970	1975	1980	1985	1985
	All traded products						
Japan	17	21	26	19	19	23	..
United States	40	40	53	57	57	54	..
France	45	60	67	65	67	74	..
Germany	39	42	54	52	57	63	..
	Manufactured products only						
Japan	..	..	32	26	28	26	23
United States	..	..	57	62	62	61	54
France	..	..	78	78	82	82	74
Germany	..	..	60	58	66	67	63
South Korea	..	..	19	36	40	49	44

1. The calculation of intra-industry trade (ITT) in a single industry is based on the standard equation $ITT = 1-[x_i-m_i]/[x_i+m_i] \times 100$, where i = industry, x = exports and m = imports. The average index for trade in all industries within a nation is calculated by weighting each industry by its share in total trade.
2. SITC = Standard International Trade Classification.
Source: United States, National Academy of Engineering (1990).

> Box 49. **Absolute and comparative trade advantages**
>
> In a Ricardian or Heckscher-Ohlin-Samuelson world, every country, by definition, must be relatively competitive in something. Competitive in this sense amounts, however, to little more than a tautology: being competitive might simply mean that anyone is bound to be less bad in something and worse in something else.
>
> The implications of the findings presented here are rather different. "Whenever strong technological interdependencies, hierarchical links between technologies and significant externalities to firms stemming from national environments are found (in terms of cross-sectoral fertilisations, spillovers, etc.), the pattern of absolute advantage in these dominant technologies, skills or capabilities will have to be taken as an autonomous determinant of international competitiveness, independent of the pattern of comparative advantage associated with relative efficiencies" (e.g. the inter-sectoral, intra-national comparison of technological lags and leads and the related differences in productivity and profitability) (Dosi *et al.*, 1990, p. 149).

become inoperative. The notion of absolute trade advantages offers a better explanation of why, since the late 1970s, some OECD countries have run long-term, almost permanent trade surpluses and others almost equally permanent deficits. Today, efficient firms increasingly operate under *sine qua non* constraints to export: technology-based economies of scale and scope are often open only to firms to the extent that they are capable of gaining and preserving strong footholds in foreign markets; consequently, such firms must export and will find institutional and political impediments to expansion particularly hard to bear. But countries which build up structural trade deficits, in particular in manufacturing, cannot remain inactive. They will feel themselves obliged to try to improve their long-term competitiveness.

Using the assumptions of accepted trade theory and addressing the problem of resource allocation and "gains from trade" within the classical static framework, the new trade theorists show that in highly concentrated, high-profit, high-wage industries (e.g. computers), public intervention may *increase* national welfare, contrary to the conclusion of traditional theory. In the OECD, their findings were recently summarised as follows. "In a real world of oligopolistic competition, the number of participant firms in some high-technology industry may be relatively small and profits can be above and beyond the 'normal' return. In such a world, there are powerful incentives for a national government to ensure, by undertaking unilateral measures, that its own national participants in the competition win a large share of these industrial sectors and profits in the integrated global market" (Yoshimoto, 1990, p. 1). Such unilateral measures may of course be offset by the threat of retaliation and lead to bilateral or multilateral trade negotiations. The equilibrium which may ensue will have little to do with those of the traditional trade paradigm.

The analysis also points to some other dimensions of policy for improving competitiveness. Once due account is taken of the systemic features of technology and the role of inter-industry diffusion of generic technology, attempts to target a few high-growth, technology-intensive export sectors for international trade specialisation can be critically reassessed. A trade specialisation and growth model presented to the Paris TEP Conference shows that for such a policy to succeed, rather stringent conditions must be met (Amable, 1990). These conditions are external (e.g. strong growth of world demand) as well as internal (effective structural competitiveness). German success in many "traditional" industries (e.g. textiles and clothing but also the wide spectrum of traditional machines for other sectors), shows that technological learning and product quality improvement, supported by technological institutions (Braunling, 1990), can represent an important component of competitive exports on a wide industrial front (see also Amable and Mouhoud, 1990; and Fouquin, 1990).

4. COMPETITIVENESS, GLOBALISATION AND SITES

The process of globalisation discussed in Chapter 10 obviously has a number of implications for the competitiveness of domestic economies. These began to be discussed during the TEP but will require close examination in the future. During the TEP Symposia (Stockholm, Tokyo and Paris in particular), some papers analysed globalisation from various standpoints, while others examined the basis of domestic competitiveness. Some experts have begun to argue that in coming years the emerging global system of worldwide intra- and inter-corporate networks will form the only effective basis for value creation and distribution; they recognise that the welfare implications are far from clear. Others consider that globalisation will have more limited effects. The relationships between globalisation and competitiveness will require careful study in the coming period. This section will discuss a small number of salient issues which emerge from the available evidence (Lanvin, 1990*b*).

The status of MNE affiliates and national technological capabilities

Changes in the forms of corporate organisation adopted by MNEs and the emergence of network firms (see Chapter 4) may involve an increase in the mobility of certain types of economic activity. Some affiliates will be upgraded, and others downgraded, as technologically sophisticated production and assembly tasks become increasingly separated geographically as international specialisation occurs within MNEs.

Experts on multinational enterprise activity consider that this is a likely scenario. Dunning and Cantwell (1991) argue that "the increasingly footloose nature of international production and of innovatory activities is likely to reinforce patterns of cumulative causation within countries. Those countries which are growing more rapidly, upgrading their industrial structures, and devoting more resources to the encouragement of indigenous technological capacity (e.g. by a well formulated industrial strategy, and an educational policy designed to meet this strategy) are more likely to attract inward MNE investment in research intensive activities, and to benefit from the dissemination outside the recipient affiliates. By contrast, countries which are losing international competitiveness are more likely to attract MNE satellites concentrated in assembly and low value-added activities" (1991, pp. 53-54). Since outward foreign direct investment by domestic based MNEs may entail a similar process with complementary effects, this can provoke cumulative causation leading to increasing but also to declining technological capacity (see Box 50).

Mergers and take-overs have their place in this context. The finance-led or finance-driven dimensions of globalisation which financial deregulation has accelerated, along with the volatility in corporate ownership

---

Box 50. MNEs and technological capabilities

MNEs and "virtuous" cycles of increasing technological capability

Inward MNE investment is likely to be attracted to innovative industries caught up in a virtuous circle, to be associated with local R&D facilities and to contribute to an increase in indigenous technological capacity. In doing so, it may increase local technological dissemination to suppliers and customers, and spur local rivals to a higher rate of innovation, by increasing competition. Indeed, such sectors are probably home to outward MNE activity which plays a similar role elsewhere. By locating research intensive production in important centres of foreign innovation, and/or assembling or manufacturing abroad, firms of the most dynamic countries more effectively sustain the basis for increased research activity at home. They are able to integrate complementary foreign technologies and to devise more broadly based technological systems in their domestic plants. Thus, the activities of domestic and foreign MNEs, and their spillover effects, may aid in the development of a country's technological capacity.

MNEs and "vicious" cycles of declining technological capability

In contrast, inward direct investment may take place in declining sectors, but it is likely to be in low value-added sub-sectors of the industry, with the higher value-added intermediate products being imported. By dint of their greater efficiency, or through deliberate strategy, the foreign MNE may be able to finance a higher level of R&D within its parent company from its increased global sales, including its greater market share in the host country. Meanwhile, local firms whose markets are cut back may lack the resources to "go global" themselves, and are consequently compelled to cut back their R&D expenditure (Dunning and Cantwell, 1991, p. 55).

it has created (OECD, 1989s), may accentuate these cumulative processes thus affecting the inter-industry linkages on which the structural cohesion of economies are founded. As discussed above, the national systems of production and innovation that are sheltered most effectively against this process are those in which the banking system recognises a responsibility for ensuring the cohesion and such growth of the manufacturing sector and has built close ties with industry. Alongside governments, banks must now include long-term criteria when considering their policies towards given industries.

Building competitive advantage from local endowments

At the interface between the interactive mechanisms and cumulative processes which command competitiveness and the process of globalisation lies the important issue of the role played by sites, e.g. local or regional (sub-national) R&D and industry clusters. The creation of private competitive advantage through the ability of MNEs to make the most of R&D and human capital resources located in several countries, thanks to world-wide technology and other key input-sourcing mechanisms (see Chapter 10), is a process in which the local or regional (sub-national) dimension becomes extremely important. Thus, regional policy and policies towards sites acquire a new and vital dimension. Such policies may now represent a way of simultaneously exploiting the competitive advantages of concentration of certain types of human resources and of addressing the problems posed by the mobility of capital and other resources within multinational firms.

A site is a geographical location that offers firms proximity and specific resources, and whose exploitation generates significant externalities. The first property eliminates geographical areas that are too large, such as the national territory *per se* of all but the smallest countries, the second, geographical areas that are too poor in certain factors, notably R&D infrastructures, highly skilled labour and advanced communication infrastructures. In the light of opportunities offered by domestic conditions (e.g. macro-economic policy environment, general characteristics of the education system, employment regulations, structure of the banking system, etc.), the competitive advantages offered by a given site may fall under four main headings (basically those of Porter's competitive "diamond", see Box 47):
– Supply of factors of production and infrastructure. Such factors include the quality of available labour (e.g. existence of specialised technical institutes), the availability of capital (e.g. local savings networks, existence of venture-capital institutions), communications and research infrastructures (e.g. transfer science laboratories, test laboratories, or technical research centres), or socio-cultural infrastructure (highly skilled and mobile workers demand certain facilities).
– Quality of the local industrial fabric in terms of sub-contractors, suppliers of intermediate goods and equipment and suppliers of professional services. In particular, full exploitation of technological opportunities requires a satisfactory division of labour between small and larger enterprises, as well as close collaboration between the creators and users of new products or processes.
– Organisation, managerial practices and the competitive environment of firms.
– Demand (propensity of local customers to become players in the innovation process through their demands and capacity for initiative).

These four dimensions mutually reinforce each other and contribute to the creation of productive resources. As a result, the site's competitive advantage can be enhanced and site activities diversified. Lack of any of these dimensions will jeopardise the site's potential for development as an autonomous unit. A site is competitive – capable of retaining its factors of production and of attracting direct investment – when the exploitation of its externalities is only really open to firms and institutions which enrich and diversify its resources. Otherwise, the site will become the satellite of a more dynamic site, or something resembling an "off-shore production platform" for multinational enterprises, with little control over its future development.

Structural competitiveness as discussed in this chapter will depend increasingly on the number and nature of such sites and the degree to which different sites within a national economy also possess interactive and cumulative feed-back relationships among themselves and do not constitute isolated "islands of prosperity" within countries or regions otherwise marked by stagnation or regression. In the face of the globalisation process, government policy must ensure that these inter-site linkages develop and that sites do not engage in disorderly competition among themselves to attract outside firms. The task may be difficult; certainly, it requires a reassessment of regional and national technology policy[16].

The OECD has undertaken a review of regionally-based technology transfer strategies. The significance of technology transfer for regional development has received increasing recognition over the past decade. A working hypothesis for the study accords to technological change the potential for raising local income and output and for increasing regional capacity for inter-regional and international trade.

The "virtuous" and "vicious" circles of cumulative causation which are at work for industries (see Box 50) also apply to sites. The premium MNEs place on proximity to high-income sophisticated markets and efficient infrastructures means that when they move abroad large firms will favour sites that offer these characteristics. As globalisation proceeds, trends towards economic and social dualism inside countries may develop, along with centrifugal tendencies so strong that they dislocate domestic innovation-related networks and other forms of interactive mechanisms.

Through analysis based on survey data and empirical research, Cantwell found that in a situation characterised by strong foreign direct investment and a rapid growth of MNEs in relation to other categories of firms, "the intensification of international competition (...) enhances a pattern of cumulative causation in which centres of innovation enjoy a virtuous circle, while certain other locations are as part of the same process locked into a vicious circle of declining growth and diminishing local research (...). Once the most successful MNEs in a given international industry concentrate research in particular locations then such favoured sites are more likely to experience a virtuous circle of cumulative success, in which their world share of innovation and production steadily rises. In an established sector this is likely to be achieved at the expense of a vicious circle of cumulative decline in research activity in certain other less attractive locations." (Cantwell, 1989, p. 161).

5. SOME POLICY CONCLUSIONS

As a result of the foregoing discussion, several conclusions can be drawn. It now appears that:

i) At the levels of development reached by OECD countries and given the role played by innovation and technology, competitiveness is now based increasingly on factors other than the cost of labour and other inputs, with competition being waged by firms through non-price as well as through price factors.

ii) Sound macro-economic policies remain a condition of competitiveness, but cost indicators such as relative labour unit costs can no longer be considered as reflecting competitiveness or as capable of predicting trade performance.

iii) While competitiveness is situated in the activity of firms, corporate competitiveness is not exclusively of their own making. It is also an expression of domestic institutional and social environments; it has a structural component and is supported by a wide range of externalities; national macro-economic policies offer some of the conditions for competitiveness, but policies that enhance national environments have important complementary effects on competitiveness, even if these effects are hard to measure.

Hence the need to refocus government policy. In the new context, many traditional forms of industrial policy, in particular government aid to specific industries or large firms are increasingly unlikely to be effective policy instruments for competitiveness, independently of the breach of international agreements (the EC or GATT) they may represent. Similarly, policies targeting the growth of industries considered "strategic" in one of the several meanings attached to the term (OECD, 1991h) require close critical analysis.

One recent assessment suggests that targeting is more likely to succeed for a strategy that involves catching up than for one that involves pushing back the frontier (Teece, 1991). With no clear model to follow, it is difficult and indeed hazardous for governments to identify the leading industries and technologies of the future. Furthermore, successful targeting requires something that many governments lack: the ability to co-ordinate various aspects of government policy such as trade, investment, technology, and anti-trust.

Today, successful policies for competitiveness will be those primarily aimed at supporting the aspects of intangible investment infrastructures and collective governance which are outside the purview of individual firms and at facilitating the formation of linkages, networks and interactive mechanisms. At present, the following areas of policy which summarize many conclusions made in earlier chapters, seem particularly relevant.

i) In business environments dominated by short-term profitability and efficiency considerations, government support for long-term generic R&D, with its highly uncertain outcome, is necessary to competitiveness as well as to welfare and growth. This entails strong support to universities and the research systems linked to them (see Chapter 1). These institutions provide the qualified personnel needed for an increasingly technological economy; they provide the base for applied and corporate research, and the national competence for participation in international R&D networks.

ii) Policies directed at the building of innovation-related networks are now essential (see Chapter 3). Actions aimed at favouring the building or consolidation of such linkages and networks have always been implicitly a fairly important component of national S&T policies. Today, this dimension of policy-making must be approached much more explicitly and made a central component of government innovation policy aimed at competitiveness.

iii) A re-examination of technology policy must go hand in hand with a new look at regional policy. In combination, technology policy and regional policy may represent key components of policies designed to preserve the structural cohesion of national economies.

iv) It is indispensable to competitiveness that human resources adapt to technical change and contribute to shaping and diffusing technology with rapidly changing technologies. This requires not only appropriate basic training and skills, but also lifelong education and retraining (see Chapter 7). This investment, which complements the educational infrastructures, cannot be supported by firms alone, as is now often the case. In particular, small and medium-sized enterprises experience considerable difficulties in providing or organising such education and training.

v) Governments have always had an important role in the provision of infrastructures. Network infrastructures possessing the potential to bring about "creative" network externalities such as broadband communications have become an essential prerequisite if many of the potential efficiency gains related to IT and the related growth in new products and services are to be realised.

vi) Deregulation and privatisation have been central elements of the structural adjustment policies followed during the 1980s. On their own, however, these measures will not succeed if vigorous domestic rivalry is not found. While it is fashionable today to accept mergers and support interfirm alliances between large firms in the name of globalisation, this may simply be a slightly adapted version of the "national champions" policy. The trade-off between policies aimed at helping firms become large, successful "global" firms and policies aimed at preserving structural cohesion and competitiveness must be examined closely. Effective national competitiveness may require governments to re-examine (as the EC has recently done) the rules and policy regarding mergers, acquisitions, and alliances that involve industry leaders. This is not contradictory with the need to start measuring world concentration and to study its effects (see Chapter 10).

vii) Finally, the strict enforcement of product, safety and environmental standards (see Chapter 9) can promote competitive advantage by stimulating and upgrading domestic demand. Stringent standards for product performance, product safety, and environmental impact can push firms to improve quality, upgrade technology and provide features that respond to consumer and social demands. Independently of other, more decisive considerations, easing this type of standard may be counterproductive in the longer run, however tempting it may be for short term competitiveness[17]. Ongoing OECD work suggests on the contrary that governments that view the "environment industry" in strategic terms and provide appropriate support, including stringent application of regulations on polluters, may be better placed to realise the ecological and economic benefits of a competitive environment sector.

NOTES

1. Business schools and most large MNEs have accepted this conclusion for over a decade. See for instance European Management Forum (1980, p. 2). "Competitiveness is not a unidimensional concept. In the marketplace buyers will weigh a price advantage against poor quality or lack of after-sales service and so on. The notion that competitiveness might be reduced to considerations of costs and productivity is thus a dangerous one."

2. The earlier evidence was surveyed by the Aalborg Institute of Technology in joint work with Freeman, see IKE (Denmark, Institute of Production) (1981). Analytical implications were first drawn for the OECD project on Science, Technology and Competitiveness (Chesnais, 1981; and Dosi, 1981). More recent evidence in presented in Chapters 4 and 5 of this report.

3. The Brookings findings are discussed by Lawrence (1980, pp. 261-63), in his contribution to the US Congress hearings on the international economy who makes the first set of arguments, while the second set is made by Bartel, in the same volume pp. 109-110).

4. See Durand and Giorno (1987) remark in their introduction that "the fall in the dollar since 1985 has not yet brought about a corresponding adjustment on the trade front, and this asymmetry has refocused attention on the question of how competitiveness should be measured,

and on the relationship between competitiveness and trade performance". Unfortunately, "the paper attempts to review only the first of these questions". Measurement problems are also discussed by Hatzichronoglou (1991, pp. 177-221). He shows that a "major drawback to using costs as a measure of competitiveness comes from the fact that they refer only to labour costs. It is not an easy matter to correctly calculate the cost of capital. (...) Other important categories of costs are also not taken into account. (...) These include R&D costs of distribution (agent costs), negotiation costs (purchasing-agent costs), and various other categories of financial charges. All these drawbacks significantly limit the relevance of cost indicators".

5. This report takes the same approach as Michalet (1985, p. 23) when he argues that orthodox theory reduces the notion of a country to an abstraction or again as Ostry (1991, p. 95) when she shows the numerous drawbacks arising from the fact that "(e)conomists have long ignored cultural, historical or institutional differences as factors of significance in market analysis. While interest in international economics has greatly increased, international forecasting models, for example, are based on the assumption that there is only one market model and thus the different 'country blocks' all have identical structural properties".

6. See Chesnais (1981). Major inputs to the OECD study later considered by the group chaired by Mr. John Ingram included Dosi (1981); Michalet (1981); Momigliano and Dosi (1983); Horn (1981); Antonelli (1982); Mistral (1983); Sciberras and Payne (1985); Dunning (1984).

7. For an overview, see R. Boyer (1986 and 1988).

8. Dertouzos, et al. (1989). The presentation at OECD was first made by Lester, 1989, who presented a similar paper outlining the MIT findings to the 1990 Paris TEP Conference on Technology and Competitiveness.

9. The issue was discussed in a critical manner in United States, President's Commission on Industrial Competitiveness (1985, pp. 193-201). This led to a major report by the US National Academy of Engineering (1987) where the issue is discussed in depth. See also MacDonald (1986).

10. For a discussion of the way in which the trajectories of numerically controlled machine tools (NCMT) were shaped by specific national environments, see Watanabe (1983) and Noble (1984). In the United States the defence industries played the major role while in Japan the automobile manufacturers made the important choices, with very different effects on overall industrial competitiveness of each economy.

11. For an overall discussion of the evidence, see Tirman (1984) and Chesnais (1990b) with evidence on France. The UK evidence is discussed in the recent report on future relations between defence and civil science and technology prepared for the UK Parliamentary Office of Science and Technology (Gummet, 1991).

12. Sauviat (1990) discusses the data on trade performance in services and the numerous difficulties it presents.

13. This is why Ricardo's theory has nothing in common with the "Ricardian goods" sometimes found in trade literature: an improper name for commodities traded on the grounds of differential endowments of natural resources (see Dosi et al., 1990).

14. The findings of the last generation of US studies which made empirical tests following this model are discussed by Pugel (1980).

15. In this sense the neo-classical (Heckscher-Ohlin-Samuelson) international trade paradigm based on factor endowment represents a step backwards with respect to Ricardo, for whom differences in labour productivity, and not "endowments", are the starting point for a postulate concerning the welfare gains to be reaped from trade through comparative advantage. Ricardo begins from a situation of absolute advantage. He argues that a country which possesses absolute productivity and cost advantages in two industries should relinquish to the country with which it trades the production and exports for the industry where it is relatively less well positioned.

16. Ezra (1990, p. 8) argues that "[n]ational technology policy may run counter to the aims of regional policy by fostering the concentration of R&D and associated development in core areas, since the usage or application of technology measures tends to be greatest in the most prosperous and growth regions. Moreover, in so far as regions are at liberty to offer their own forms of assistance, prosperous regions can outbid the not so prosperous".

17. This argument is made by Porter (1990a). Ongoing work by the OECD on the "environment industries" (see Chapter 9) has begun to examine these issues.

Chapter 12

NEW TECHNOLOGY, LATECOMER INDUSTRIALISATION AND DEVELOPMENT

ABSTRACT

This chapter assesses the combined impact of the new technologies and globalisation on newly industrialised economies (NIEs) and developing countries. A key concern of late industrialising countries is how to accelerate the learning processes required to gain technological mastery. Successful firms in NIEs have combined the acquisition of foreign technology with a consolidation of their own in-house knowledge and expertise. In the early period of rapid industrial growth, the largest part of industrial production was concentrated in consumer non-durable industries in which technology requirements were low. Subsequently, capital goods sectors were developed by NIEs in order to adapt and improve imported capital goods. Foreign direct investment (FDI) also played a prominent role in development, and foreign firms occupied an important place in host country industrial structures.

The current dynamics of globalisation may drastically increase inequalities of access to sources of foreign investment, the acquisition of foreign technology and access to foreign markets. Overall data shows that the transfer of technology to developing countries has substantially declined as a result of the recent concentration of FDI flows within the OECD region. The present economic environment is considerably less conducive to development than in earlier periods, and export-led growth strategies targeted towards the major OECD markets appear increasingly problematic. Developing countries are experiencing a substantial deterioration in their capacity to use technological change for modernisation. On account of the particular resource and skills requirements of the new technologies, the gap separating rich and poor nations is likely to increase.

Examining corporate strategies and government policies may help to explain why certain countries have been more successful than others in utilising technology. Almost everywhere, the state has been an important actor in the economic development of these countries. By supporting firms in their export strategies, some governments have helped them to earn the foreign exchange needed for buying capital goods abroad. In some countries, the saving ratios obtained through government policies have also facilitated high rates of capital formation. In several NIEs, governments have played an important role in building scientific and technical institutions. However, many important dimensions of corporate strategy, including the building of industrial conglomerates and the capacity to establish flexible supplier/subcontractor networks are equally decisive.

NIEs have no choice but to adopt computer-based advanced manufacturing activities in order to raise labour and total productivity and acquire organisational flexibility. As manual skills become less important, NIEs must develop formal education; this can only be done through appropriate public investment. The state also has a role to play in fostering the diffusion of these technologies. Even average-level developing countries are confronted with an erosion of competitiveness as exports based on traditional resources and industries decline because low labour costs lose their effectiveness. User-oriented strategies of technology diffusion offer the most sensible solutions.

The situation of the least developed countries and their capacity to acquire and use technology have been reassessed recently. This chapter ends with a description of the new policy orientations addressed at both donor and recipient countries.

INTRODUCTION

Decades of uninterrupted and intensive internationalisation in numerous and varied forms have created a world economy characterised by a deep, almost total economic interdependence of countries and continents. Thus, the opportunities, but also the acute problems, currently confronting the newly industrialised economies (NIEs) and the developing countries must be considered in the light of the issues dealt with in this report.

Almost all these countries are keenly aware today of the need to acquire new technologies, to make organisational changes and to increase their educational expenditures in order to improve their international competitiveness. However, they are separated from OECD Member countries by a large and often growing gap in their capacity to use technological change as a motor for growth and structural transformation. While some countries in Asia and Latin America, in particular the East Asian newly industrialised economies (EANIEs)[1], have clearly improved their capacity to acquire, absorb and adapt a variety of important production technologies, this is certainly not the case for the great majority of developing countries. The situation of the least developed countries is particularly dramatic; their capacity to use and adapt technology, in particular the new information technologies (IT), has substantially deteriorated, while their traditional export markets have been strongly eroded by the fall of traditional exports for a number of reasons including the substitution effects of new materials technology and biotechnology[2].

Since the early 1980s, the main indicators of the volume of international technology flows – FDI, capital goods imports, payments for licences and know-how, and official technical assistance – show an unprecedented shrinkage of technology flows to the developing countries, in particular the least developed ones. The only exceptions are China and the EANIEs. Since this state of affairs cannot be divorced from the processes discussed earlier and since it conditions, to a fairly large measure, the degree to which developing countries can meet the challenges of technological change, this chapter begins by reviewing the evidence regarding overall patterns of FDI and technology transfer accruing to these countries. It then describes the foundations on which successful latecomer industrialisation was built, before looking at the source of the problems facing the NIEs today. The situation of the developing countries and the particular problems confronting the least developed countries are then examined. The discussions focus on issues related to the diffusion of new information technologies[3]. These technologies are both an important factor in overall economic development and a good case study of the problems posed by new technologies to the various categories of countries considered here.

1. SOME MAJOR TRENDS IN INVESTMENT AND TECHNOLOGY TRANSFER

An important strand of analysis, approached from many standpoints in this report, concerns the way selection mechanisms and virtuous – but also vicious – circles in tangible and intangible investment accumulation and in technological learning occur in technological change. The footloose character of private capital and the great flexibility attained by MNEs in their activities means that selection and exclusion mechanisms can now occur on a large scale in the world arena.

Basic shifts in international investment patterns

Throughout most of the post-war period, international investment was an important vehicle for the international diffusion of technology. Until the mid-1970s, the United States and to a lesser extent the United Kingdom were by far the largest sources of FDI and the main home countries for MNEs (see Chapters 4 and 10). These firms were under strong pressure to invest exports, in order to extend the life cycle of mature technologies by means of global strategies of planned obsolescence, recover the costs of R&D (e.g. through highly sophisticated transfer-pricing techniques), and penetrate closed markets. While most of these technology flows remained confined to the OECD region, and to a lesser degree to COMECON countries, at least some of them ended up in Third World growth poles.

During the 1980s, the situation was reversed significantly (see Table 58). The United States became the world leader with respect to the inflow of FDI and an increasing share of international investment flows were concentrated within the OECD, in particular among the three poles of the Triad (see Chapter 10; and UNCNC, 1991). In spite of a notable increase of average annual flows to the developing countries between 1980-84 and 1985-89, their share in world-wide inflows between these two periods fell from 25 per cent to 19 per cent.

Furthermore, this overall fall was accompanied by an increasing disparity among countries and groups of countries. Ten developing economies maintained a share of about three-quarters of total inflows to developing countries throughout the 1980s. The ten and their respective shares of the $16 billion in average

annual inflows to developing countries over 1980-89 (excluding tax havens) are: Singapore (12 per cent), Brazil (12 per cent), Mexico (11 per cent), China (10 per cent), Hong Kong (7 per cent), Malaysia (6 per cent), Egypt (6 per cent), Argentina (4 per cent), Thailand (3 per cent) and Colombia (3 per cent). While the shares of Africa, taken as a whole, and East, South and South-East Asia in total world-wide inflows remained stable over the periods 1980-84 and 1985-89, the share of Latin America and the Caribbean declined from 12 to 7 per cent. For the non-oil exporting sub-Saharan African countries, FDI inflows have remained below $500 million since 1981, with some year-to-year fluctuations.

More generally, the less developed countries accounted for only 0.7 per cent of average annual inflows to developing countries in 1985-89, as compared with 1.5 per cent in 1980-84. Total FDI in this group only reached about $200 million in 1989. The marginalisation of less developed countries, most of which are located in Africa, is likely to continue since flows to developing countries are highly concentrated in the newly industrialised economies and a small number of resource-rich countries (essentially those rich in oil).

Table 58. **Inflows of foreign direct investment to all countries and to developing countries, by region, 1980-84, 1985-89 and 1988-89**

Host region and economy	Annual average inflows ($ billion)			Shares (Percentage)		
	1980-84	1985-89	1988-89	1980-84	1985-89	1988-89
All countries	49.70	119.0	173.0	100.0	100.0	100.0
Developing countries	12.50	22.20	29.20	25.2	18.6	16.9
Africa	1.20	2.60	3.20	2.4	2.2	1.9
Latin America and Caribbean	6.10	8.30	10.00	12.3	7.0	5.8
East, South and South East Asia	4.70	10.70	15.20	9.4	9.0	8.8
Oceania	0.13	0.14	0.20	0.3	0.1	0.1
West Asia	0.37	0.40	0.54	0.8	0.3	0.3
Other [1]	0.04	0.03	0.05	0.07	0.03	0.03
Ten largest host economies	9.01	14.31	19.26	18.1	12.0	11.1
Argentina	0.44	0.73	1.09	0.9	0.6	0.6
Brazil	2.10	1.59	2.53	4.2	1.3	1.5
China	0.53	2.49	3.29	1.1	2.1	1.9
Colombia	0.40	0.56	0.39	0.8	0.5	0.2
Egypt	0.56	1.23	1.40	1.1	1.0	0.8
Hong Kong	0.68	1.65	2.04	1.4	1.4	1.2
Malaysia	1.13	0.83	1.28	2.3	0.7	0.7
Mexico	1.50	2.02	2.42	3.0	1.7	1.4
Singapore	1.39	2.50	3.29	2.8	2.1	2.0
Thailand	0.29	0.72	1.40	0.6	0.6	0.8
Least developed countries	0.19	0.15	0.17	0.4	0.1	0.1

1. Includes Malta and Yugoslavia.
Source: UNCTC estimates, based on International Monetary Fund balance-of-payments tape, retrieved on 10 January 1991, OECD estimates and UNCTC, Directory of Transnational Corporations, New York, UNCTC (1991).

Other channels for the transfer of technology

The other two private sector channels of technology transfer – capital goods exports and direct payments – indicate a concentration of technology flows to the newly industrialising economies of Asia (see Table 59); at the same time, East Asian newly-industrialised economies (in particular the Republic of Korea and Taiwan) are becoming technology exporters. Official technology assistance flows – the smallest of the four technology transfer channels – have been more evenly distributed among various developing country recipients than have private flows. For many developing countries, technical co-operation grants form virtually the only source of technology transfer. Given the growing importance of technology flows to economic development, the uneven distribution of aggregate technology flows among developing countries indicates the presence of a substantial barrier to improved development prospects for a large number of countries.

Table 59. **Structure of technology flows to developing countries, 1988**

$ billion, current prices

	Capital goods imports	Foreign-direct investment inflows	Technical co-operation grants
All developing countries	144	28.7	12.6[1]
Africa	17	2.1	4.9
Asia	87	14.9	2.9
Latin America and the Caribbean	36	11.4	2
cf.			
Least developed countries	4	0.1	2.6

1. Grants not allocated to individual countries are included in total, but not in regional group.

Sources: UNCTAD, "Transfer and development of technology in a changing world environment: the challenges of the 1990s", (TD/B/C.6/153) Table 2; updated data from UNCTAD Secretariat; International Monetary Fund balance-of-payments tape.

2. TECHNOLOGICAL CAPABILITIES AND LATECOMER INDUSTRIALISATION

Since the early 1960s, the NIEs have achieved extremely high rates of growth in gross national product (GNP), exports of manufactures and per capita income, with the result that, in 1987, Hong Kong and Singapore both had per capita GNP greater than Spain, Ireland, New Zealand, Greece and Portugal. This section examines how these countries have used technology to strengthen their industrial capabilities and to sustain a process of transformation from agriculture to industry. The analysis is centred on a limited number of countries: the first-tier Asian NIEs (Republic of Korea, Taiwan, Singapore and Hong Kong), two quasi-continental economies (India and Brazil), and the second-tier Asian NIEs (Malaysia, Thailand, the Philippines and Indonesia)[4]. The first group (sometimes referred to as the "four Asian dragons" or the "Gang of Four") offers the most widely analysed records of success; the literature on the last group is more limited, but an extensive study on technological capabilities in Thai industry was recently completed (Thailand, TDRI, 1989).

It is useful to consider what the industrialisation of the NIEs has in common with and what sets it apart from other developing countries. Industrialisation refers to the simultaneous occurrence of high growth of industrial output – with the industrial sector accounting for a progressively larger share of GNP, largely at the expense of agriculture – and a transformation of the structure of industrial production itself. The rates

of GDP growth of the NIEs over the last few decades have generally been above the average for developing countries and for the world economy (see Table 60). That growth was largely attributable to a rapid increase in industrial production. The structure of the

Table 60. **Manufacturing value added in the least developed countries, 1975-87**

	Manufacturing value added (MVA)		Share of GDP in percentages
	In millions of US dollars	Growth rate in per cent	
1975	6 240	–	8.5
1976	6 316	1.21	8.3
1977	6 340	0.39	8.1
1978	6 338	−0.05	7.8
1979	6 168	−2.67	7.6
1980	6 470	4.90	7.8
1981	6 430	−0.62	7.6
1982	6 449	0.29	7.4
1983	6 816	5.68	7.6
1984	7 035	3.22	7.7
1985	7 248	2.98	7.7
1986	7 468	3.08	7.6
1987	7 748	3.75	7.6

Note: Data for 1983-87 concerning MVA is influenced by one country (Bangladesh).

Source: UNCTAD (1989c, p. 32).

manufacturing sector also grew more rapidly and changed more dramatically in the NIEs than in other developing countries. Using an index of structural change in manufacturing constructed by UNIDO, Table 61 shows the indices for several NIEs as well as the averages for developing countries and the world: for Korea, the index was 31.37, for Singapore, 48.32, for Brazil, 30.03, and for India 20.89. The developing country average was only 13.83 and the world average 10.6. Structural change was generally most pronounced in the manufacturing sectors that grew the fastest. Most NIEs have also experienced unusually high rates of growth of manufactured exports over recent decades.

Technology is by no means the sole reason for the rapid industrialisation of the NIEs. What requires explanation in particular is why certain developing countries have been far more successful than others in utilising technology to move from agricultural to industrial societies. There may be lessons to be learned from the NIE experience with regard to the determi-

nants of technological dynamism and the conditions under which technological change and industrial development become mutually reinforcing processes.

Conventional wisdom about growth in the NIEs

No model can capture the diversity of NIE experiences. Nevertheless, neo-classical economists have sought neat explanations which would conform to their notions of what constitutes correct policies and can be comfortably integrated within a general theory. Thus has emerged what Bradford (1986) has referred to as the "new orthodoxy" and Hirschman (1986) calls "mono-economics". This view draws a sharp contrast between the approaches followed by Latin American NIEs and those followed by Asian NIEs, stylising (or caricaturing) them as inward-oriented and outward-oriented growth strategies respectively. The story goes that the latter outperformed the former because they "got prices right". In other words, they gave greater freedom to the operation of market forces.

Detailed empirical studies of the experiences of countries like Korea, Taiwan, and Singapore seriously challenge this version of success stories (S. Smith, 1991). Evidence shows that the governments of these countries were highly interventionist, although in different ways. The Korean government, for example, used subsidised credit and highly selective rationing of credit to favoured sectors and firms as a way of directing investment into strategic industries. The Taiwanese government relied more heavily on fiscal incentives, also administered in a highly selective way to promote certain priority industries. Moreover, public enterprises in Taiwan have played a much larger role than in Korea in fixed capital formation. A more rigorous demonstration of the fallacy of the "getting prices right" argument is contained in an exercise performed by Bradford, using data from the United Nations International Price Comparison Project (Bradford, 1987). The most rapid structural changes in manufacturing sectors of NIEs have occurred when investment goods prices were low relative both to the international prices of such goods and to the prices of other goods in their own economies (in particular, consumption goods), suggesting a certain degree of price distortion. Most East Asian first- and second-tier NIEs fall into this category, as do some Latin American NIEs. Obviously, the under-valuation of investment goods has resulted in above average rates of capital formation, as reflected in the ratio of investment to GDP.

A variation of the "getting prices right" argument is the "stages theory of comparative advantage", which views the shifting industrial structures of the NIEs as merely the reflection of shifting comparative advantages enjoyed by relatively open economies, i.e. economies with few price distortions (Balassa, 1977). By this reasoning, countries moved from producing mostly pri-

Table 61. **Structural change and industrialisation, 1965-80**

Transitional economies	Index of structural change in manufacturing[1]	Average growth rate of value added in manufacturing
European NIEs		
Spain	24.73	6.78
Yugoslavia	12.01	6.94
Portugal	21.61	7.18
Greece	13.56	7.00
Asian NIEs		
India	20.89	2.59
Korea	31.37	18.99
Taiwan	n.a.	n.a.
Hong Kong	9.87	6.05
Singapore	48.32	11.41
Next-tier NIEs		
Philippines	10.95	5.45
Thailand	17.69	7.98
Malaysia	15.86	8.12
Colombia	10.90	6.36
Natural resource NIEs		
Brazil	30.03	9.50
Mexico	14.83	7.09
Argentina	15.90	3.12
Indonesia	19.52	10.20
Global averages		
Developed countries	10.90	4.66
Developing countries	13.83	6.55
World	10.60	4.85

1. The index of structural change is derived from sixteen manufacturing branches. It is a measure of the degree of correlation between the value-added shares in 1965 and 1980. If the correlation is high, then there is little structural change and the index is low. But if the correlation is low, then there is a lot of structural change and the index is high. Both expanding and shrinking branches contribute to the index.
Source: UNIDO (1985, pp. 31-40 and country tables, pp. 35ff).

mary commodities to light manufactures to more skill- and capital-intensive manufactures as their factor endowments changed in the course of development. By this logic, unskilled labour scarcity in the first-tier Asian NIEs is now forcing the transition to skilled labour-intensive and capital-intensive activities. On the face of it, the argument has a certain plausibility, which is perhaps why it is so frequently repeated. In addition, it also appears to conform nicely to the precepts of neo-classical trade theory. However, it does not stand up to close examination, for it ignores the significant role played by explicit policy interventions in creating comparative advantages that would have developed far more slowly in response to market forces alone (Amsden, 1989). Moreover, this analysis fails to explain satisfactorily the fact that many of the heavy industries in the NIEs that eventually became success- ful exporters were built up initially as import-substitut- ing industries under considerable government protection.

The initial manufactured products export boom of countries like Taiwan, Korea, and Singapore was, in fact, heavily import-dependent. Local value added was limited. The rapid growth of light manufactured exports created conditions for the development of industries to supply intermediate inputs and capital goods. However, many intermediate goods require very high volumes to be produced economically. Thus, even with protected domestic markets, exports still proved critical to efficient production. This is why the Korean government, for example, subsidised exporters of capi- tal-intensive goods until they became internationally competitive on their own. The main function of the protected domestic market was, in fact, to provide val- uable learning economies for the efficient operation of complex process plants and other production facilities.

Beyond conventional wisdom: the role of technological capabilities

Following work by Chenery (1986) a number of macro-economic studies have shown the inadequacy of conventional explanations of economic growth and structural transformation. They establish a statistical correlation between high growth rates in productivity and exports and technological effort, measured by some aggregate such as patenting activity or R&D expenditures as a share of GDP (Fagerberg, 1988b). But they stop short of explaining the concrete contribu- tions that technology, and the acquisition of technolog- ical capabilities, have made to the rapid industrial development of the NIEs.

The notion of technological capabilities attempts to capture the great variety of knowledge and skills needed to acquire, assimilate, use, adapt, change and create technology. It goes well beyond engineering and technical know-how to include knowledge of organisa- tional structures and procedures as much as knowledge of behavioural patterns, e.g. of workers and customers. Firms need certain complementary assets and capabili- ties in order to create, mobilise and improve their tech- nological capabilities, among which may be noted: organisational flexibility, finance, quality of human resources, sophistication of the support services and of the information management and co-ordination capabilities.

The debate has so far focused on producers' capa- bilities, grouped in four broad categories[5]:

- the knowledge and skills required for the process of production, where shop-floor experience and "learn- ing-by-doing" plays an important role;
- the knowledge and skills required for investment, i.e. the establishment of new production facilities and the expansion and/or modernisation of existing ones;
- the vast area of adaptive engineering and organisa- tional adaptations required for the continuous and incremental upgrading of product design and per- formance features and of process technology;
- and, finally, the knowledge required for the creation of new technology, i.e. major changes in the design and core features of products and production processes.

Today, knowledge about user needs has become quite as important. In practice, constraints on the demand side develop alongside the barriers to produc- tivity enhancement in the economics of production. More than ever before, not only innovation but even successful learning require knowledge about user needs as much as knowledge about the process of produc- tion (Freeman, 1982; Lundvall, 1988; and Perez, 1990 and 1989). The knowledge of how a particular technol- ogy can be utilised requires significant "learning-by- using" experience. Without sophisticated users who can conceptualise a need and translate it into a requirement for products and services, there will not be sufficient pressure to improve the technological capa- bilities of producers (see Chapters 3 and 11).

In order to develop viable forms of user-producer interaction, clear rules and long-term relationships are essential. Substantial intangible investments are required in order to establish channels of information as the necessary complement of investing in physical capital. It is a time-consuming process often involving quite heavy costs. The fundamental changes which are now occurring in the nature of export markets must be discussed in this perspective. If "differentiated mass production" rather than "flexible specialisation" (Piore and Sabel, 1984) is the dominant trend, this will have far-reaching implications for NIEs. As even those firms that specialise in a single standardised product must be able to adapt to variations in demand volume and in delivery times, strength in mass producing stan- dard price-sensitive products may matter less and less, while the capacity to quickly identify and exploit new high value-added market opportunities and to customise products and services accordingly may gain in importance. This will certainly require major changes in the prevailing approach to technological capability formation, with a shift to product design and market development capabilities, for which even the most successful EANIEs are still badly equipped.

Spatial proximity and a common cultural background are important, particularly when both technology and user needs are complex and changing. "Organisational proximity" through vertical integration may overcome geographic and cultural distance. This is one reason why, particularly for NIEs, a focus on large, multiple-product firms would seem to make sense (Amsden, 1989). Yet vertical integration can have important drawbacks, as it tends to cut off the integrated units from interactions with outside, independent users and producers (Imai and Itami, 1984). The alternative would be to rely on "organised markets", defined as an intermediate form of economic institution which combines both the formal characteristics of a market – product, seller, buyer and a price – with "(...) elements of organisation – as exchange of qualitative information, co-operation, dominance and trust" (Lundvall, 1991, p. 449).

Firms' technological capabilities in NIEs

Firms in NIEs have acquired their technological capabilities, first and foremost, by making judicious use of foreign technology sourcing as a way to acquire quickly the needed knowledge. They relied on customer firms to provide specifications and concentrated on refining productive capacity to produce to specification at low cost. Korean and Taiwanese firms used Original Equipment Manufacturer (OEM) agreements and Singapore and second-tier NIEs relied largely on FDI as their means of entry into world markets. The greatest reliance on foreign technology sources is generally found in sectors with the most demanding technological requirements. FDI, technology licensing, and other forms of foreign technology sourcing are much more important in the electronics industry than in the textile industry (Westphal et al., 1985). What is crucial is how effectively a firm combines foreign technology elements with its own experience and knowledge in order to strengthen its internal capabilities. It is principally this ability to learn how to utilise technology to strengthen competitiveness which distinguishes the successful NIE firms from the less successful ones.

Production capability

In the early period of rapid industrial growth, most NIE industrial production was concentrated in consumer non-durable industries with relatively low technology requirements, and production capability was thus restricted to the efficient operation of labour-intensive production processes. At least from the early 1960s, many light manufacturing firms in Korea and Taiwan were encouraged, if not pressured, to export. As a result, they were forced to maintain satisfactory product quality to meet standards in major markets like the United States and Western Europe. By focusing on the "low end" of mass consumer goods markets,

---

Box 51. Mastery of industrial production and economic development

The mastery of industrial production rests on a number of factors which include production management, production engineering, and repair and maintenance of physical capital. The first involves organisation and control of the production process, as well as its interactions with upstream, downstream, and ancillary activities. Production engineering includes raw material control, production scheduling, quality control, and trouble-shooting. Adaptive engineering is an important activity at the borderline of production capability. It involves making minor adjustments and incremental improvements in the production process. There are many possible reasons for such adjustments, such as the need to adapt the process better to local environmental conditions, e.g. to accommodate a locally available raw material.

Firms differ widely in their mastery of production capabilities, and different techniques require different technological capabilities for effective use. For example, one textile firm may use a highly automated loom and another a traditional mechanical loom. The skills required to operate the two types of machinery are worlds apart, and each type of machine functions best in a particular environment. An automated loom requires different work scheduling, inventory, maintenance, and other procedures from a mechanical loom. Material inputs – e.g. yarn properties – may also be different, and the firm needs to be able to evaluate the material requirements. Thus, for a given product, more than one production technique can generally be used. But each requires a particular set of technological capabilities to be operated most efficiently. The sets may overlap in part, but knowing how to operate one technique efficiently does not necessarily enable a firm to operate another just as efficiently.

---

these firms acquired valuable experience in mass production methods. The more successful were able to learn from that experience how to upgrade product quality so as to be able, eventually, to move into higher value-added segments (though not yet the "high end") of the market.

In industries like textiles and garments, production capability was not a serious constraint, since the production technology was largely embodied in capital equipment (spinning machines, looms, sewing machines) available on the world market (Pack, 1987). Still, installing, operating and maintaining that imported equipment is not always a straightforward matter, and NIE firms were obliged to rely on foreign equipment suppliers or independent consultants to provide them with the necessary know-how. Once foreign technical assistance ended, experience differed: Korea's major steel producer (POSCO) had relatively few difficulties carrying on operations efficiently, in part because much of the process technology is embodied in equipment; on the other hand, its major shipbuilder (Hyundai Heavy Industries) encountered greater problems, due to the fact that the process of shipbuilding is largely embodied in worker skills (Amsden, 1989).

Imported equipment designed for use in industrialised countries can seldom be used optimally by NIEs or other developing countries without substantial adaptations. The need to make such adaptations and improvements has been a major stimulus to the emergence of producers of capital goods in NIEs. As a result, Hong Kong, Korea and Taiwan have emerged as sizeable exporters of textile machinery to other developing countries (Chung-Hua Institution for Economic Research, 1987, p. 125). Blending different types of technological capabilities is of utmost importance for latecomer industrialisation. Production capability in knitting and weaving, for example, requires sound adaptive engineering (minor change) capabilities and, when combined with reverse engineering, it makes possible the complete local fabrication of textile machinery. Thus is born a new industry which in turn requires the development of investment, production and change capabilities.

In mature industries like textiles, pressures to upgrade developed in the first-tier Asian NIEs by the late 1970s, as lower cost producers made it difficult for them to maintain market share at the low end (Amsden, 1989, p. 261). However, NIE growth could not be sustained indefinitely through upgrading of products within traditional industries. Diversification became a dominant impulse during the 1970s (see Table 61). (Hong Kong is the major exception to this rule: the island already had a highly developed textile and garment industry by the mid-1960s, and that industry continues to weigh heavily in the industrial structure; it accounted for 40 per cent of MVA in 1986.) The diversification occurred both within firms, as in the conglomerates formed in Korea, and by the formation of new firms, as often occurred in Taiwan. In Singapore, Malaysia and Thailand, entry of new foreign investors was a predominant means of industrial diversification.

Investment capability

The creation of new industries and the entry into new product markets creates a tremendous demand for investment capability. This covers two major sets of activities: pre-investment feasibility studies and project execution. Project execution in turn involves several subsidiary activities, including project management, project engineering (detailed studies, basic engineering and detailed engineering), procurement and start-up operations.

Where foreign technology sourcing is involved in setting up a new production facility, the local firm need not possess the full range of investment capabilities enumerated above. In effect, it hires some of them from foreigners. Even then, it must possess a number of other capabilities if it is to be able to source technology efficiently, i.e. it must be able to search out and then choose the most suitable technology, negotiate its procurement (in the appropriate form), determine the modes of transfer, etc. Finally, the firm must supervise the installation and start-up of the facility. Even if the firm relies on foreign engineers for installation and start-up, it must become an active participant in the process if it is to absorb the principles of the technology (see Box 52). There may be limited "learning spillovers" when diversification involves investment in a process or product closely related to equipment already used by the firm, but this is rarely the case. Newly created firms, of course, do not have the possibility of building upon prior investment capability.

The role of conglomerates

In addition to macro-economic structural factors, elements of industrial structure also matter for the acquisition of technological capabilities[6]. Two factors that have been of particular importance for the NIEs are conglomerates and flexible supplier/subcontractor networks. Account must also be taken of the role of supporting institutions (see Box 53).

The Korean *chaebol* illustrate the positive role played by conglomerates in latecomer industrialisation. Faced with limited prospects for further growth in some of their mass consumer markets, the *chaebol* chose to diversify in order to capitalise on their manufacturing experience in other markets. This allowed them to transfer substantial resources (in technology, human resources, etc.) and at the same time to create synergies in their distribution networks. By avoiding the dangers of specialisation which had been recognised by the Korean government, the *chaebol* were free to undertake risky investments (using subsidies provided by the state), notably in R&D (Amsden, 1989, p. 151).

Even though conglomerates clearly played an important role during the "catch up" phase of Korea's industrialisation, they can also exert a negative impact on the economy by stifling the emergence of local SMEs. The latter have more entrepreneurial spirit and greater organisational flexibility than the *chaebol* and can be an important source of technological innovation, especially product innovation. Aware of the need for innovation, the *chaebol* have come to recognise that the more they advance technologically, the more they need to devote greater effort to R&D and focus their activities in fewer areas. As the leading *chaebol* target high-growth, high-technology markets like electronics, for example, they are beginning to shed certain of their more traditional, low technology activities. Thus, while they will remain conglomerates, their structure may evolve towards that of the major Japanese conglomerates. These generally have more tightly integrated structures in which their different products are technologically related; they are thus able to reap sizeable economies of scope in R&D as well as in production.

The role of supplier/subcontractor networks

The development of supplier/subcontractor networks – as an alternative to vertical integration – is a crucial factor in ensuring the diffusion of technology through interaction between component producers and their industrial customers (Imai, 1988a). A major feature of Japanese industry, this system has been transplanted to Korea and Taiwan. The Brazilian conglomerates, instead, have tended to opt for the internalisation of component production and have thereby forfeited significant economies of scale[7]. The main form of technology transfer from principal to subcontractor has been in the area of training in inspection standards and tight quality control, the organisation of just-in-time inventory management systems, and close co-operation. Transfer of personnel is an important source of improved managerial methods and technical expertise, while pressures exerted by the principal serves to force subcontractors to improve efficiency and quality.

The role of the state and structural influences on the formation of technological capabilities

Various positive or negative "externalities" shape the environment within which firms operate; they help explain why a higher proportion of firms succeed in certain NIEs than in others. These externalities may be political, structural, or institutional, but a certain number of them seem particularly important for explaining the different performances observed in the various groups of NIEs.

Relations between government and business

One important feature of the success of EANIEs has been the effectiveness of the State as an agent of industrial transformation. This clearly sets these countries apart from the Latin American NIEs. In the latter, the state has generally been beholden to powerful vested interests, whose privileged economic positions it has helped to preserve and even reinforce without extracting any quid pro quo (UN ECLAC, 1990). Indeed, in Latin America the State has generally exacerbated the "rent-seeking" activities of firms, first by creating conditions which have made such activities profitable and then by demanding a share of the rents. In some cases, these interests were inimical to industrial development, while in others they sought to carve out specific industrial sectors or activities as their exclusive profit-making preserves. With a heavily protected domestic market serving as their principal source of profits, firms faced little external competitive pressure to improve their efficiency. Yet those economies were not entirely devoid of technological effort. What was lacking was an institutional structure and policy regime which systematically encouraged and rewarded such efforts.

In the EANIEs, sweeping agrarian reform measures enacted after the second World War made for the relative autonomy of the state from social classes committed to the pre-industrial order. This was the case in Korea and Taiwan. Hong Kong and Singapore clearly had no comparable agrarian question. Once the State had consolidated power, there was no guarantee that it would direct its energies toward industrialisation, nor that, even if it did, its industrial development policies would prove effective.

Many newly independent developing countries shared a political commitment to industrialisation, one based moreover on a firm social and economic rationale. Yet not all governments possessed the same capacity to implement effective policies. What appears to distinguish the EANIEs is the determination of the government to extract certain "performance guarantees" from firms in exchange for state support, whatever form that might take (Amsden, 1989). In particular, export requirements were an important means of imposing discipline on firms that received subsidised credit, tax incentives or other concessions from the government. These subsidies were administered on a highly selective basis to direct both investment funds and other resources, as well as technological effort, into certain industries and activities.

The difference between East Asia and Latin America was not between protected and unprotected domestic markets. In East Asia, governments frequently combined import substitution with export promotion. However, firms were not left to hide behind tariff walls or import bans indefinitely. They were to "pay back" the government for the special treatment they received by demonstrating their capacity to export, thereby helping to earn much-needed foreign exchange for the imported capital goods and intermediates needed to sustain the industrialisation process. Even though at any given time there were products in various stages of import substitution, the rapid transformation of the industrial structure was continually creating new demands for capital goods and intermediates that could not yet be supplied locally. The greater capacity which high export growth has afforded to the East Asian NIEs to import new capital equipment – and thereby acquire rapidly new production capability – has been an important source of the superior productivity performance of East Asian over Latin American NIEs.

Savings rates, income distribution and foreign indebtedness

Another frequently cited determinant of the differential performance of NIEs is the large discrepancy in savings rates between East Asia and Latin America, which in turn is related to the wide disparities in income distributions between the two regions (Felix,

1989). The savings rate is critical to the rate of capital formation, and a high rate of capital formation is critical to the structural transformation of the economy. This transformation induces the demand for technological capabilities, especially investment and production capabilities. While savings may come from either domestic or foreign sources, reliance on foreign savings to compensate for a low domestic savings rate is not a viable long-term strategy, as the experience of the most heavily indebted countries has shown. The behaviour of East Asian NIEs contrasts sharply in this respect with that of their Latin American counterparts (World Bank, 1989), whose poor savings rates also reflect marked social inequalities.

The wide income disparities in Latin America have also contributed to the narrowness of the domestic market. The limited purchasing power of the average citizen, even in countries with large populations such as Brazil and Mexico, has severely constrained the prospects for efficient import substitution in industries where economies of scale are important. In East Asia, import substitution in intermediates was linked to large volume exports of final consumer goods. In Latin America, by contrast, second stage import substitution has generally been linked to first stage import substitution of consumer durables for a relatively small middle and even smaller upper class. Thus it is not surprising that few such industries have been able to achieve efficient production scales.

All the NIEs resorted to heavy foreign borrowing to offset their low domestic savings rates. The EANIEs differed from their Latin American counterparts, however, in that the state was able to ensure the productive investment of funds borrowed from abroad by maintaining tighter accountability. Korea made efforts to replace foreign saving as the main source of capital accumulation. Whereas in 1962-66 foreign saving outweighed domestic saving in its contribution to investment, in 1984 domestic saving accounted for 27.6 per cent of GNP and foreign saving for only 2.3 per cent (Hong, 1987). In the case of Brazil, gross domestic saving remained virtually constant as a share of GDP, at 22 per cent, from 1965 to 1988 (World Bank, 1989). Overall net saving of the Brazilian economy actually fell from 1975 to 1985, due to the massive debt servicing requirements and the resulting net resource transfer, from 21 per cent to 18 per cent of GDP (UN, 1986a).

The heavy debt servicing burdens of the Latin American NIEs has proved to be a serious impediment to productive investment during the last decade. The Latin American industrial base has not been upgraded or expanded appreciably during this period, and in many industries it has lapsed into obsolescence (UN ECLAC, 1990). The Latin American debt crisis has not only stunted firms' capacity to invest in productive assets; it has also prevented them from undertaking critical investments in the training of personnel and in R&D, which are necessary even for successful imitative learning. A statistical exercise designed to "explain" why growth rates differ across countries found that a large part of the difference in growth rates between Latin American NIEs and EANIEs over the last two decades can be attributed to differences in their innovative activity, as reflected in R&D expenditures as a proportion of GDP (Fagerberg, 1988b). Worse still, the state's capacity to invest in education has been severely constrained by the fiscal crisis. For example, between 1972 and 1987 the percentage of the central government budget allocated to education fell from 16.4 per cent to 8.7 per cent in the case of Mexico and from 20 per cent to 6 per cent in the case of Argentina (World Bank, 1989).

The characteristics of the indigenous scientific and technological (S&T) infrastructure

While the importance of this type of infrastructure can be questioned during the early phases of industrialization, there is no doubt that indigenous S&T capabilities have now become essential for sustaining the industrialisation process. All the NIEs recognise this, but not all are equally capable of making the necessary adjustments and improvements in their S&T infrastructure in a timely manner. Their different capabilities to adapt to the new environment depend, in part at least, on how their S&T institutions have evolved to date.

Both India and China have a long-standing tradition of basic and applied research which, at least in certain military-related fields, is very near the frontier. Yet, despite their scientific expertise, they have been slow to acquire proficiency in commercial application of new technologies. In short, there would seem to be seizable barriers to diffusion of scientific knowledge into the industrial economy. Such diffusion is by no means an automatic process, but involves a number of intermediate links (or "bridging institutions") which may be only imperfectly formed in those economies. One weak link in both China and India has been their inefficient engineering and capital goods sectors. Another shortcoming of the S&T infrastructure in the quasi-continental economies (as well as in Brazil) has been the high degree of concentration of R&D activities in government laboratories and state enterprises, and a relatively narrow focus of such R&D on defence-related applications. This may partially explain the advanced state of aeronautics and space-related research, as well as of nuclear-related research in China, India and Brazil. Spillovers to the civilian sector have been limited.

In the Asian first-tier NIEs, by contrast, a much higher percentage of R&D focuses on commercial applications. Taiwan's R&D sector, for example, includes a number of quasi-public research institutes with close links to the private sector, and scientists are even encouraged to establish their own firms to commercialise the results of their R&D. In South Korea a reorganisation of the R&D infrastructure is under way: the government is encouraging private firms to set up research institutes to facilitate industrial restructuring and technological development. At the same time, government research institutes are concentrating more on improvements in production technology to support the needs of private sector firms.

The educational level of the population also varies widely across countries. The Asian NIEs, in general, have very highly educated populations and educational access is fairly evenly distributed, at least at primary and secondary levels. Moreover, educational institutions have been quite flexible in adapting to the demands generated by the restructuring of the economy and the changing skill composition of the workforce. For example, the percentage of tertiary students in engineering in South Korea is at least double the figure for the Latin American NIEs and India (Lall, 1990). In countries like India and Brazil, on the other hand, educational opportunities are far more unevenly distributed and the average educational level is below that in the Asian first-tier NIEs. In Latin America, moreover, the fiscal constraints of a number of countries have caused the quality standards of educational institutions to deteriorate.

As technological innovation has become more widespread and systematic in certain NIEs, specialised financing institutions have evolved to enable firms to undertake high-risk investments in R&D and engineering. In short, venture capital markets have developed in certain Asian NIEs and Brazil, but they function imperfectly. In South Korea, for example, a large share of loans and equity investments by venture-capital firms are channeled to the more established large-scale conglomerates which are relatively good risks and have substantial collateral. In Taiwan, venture capital is more widely accessible to small start-up firms, and informal credit networks are more developed than in South Korea. The venture-capital firms themselves are entrepreneurial start-ups in Taiwan, whereas in South Korea they tend to be firms started at the government's initiative.

3. INFORMATION TECHNOLOGY AND THE NEWLY INDUSTRIALISED ECONOMIES

The current world-wide restructuring of industrial manufacturing based on the diffusion of new IT and of related organisational changes, have important implications for NIEs, both as users and as suppliers. As users, NIEs must determine a desirable, and feasible, rate of adoption of the new technologies within existing industries. To a large extent, these decisions are made by private firms, but the government clearly can shape the environment in a way which either accelerates or retards the pace of adoption. As technology suppliers (or, more accurately, as suppliers of products embodying new technologies, such as controllers or microprocessors), NIE firms are confronted with very difficult decisions. Should they continue to pursue imitative strategies and remain locked into subcontracting or OEM arrangements? Or should they, at considerable expense, generate the product, design and market development capabilities to compete on their own in world markets? Here also, governments may be able to influence the probability of firms' success in moving beyond imitation.

Adaptation and diffusion of advanced manufacturing technologies

Faced with the need to adapt to technological change, firms in traditional industries have had to choose among automating their activities, using computer-based advanced manufacturing technologies (AMTs)[8], searching for cheaper sources of labour abroad, or abandoning some industries altogether in favour of less labour-intensive ones.

Factors affecting the choice of technology

The choice is governed by factors such as firm size and structure (large, multi-product firms have greater scope for diversification than small, specialised firms), financial resources, and the existence of economies of scale which justify investment in automated production equipment. Because the average NIE firm is much smaller than the average US, Japanese or European firm[9], automation possibilities are reduced. In addition, unlike NIEs, OECD firms have a long tradition of automated production and thus have accumulated a vast body of empirical and theoretical knowledge. Therefore, NIE firms typically have to engage in "quantum jump" decisions, which would move them from manually operated or at best semiautomatic equipment to computer-based programmable automation. Obviously, this creates substantially higher uncertainties and risks than those faced by OECD countries.

If cost considerations alone dictated the automation decision, it is doubtful whether many NIE firms would risk the investment. NIE firms have in fact frequently preferred to move their production offshore to lower wage countries rather than to invest in automation at home. This is the case, for example, with many NIE textile and garment firms. At least for garments, automation technologies are still relatively underdeveloped. Firms thus have other options to lower their labour costs, at least in the short run.

Costs are not all that matter, however, in most industries. Especially at the higher end of the textile and garment industries, quality, speed-to-market and the flexibility to respond to rapidly changing fashion trends are of at least equal importance. Thanks to AMT, this can be achieved without excessive tradeoffs in terms of higher costs. Consequently, NIEs are under strong pressure to invest in flexible automation systems in order to penetrate markets where product differentiation is what matters. While these investments may prove necessary, they are by no means always sufficient. Frequently a successful product differentiation strategy requires substantial complementary investments (mostly intangible ones), for example in advertising, marketing and distribution channels. In some markets, proximity to customers constitutes a decisive competitive advantage, since it allows quick response to changing requirements and permits close interaction with the user in the design and development of a new product. In such cases, NIE firms whose major markets are in the OECD region not only have to invest in production automation but must do so overseas. Such outward FDI is driven, first and foremost, by the increasing trade restrictions on NIE exports to the United States and by the uncertainties surrounding access to the European market after 1992.

In short, NIE firms considering whether to invest in AMTs are confronted with enormous challenges which, by any standard, surpass the challenge faced by their OECD counterparts. They must simultaneously deal with cost pressures at home and with intensifying pressures from their major trading partners. They must also learn, at very short notice, how to manage investments in AMTs and how to co-ordinate global investment networks, activities in which they have virtually no prior experience.

Changes in industrial organisation

Major changes in organisation generally must accompany the introduction of AMTs for their full benefits to be realised (Hayes and Jaikumar, 1988). NIE firms need to possess or to acquire organisational flexibility as a precondition for achieving production flexibility, but they have faced major problems in undertaking the organisational changes required. In most NIEs, firms are organised in accordance with rigid hierarchies, which tend to result in a number of informational diseconomies, strictly segmented job functions and little contact between engineers and workers on the shop floor (on these issues, see Chapters 4 and 7). The hands-on engineers common in Japanese enterprises (and also in the Korean *chaebol*) are all too rare in many NIE firms, especially in Latin America and India. Important learning opportunities are thus foregone.

Another important issue related to organisational restructuring is the appropriate degree of product specialisation/diversification of the firm. As economies of scope have gained in importance, there may be justification for somewhat greater differentiation of product offerings, as long as those products draw upon a common technical and skill base. Thus, the Korean *chaebol* are confronted with the question of whether and how far to streamline their operations to concentrate on core businesses (Ernst and O'Connor, forthcoming).

New skill requirements

Highly specialised manual skills are becoming less important with the diffusion of AMTs; broad, generic skills are gaining in importance[10]. Consequently, on-the-job apprenticeship is less useful as a means of skill formation; formal education becomes critical (see Chapter 7).

Most NIEs are well aware that formal education, especially in engineering and the natural sciences, is critical to an effective utilisation of ITs. Yet only the East Asian NIEs have invested heavily in upgrading educational levels. At least since the early 1980s, Latin American NIEs have been strapped with heavy debt service burdens that have diverted resources from human capital formation. In 1987, Mexico had only around 6 200 students enrolled in US universities, Brazil 3 000, and Argentina 1 700. In contrast, China had 51 800, India 21 000, Korea 20 500 and Malaysia 19 500. With the exception of Argentina, the Latin American NIEs also have relatively low secondary enrolment rates as compared with the first-tier Asian NIEs. In the context of new production technologies and a work organisation which requires a highly flexible and adaptable work force, learning how to learn is obviously a crucial capability. To the extent that education, particularly at the secondary level, is largely concerned with developing this skill, a country with a low secondary level enrolment will be severely handicapped in reaping the benefits of AMTs.

Financial constraints

Adopting AMTs normally involves huge capital outlays. This is not to imply that the capital input per unit of output is greater than with the older technologies. Some of these technologies may be in fact capital-saving as well as labour-saving, if savings in working capital and the potential for more efficient capacity utilisation are considered (Freeman and Soete, 1987). But where capital markets are not operating efficiently, firms will find it difficult to borrow against future potential earnings to pay for the purchase of new equipment. Small firms without adequate retained earnings to finance such purchases on their own may thus be effectively prevented from undertaking investment in AMTs.

Beyond the initial capital costs, substantial follow-on costs have normally to be met. A large part of these follow-on costs involve investments in intangible assets, whose share in total investment is rapidly increasing[11]. Intangible investment covers complementary investments in software development and the training of workers to make effective use of the new production methods (see Chapter 5). Firms may also need to hire consultants to advise them on how to reorganise their production process, to set up a management information system, to implement a quality control programme, etc. Small firms in particular are at a distinct disadvantage in undertaking these investments. Thus, without some form of investment support to such firms, they face the prospect of lagging behind their larger OECD or NIE competitors in the introduction of process innovations.

The role of externalities

Especially for small firms, a number of supporting institutions are required in order to foster a greater diffusion of AMTs. With new communications technologies, information search is perhaps less costly than it once was, but even now, to attain significant economies of scale, it is desirable for a single institution – a government technical information centre or an industry association – to undertake searches on behalf of many firms[12].

Beyond technical information, technical training is of crucial importance. In Singapore, for example, the government has established a training centre where firms can send workers to learn the use of CNC (computer numerically controlled) machine tools, computer-aided design (CAD), and other new technologies. In Taiwan, the China Productivity Centre (CPC) has been instrumental in promoting effective use of automation (Dahlman, 1989). The Korea Productivity Centre (KPC) performs a comparable function in promoting micro-electronics based factory automation by local firms.

The new ITs are valuable not only for their applications to production within the firm, but also for the advantages they offer for logistical co-ordination among various firms. To facilitate such co-ordination, substantial infrastructural investments are required, particularly in the communications network, to make possible the rapid and reliable transmission of data in a local area or a wide area network. EANIEs that have been able to undertake the necessary investments, e.g. in fibre optic transmission systems, are in a much better position to exploit the full potential of such information networks. The heavily indebted NIEs of Latin America have so far been unable to afford such investments.

The role of the state

The significant externalities involved in the broader diffusion of new information technologies through the industrial system suggest an important role for the state in fostering such diffusion. In OECD countries, government diffusion-oriented programmes have by no means been universally successful (OECD, 1989a). The rather mixed results have led some to conclude that diffusion should be left to the decisions of firms mediated through markets. As already noted, however, market forces alone would most probably lead to under-investment in the new technologies in the NIEs, especially by small firms. The State can offset some of the costs of adopting IT-based innovations by offering to subsidise training costs (or the costs of hiring consultants). Alternatively, it can take upon itself the task of gathering and then disseminating information through training centres and research institutes which focus on the application of new technologies to specific production sectors. Examples are the government-sponsored Korea Institute of Metal and Machinery (KIMM) or Taiwan's Industrial Technology Research Institute (ITRI).

Even with government efforts to promote the diffusion of AMTs, particularly in EANIEs, the extent of diffusion is limited by comparison with the OECD countries. For example, in 1985 Korea had the largest number of numerical control machine tools in use of all the NIEs, with 2 700 units. This compares with a figure of 120 000 in Japan. In the case of industrial robots, all the NIEs combined had roughly 400 in use in 1985, while the OECD countries had more than 100 000. Finally, in the case of CAD systems, the United States, Germany, Japan, the United Kingdom and Sweden combined had more than 90 000 "seats" in 1985; the NIEs combined had roughly 2 100 (Edquist and Jacobson, 1988). Without strong initiatives on the part of NIE governments to encourage more rapid diffusion, the gap is more likely to increase than to narrow.

NIEs as exporters of IT-intensive products

Since the early 1970s, EANIEs have been large exporters of engineering products, many of which now embody the new micro-electronics technologies. Figures on growth in overall engineering product exports are suggestive. Between 1970 and 1986, the average growth in these exports for OECD countries, COMECON countries and the NIEs was 13.6 per cent per year. The United States and the United Kingdom were below that average at 11 per cent and 10.1 per cent respectively. Japan was above it at 20 per cent. The performance of certain NIEs far exceeds even that of Japan. Korea's engineering exports grew at 39 per cent a year, Singapore's at 27.5 per cent. Hong Kong's grew at roughly the same rate as Japan's. As for Taiwan, between 1978 and 1987, its exports of non-electrical machinery grew by 20 per cent a year and its exports of electrical machinery by 23 per cent (Taiwan Statistical Data Yearbook, 1988).

Not all of these exports represent engineering products incorporating new technologies. Nevertheless, of total NIE machinery exports, electronic and electrical machinery represents a large and growing proportion. In the case of Korea, such machinery represented 60 per cent of total machinery and transport equipment exports in 1986. For Singapore, the figure was 77 per cent, for Hong Kong, 91 per cent. Taiwan's 1987 electrical (including electronic) machinery exports were five and a half times as large as non-electrical machinery exports (Taiwan Statistical Data Yearbook, 1988).

These aggregate figures do not convey a picture of the NIEs' export structure that is sufficiently detailed to show important trends. It remains unclear to what extent the growth in exports reflects an expansion of assembly-type activities, the introduction of more sophisticated production processes, expanded production of low-end consumer electronics goods and standard computer peripherals, or a shift toward higher-end consumer goods and more sophisticated office automation equipment. To shed more light on the prospects for NIEs to improve upon existing capabilities as suppliers of new information technology products, a micro-economic analysis of industry dynamics in those countries is required[13]. This section merely hints at broad outlines of the picture.

Building upon acquired technological capabilities

Most EANIEs have gone through a protracted learning process, during which they have amassed significant and diverse technological capabilities. Much of that learning occurred prior to the accelerated growth in the applications potential of AMTs and the explosion in the use of ITs. The question is whether the rapid technological changes in the world economy will allow NIEs to benefit from some of their previous learning.

With the transition to numerical control machines, of which the Japanese are the main suppliers, remarkable technological progress has been

achieved in the machine-tool sector, in which several EANIEs were exporters[14]. The only EANIE firms that will be able to challenge this competition are those which, like Korean firms, have the requisite prior experience in CNC tool production (Lee, 1988).

EANIE firms that have already entered technology-intensive sectors like electronics are faced with the prospect of another sort of transition, namely, from traditional, relatively labour-intensive assembly operations to higher value-added activities such as the introduction of automated assembly and testing, design and production of sophisticated components, software development and system integration. The transition to any one of these activities will place new demands upon the technological capabilities of firms. However, even industries whose products and processes have traditionally involved a relatively low technology content are being transformed by the diffusion of micro-electronics based innovations[15].

The transition to advanced electronics: Korea and Taiwan

Since the early 1980s, Korea, Taiwan and Singapore, whose electronics industries had in large part based their growth on the supply of efficient, inexpensive labour, have had to automate their plants (Ernst, 1983). These NIEs therefore had not only to invest in advanced equipment, but also to choose equipment carefully (or even to design semi-automated equipment better suited to local conditions). They also had to exhibit good organisational capabilities and flexibility, because of the radical impact which the implementation of automated techniques can have on the internal organisation of firms.

The approaches adopted by Korea and even Taiwan with regard to the fabrication of very large-scale integrated circuits (VLSIs) and, in particular, direct random-access memory chips (d-RAMs) illustrate the strategies adopted for moving into more sophisticated areas (Ernst, forthcoming). The decision to enter this highly competitive market, which is prone to rapid and excessive demand fluctuations (Ernst, 1983, pp. 25-27), called for a massive investment by the Korean *chaebol* in highly sophisticated capital goods, in acquiring a knowledge of foreign markets (essential because of the weakness of domestic demand) and in obtaining manufacturing licences. To gain user knowledge, they relied on second sourcing arrangements and efforts to be placed on the "preferred supplier list" of leading Japanese and US computer companies. In approximately half a decade, three Korean d-RAM producers, i.e. Samsung, Hyundai and Goldstar, have carved out a sizeable world market share in d-RAM devices one generation behind the state of the art. Given the pace of technical change in this sector, firms have had to invest substantially in R&D, frequently carried out in collaboration with the major international firms (e.g. the agreements between IBM and Siemens and between Goldstar and Hitachi). Only the large *chaebol* seem to have the necessary resources to embark upon such a course.

However, a number of Taiwanese firms have recently decided to enter into d-RAM production. With few exceptions, Taiwanese firms are much smaller than their Korean counterparts and thus are unlikely to be able to finance their own internal R&D efforts on the requisite scale. One of the first entrants, the highly successful personal computer manufacturer Acer, has overcome this short-run problem by entering into a joint venture with Texas Instruments, one of the few remaining US d-RAM producers. The latter's design and process technologies make up a significant share of its equity contribution to the venture. For R&D on 16M d-RAMs and certain other memory chips, the government-sponsored Industrial Technology Research Institute (ITRI) has entered into a "technology development alliance" with several major chip manufacturers[16]. The main factor behind the large-scale entry of Taiwanese firms into d-RAM production is the leading role this country has come to play in the world microcomputer industry. In Taiwan, the emergence of a vast domestic (indirect export) market was the major inducement to d-RAM investments, although firms will undoubtedly be required to export directly a large share of output if they are to achieve sufficient economies of scale and move quickly down their learning curves.

External constraints on the adjustment process

The two major types of external constraints which NIEs face in the present global environment are restrictions on access to technology and to major OECD markets[17]. The narrower the gap becomes between certain NIE firms and the OECD-based leaders in specific technologies, the more reluctant the latter become to allow those NIE firms access to their technologies. This reflects concern over what has come to be known as the "boomerang effect". To some extent, EANIE firms have exploited effectively the rivalry between US and Japanese firms, as well as the weakened competitive position of small US high technology firms, to gain access to advanced technologies on favourable terms. However, with fewer and fewer small, independent US firms in fields like semiconductors, there are fewer potential technology sources.

Moreover, governments in some OECD countries have been actively restricting technology access. There appears to be a trend in the United States, for example, towards restricting access by foreigners to the research results of some government-sponsored R&D projects. Foreign participation in R&D consortia has long been restricted if not prohibited in many OECD countries. Moreover, a growing proportion of the basic research undertaken in US universities is sponsored by large domestic corporations which have an interest in limiting access by potential competitors – whether domestic or foreign – to the results. Governments of OECD countries are also endeavouring to protect the proprietary technologies of their firms by developing stricter intellectual property regimes (e.g. the negociations on IPR issues in the Uruguay Round). Even if the NIEs acknowledge the need for such measures, stricter IPRs would obviously create a further barrier to broader diffusion of the new technologies (see Chapter 2).

Despite the fact that most NIEs have instituted stricter IPRs in recent years and adopted certain market liberalisation measures to appease their trading partners, trade frictions still abound. "Managed trade" is coming to affect an increasing number of the NIEs' manufactured exports, but in certain high technology industries like electronics, its effects are not always those sought. For example, managed trade between the United States and Japan in computer memory chips (d-RAMs) undoubtedly worked to the advantage of new Korean entrants into the market. Now, even efforts of the US government to enforce a floor price in the US market cannot prevent cut-throat competition and a d-RAM price war in the world market. Moreover, US computer and other electronic equipment manufacturers have been hardest hit by managed trade in chips, and no doubt they would also be hard hit if similar restrictions were placed on other elements of computer hardware which they source internationally.

Another major concern to EANIE firms are the rules they are likely to face after 1992 for selling in the European Community (EC) market. Already, severe import restrictions are in place, in particular against colour TVs and VCRs. They are estimated to have cost the Korean electronics industry $300 million in lost exports. In the year to the end of September 1989, Korean electronics shipments to the EC shrank by 11.3 per cent to $1.72 billion[18]. In addition, strict local content requirements have been imposed on integrated circuits sold in the EC, which effectively require foreign manufacturers to undertake chip fabrication within the Community in order to qualify for unrestricted market access. Japanese firms have been most immediately affected. However, the major Korean, and even Taiwanese, groups have already undertaken investments in European production of consumer electronics (a strategy denied to small and medium-sized NIE firms).

As international market access is increasingly restricted, an external constraint soon becomes an internal constraint. With rising R&D cost burdens and high fixed capital requirements, firms need to increase market shares as rapidly as possible if they are to recover their investments. While domestic markets in the NIEs may provide important learning economies, they are not likely to prove sufficient for amortising the large investments required. Ensuring relatively free international market access, accordingly, is a virtual survival requirement for many NIE firms.

Prospects for the second-tier Asian NIEs

The industrial restructuring occurring under intense cost pressures and intensifying trade frictions in Japan and the first-tier Asian NIEs has caused the relative decline of a number of labour-intensive industries and activities. Rather than abandon those industries and product markets altogether, firms have begun to shift much of their labour-intensive production to the low-wage countries of South-East Asia. Additionally, changing comparative costs have created profit opportunities for local firms in South-East Asian countries, which are now better able to compete with first-tier NIEs in many labour-intensive industries. This situation has created new opportunities for second-tier NIEs to acquire technological capabilities unavailable to the vast majority of developing countries. The question is whether the process of economic adjustment occurring in Japan and the first-tier Asian NIEs will allow the second-tier Asian NIEs to move up, or whether internal constraints will prevent them from taking advantage of the opportunities.

A favourable situation for foreign direct investment

The last decade has witnessed a drastic reduction in the role of the United States as a source country for FDI, due to macro-economic imbalances which have transformed the country into a massive net capital importer. Meanwhile, the enormous trade surpluses of Japan and certain EANIEs have made them the new large-scale capital exporters[19]. While the bulk of this outward FDI has gone to other OECD countries, the main recipients among developing countries have been the second-tier Asian NIEs (Summerville, 1990, p. 8). Much of that investment is in products based on relatively mature technologies, as well as in relatively labour-intensive industries and assembly operations. Most has been channelled into wholly foreign-owned subsidiaries and only rarely into joint ventures. According to the results of an extensive study of technological capabilities in some key Thai industries (including electronics, biotechnology and agro-industry, and metal and materials processing), the major impact of the inflow of FDI has been to strengthen production capability, with relatively little effect on investment and on minor and major change capabilities (Thailand, TDRI, 1989).

As a result of the massive influx of FDI occurring in ASEAN (Association of South-East Asian Nations) countries, demand for parts and components has sub-stantially increased. This could lead to the establishment of local supplier networks. But local suppliers may remain largely confined to the low-technology, small-scale parts and components sector, with foreign firms bringing in their own subcontractors to supply the more sophisticated and the higher volume parts and components. At present, most local firms lack the investment capability to undertake large-scale production, and very often they lack the production capability to produce more sophisticated components in accordance with the quality and reliability standards required by major foreign customers[20].

In Malaysia, for example, where there are a growing number of foreign (mostly Japanese) manufacturers of video cassette recorders (VCRs), the Japanese subcontractors of these firms have been persuaded to set up local operations to supply their demand for sophisticated components. Local companies have largely been confined to supplying plastic parts, printing and packaging materials.

On the other hand, one of the largest semiconductor assembly plants in Malaysia – a subsidiary of the Intel Corporation – decided several years ago to contract out the manufacture of much of its tooling requirements to local engineering firms rather than continue to make the tooling in-house. The tools involved are precision dies, jigs, and fixtures for use with its automated assembly and testing equipment. It provided technical support to the most capable vendors, including training in statistical process control (SPC) and the use of CNC machines. It also entered long-term supply contracts with the selected engineering firms to lessen the risk to them of investing in expensive equipment.

The pattern which seems to be emerging in the second-tier Asian NIEs resembles that described by Imai (1990), in which Japanese (and first-tier Asian NIE) firms concentrate in or near their home base their core factories (which produce critical technologies) and factories which produce closely related components or subsystems, while dispersing the production of less critical components or subsystems to the second-tier Asian NIEs. As a result, certain second-tier Asian NIEs have become major producers and exporters of consumer electronics and computer peripherals. A very high proportion of that production and those exports are controlled by wholly foreign-owned firms, but in a few cases local firms have qualified as OEM vendors for foreign electronics firms. The OEM electronic parts and components business may well continue to grow in the second-tier Asian NIEs in the coming years.

Supplying IT-intensive products

There would appear to be some prospect for SMEs in the second-tier Asian NIEs to supply certain complementary inputs into the many electronic systems which are being developed and manufactured throughout the region, whether by OECD firms or by first-tier NIE firms. Something similar to what occurred in Taiwan might be repeated in one or more second-tier Asian NIEs, although probably on a much smaller scale; as demand for the "core" product rises, demand for their products will rise with it. As yet, however, indigenous product design and development capabilities as well as manufacturing capabilities remain too weak to enable many firms to enter on their own into markets for more sophisticated information products. To emerge as competitive world or regional suppliers of IT-based products, local firms need to be innovative and quick to respond to new opportunities[21]. In addition, they must find ways to finance the commercialisation of their products. Given the high risk, this implies some sort of risk capital facilities, either from private capital markets or from government-supported financial institutions. Firms must also be able to draw upon a pool of skilled engineering and technical labour. Finally, the broad policy and institutional environment should encourage and reward risk-taking and adaptability on the part of firms.

While greater government encouragement of local investors could act as a catalyst to the emergence of a sector supplying IT products in these countries, excessive government direction could prove extremely costly. Competent States played an important role in facilitating the process of industrial transformation in South Korea and Taiwan. That occurred, however, in a period when technological change had not yet drastically accelerated. Even strong states in the first-tier Asian NIEs are increasingly wary of providing detailed direction to investors. The risk of government misdirection is exacerbated in countries where the state has not been very effective in directing economic activity. The second-tier Asian NIEs fit this description, and their record in undertaking major investments in the industrial sector is not particularly outstanding (Yoshihara, 1988).

The application of new information technologies

The second-tier NIEs have become major exporters of manufactures over the last several years, and those exports are anticipated to continue to grow strongly in the future. There is thus a rapidly growing demand for engineers and technicians capable of operating and maintaining automated equipment (technology transferred by Japanese investors). Already, the surge of investments in Thailand has created a severe shortage of skilled personnel. Malaysia is somewhat better supplied with trained and experienced electronics engineers and technicians, but it too faces imminent shortages. The foreign investments from the first-tier Asian NIEs generally involve less sophisticated production processes and fewer AMTs and may be more easily absorbed. However, they may also involve outmoded technology and thus make it more difficult for the second-tier NIEs to upgrade quality and to introduce greater flexibility into their operations. Besides, despite the relative abundance of unskilled and semi-skilled labour, there are specific skills which are already in short supply; selective automation may be the only solution to the problem.

Clearly, second-tier Asian NIEs face a far more difficult challenge than first-tier NIEs in applying IT and related organisational innovations. Most of their firms are financially less capable of undertaking the necessary investments and technically less competent to assimilate the new technologies. For most firms, size still creates a major constraint for the adoption of AMTs. This applies to a wide spectrum of sectors, ranging from textiles and clothing manufacturing to metal-working[22]. Moreover, many firms are hesitant to purchase sophisticated computer-controlled machine tools for lack of personnel trained to operate and maintain the machinery properly (Thailand, TDRI, 1989). Because most foreign investors ask their subcontractors to relocate to South-East Asia, local suppliers have limited exposure to AMTs. This precludes many valuable learning opportunities. However, when foreign firms have been willing to engage in a systematic transfer of the know-how required for making sophisticated parts, local subcontractors invariably demonstrated their ability to absorb even sophisticated production know-how. Governments should find a way to induce foreign investors, wherever possible, to increasingly localise their sourcing of parts and components.

4. INFORMATION TECHNOLOGY AND OTHER DEVELOPING COUNTRIES

The situation for developing countries and in particular the least developed countries appears considerably more serious, because of structural constraints, weaker physical infrastructure, and underdeveloped human resources. The current trends in globalisation show that these countries are losing ground on a variety of fronts (see Section 1). They risk being unable to overcome the rapidly increasing distance that separates them even from the second-tier NIEs and thus being closed out of world markets.

Even in traditional industries, of which textiles and clothing are prime examples, the need to improve the competitiveness of domestic productive capacity by advanced countries has led to the internationalisation of production and to a rapid increase in knowledge-intensiveness of production. As a result the acquisition and adoption of the new technologies can no longer be regarded as a "cost" of modernisation but *needs to be seen as an investment in competitiveness* (Mytelka, 1990). Furthermore, the growing uncertainty that results from the rapid pace of technological change and the extensive internationalisation of capital and production has placed a premium on flexibility and innovativeness in the adjustment process. At no time, therefore, has the building of indigenous technological capabilities in developing countries assumed as much urgency as it has today. Low labour costs become less effective as a basis of competitiveness as automation advances. The insufficient educational levels make the succesful introduction of new technologies even harder. However, the evidence suggests that a user-oriented strategy that introduces IT-based innovations where they can be most effective can still improve their situation.

External constraints on the adjustment process

Most developing countries continue to rely heavily if not exclusively on their agricultural and primary commodity sectors to generate desperately needed foreign exchange. They face enormous difficulties both in moving in the direction of manufactured goods and in becoming major exporters. Many still encounter severe difficulties in feeding their populations, so that the generation of surpluses from agriculture for transfer to industrial investment is virtually impossible. Savings rates are exceedingly low, and reliance on foreign saving to supplement domestic is problematic, as many of these countries are already heavily indebted.

Access to markets and to technology

The global economic environment is far more hostile today than it was when economies such as Korea and Taiwan underwent their rapid industrial transformation. Protectionism is on the rise. Even some of the least developed countries have not been spared: a surge in garment exports from Bangladesh in the mid-1980s led to the imposition of quotas by the United States and several other major importers. Should other least developed countries succeed in boosting their manufactured exports significantly, they could expect to face increasing trade restrictions in the OECD countries. As the Asian NIEs have demonstrated, there are creative ways to circumvent or to minimise the impact of trade barriers. Yet even they have had to bear the costs of protectionism in terms of diminished export and GNP growth rates. The least developed countries will have even less room for manoeuvre, given their relatively weak bargaining position and limited resources.

In this environment, an export-led growth strategy targeted on the major OECD markets appears increasingly problematic (Athukorala, 1989). As noted, the export growth in the EANIEs has been an important source of technological learning. Not only have interactions with foreign customers provided valuable technical information, the capacity to import capital goods has also proved an important source of technological capabilities. Diminished export capacity of developing countries would clearly mean diminished capacity to import capital goods (and the technology embodied in them) from the OECD countries or other capital goods exporters. In addition, most developing countries have seen a decline in FDI inflows. Foreign capital in their economies remains largely confined to enclaves of mining, forestry, and agricultural export production. If anything, the advent of ITs has worsened their prospects of being integrated into global production networks.

The erosion of competitive advantages

Even in NIEs, the density of IT is only a fraction of what it is in OECD countries. Very little data exists on diffusion patterns in the rest of the developing world, which is now being left far behind in the race to apply AMTs to upgrade production capabilities[23].

The growing applications gap for IT has far-reaching implications for the international competitiveness of developing countries. Since the late 1970s, OECD countries have increasingly applied ITs in a broad spectrum of industries, especially in those industries and stages of production which had been labour-intensive and in which, consequently, OECD countries had been losing competitiveness. The crucial question for developing countries is to what extent IT applications have restored, or have the prospect of restoring,

The textile and garment industries represent a classic case of labour-intensive industries in which cheap labour has traditionally conferred competitive advantage. In the case of spinning and weaving technologies, the diffusion of the newest innovations is generally much greater in OECD countries than in developing countries (Toyne *et al.*, 1984; and Bowring, 1989). Of the two sectors, weaving is more labour-intensive. In weaving, shuttleless loom technology represents the most advanced presently available, incorporating sophisticated electronic controls. Taking shuttleless looms as a percentage of total installed loom capacity in 1984, the figure for Germany and France was 46.3 per cent, for Italy 44.3 and for the United States 31.2 per cent. In contrast, the figure for Indonesia was 4.2 per cent, for Pakistan 1.6 per cent, for India 0.7 per cent, and for China 0.3 per cent (Japan Chemical Fibre Association, 1988). As might be expected, the NIEs fell in the intermediate range between these two extremes. The widespread diffusion of AMTs in the OECD textile industries has taken place principally in the last decade.

In 1973, developed countries accounted for approximately three-quarters of world textile exports. By 1982 that share had fallen to two-thirds. Since 1982, however, there has been only a marginal decline in the developed country share, to 63 per cent in 1987 (GATT, 1989). Figures on textile imports into the OECD tell a similar story. Imports from the four leading Asian suppliers (China, Taiwan, Korea and Hong Kong) rose from 7.3 per cent in 1975 to 10.2 per cent in 1981. From 1981 to 1987 they increased only marginally to 11.3 per cent. In the case of the five ASEAN exporters (excluding Brunei), the OECD import share roughly doubled from 0.7 per cent in 1975 to 1.3 per cent in 1981, then crept up to 1.7 per cent by 1987. Finally, for the two leading South Asian exporters (India and Pakistan), the OECD import share in 1975 was 2.9 per cent, in 1981 3.8 per cent, and by 1987 4.2 per cent. What cannot be determined from such figures is whether the slowing growth of developing countries' world textile market shares is a function principally of more advanced technology diffusion in the OECD industries or of effective protectionism in the major OECD markets. Undoubtedly the latter has played an important supporting role.

In the case of the garment industry, the impact of AMTs has been far less dramatic. For most clothing items, assembly remains a highly labour-intensive process. Therefore, developing countries continue to play a prominent role in world clothing trade, and even protectionism has not been completely effective in stabilising market shares of OECD producers. It is possible, however, that in the next decade a growing stream of AMTs will make possible increasing automation levels in clothing assembly. An analysis of patent activity in the textile and clothing industries in the mid-1980s indicates that the bulk of activity in the two most innovative countries (Japan and the United States) was focused on clothing assembly operations (Kurt Salmon Associates, 1987). If the threat to developing countries' competitiveness in clothing is not imminent, it may not be too far in the future.

the competitiveness of OECD countries in those industries. Existing information suggests that developed countries have largely managed to use the IT based technology to reduce their import dependency on developing countries (see Box 54) with the help of trade protection.

This pessimistic assessment should be tempered by the observation that developing countries are able to adopt the same AMTs that are available to their OECD competitors, as the experience of the semiconductor assembly industries of the second-tier Asian NIEs has shown. Those industries have undergone extensive automation over the last decade. Initially, it was expected that automation of chip assembly would eliminate the cost advantage of offshore locations like Malaysia, Thailand and the Philippines[24]. In fact, assembly plants in those countries have been heavily automated (admittedly, largely by the subsidiaries of multinational semiconductor firms). The work forces in those countries have proven quite capable of mastering the new AMTs and their composition has shifted markedly in response to changing needs. For the most

part, these countries have been able to supply the higher skill requirements, although Thailand is currently faced with a serious shortage of electronics technicians.

The persistence of internal constraints

Constraints to industrial applications

While raising agricultural productivity is a necessary component of industrial transformation for all developing countries, it is only a first step. Sooner or later, countries must seek to raise the share of non-agricultural activities in GDP, or they will be unable to provide productive employment for the agricultural surplus population. Creating industrial employment opportunities is not primarily a matter of technology, appropriate or otherwise. The two principal constraints to greater industrial employment generation in most developing countries are the familiar capital constraint (of which the foreign exchange constraint can be considered a relative) and the skill constraint. The critical question here is whether either or both of these constraints can be relieved through the diffusion of ITs.

For capital-scarce countries, a primary concern is how to maximise the use of existing productive capacity. Pack (1987) has shown, for the textile industry, how widely the technical efficiency of production can vary within a single country and industry, even when basically the same technology is used. Factor productivity can be raised if the majority of firms in an industry can be brought closer to available best practice. Achieving that goal will depend very much on the constraints faced by firms in a particular context. Improvements in factory organisation are likely to be important and may involve a reorganisation of the production process, work scheduling, inventory management, etc. The organisational changes required to eliminate gross technological inefficiencies are likely to be rendered more effective if complemented by relatively modest investments in ITs. Since many of these are modular, investments need not be for large indivisible systems. For example, a small firm interested in installing a new inventory management system to reduce working capital requirements may be able to do so on a microcomputer. The purchase of a microcomputer and related peripherals and software may represent a barrier to adoption by some users, but if the hardware is available at world market prices, that barrier should be relatively low. For users who are really so small that they could not afford to purchase the hardware and related software, the government obviously needs to subsidise such investment. Where inefficient utilisation of existing plant and equipment is a major cause of low productivity, substantial improvements in industry performance can be achieved with relatively small investments in ITs combined with appropriate organisational innovations by laggard firms.

Weakness of infrastructures and lack of skilled workers

The relatively sanguine experience of the second-tier Asian NIEs with assembly automation may not be easily repeated by other developing countries. Once it has become cost-effective to introduce automated assembly technologies widely, cheap unskilled labour no longer attracts investment. Furthermore, few other countries have the same capacity as the second-tier Asian NIEs to provide the pool of skilled engineers and technicians needed to operate and maintain the automated equipment. Finally, AMTs tend to be more infrastructure-intensive, requiring not only a reliable electricity supply, but also a reliable telecommunications network linking offshore assembly plants to their host factories in the industrialised countries or the NIEs.

The vast majority of developing countries have been by-passed by FDI, particularly in industries that have witnessed a broad application of AMTs. An important reason has been the very low educational and skill levels of their work forces. Implementing AMTs, after all, requires an educated, adaptable and highly motivated work force able to learn new skills quickly. In the second-tier Asian NIEs such workers can generally be found. But this is not the case in most developing countries. The low wages in those countries do not necessarily translate into low labour costs, since they coincide with low educational and skill levels and low productivity.

Sophisticated capital equipment is, in many cases, only as effective as the workers who operate, maintain, and service it. If workers lack the skills necessary for utilisation of that equipment, its productivity is likely to be low. Traditionally, late industrialists have overcome skill constraints by combining foreign technical assistance with intensive training of nationals in the use of imported capital equipment. Relying on foreign expertise can be costly, especially if foreign experts cannot be quickly replaced by local counterparts. How quickly the substitution occurs depends in turn on the speed and efficiency of the learning done by domestic engineers, technicians, and machine operators. Prior education and training as well as experience clearly condition the speed of learning.

ITs and related organisational innovations may enable developing countries to shortcut some of the learning processes required for the effective use of sophisticated capital equipment or the operation of complex production processes. Machines which incorporate ITs acquire a certain amount of "intelligence", in the sense that they "learn" certain routines and can be adapted to changes in their environment. The degree of "intelligence" varies widely depending on the machine – and in particular on the sophistication of the software used to control it. The important point is that machine "learning" can, within fairly narrow limits, replace human learning in the performance of certain production tasks. The relevant question for the developing country user is how much learning is necessary for effective use. At the very least, a computer literate work force is likely to be essential, since some basic programming skills are virtually indispensable for using micro-electronics based equipment. How much programming is needed depends on how much software is embedded in the machine's hardware. Basic electronics hardware maintenance skills are also essential if the electronic capital equipment is to be kept operational. In short, a substantial knowledge and skill base is required to reap the productivity-enhancing potential of ITs[25]. Many developing countries thus seem to be caught in a low productivity trap: their inability to attract significant foreign manufacturing investment limits their acquisition of production capabilities and, consequently, growth in their labour and total factor productivities.

Weak inducement mechanism for technological change on the demand side

The ability to absorb imported technology and the extent to which local technologies are generated or upgraded are also affected by the size and structure of the user consumer base (Mytelka, 1990). Domestic markets in the less developed countries are excessively small either because of low levels of per capita income (Nepal, Ethiopia, Malawi, Tanzania), small population size (Botswana, Democratic Yemen, Mauritania, Vanuatu) or a combination of both (Bhutan, Guinea Bissau, Lesotho). But the disadvantages of small size of domestic markets is exacerbated by the type of industrialisation model adopted in most least developed countries.

Since the productive population consists overwhelmingly of peasant farmers and artisans with low purchasing power and hence limited effective demand for new or improved technologies industrialisation through large-scale production has been particularly problematic. Consequently the dictates of the "market" for manufactured goods has oriented production toward urban consumers whose tastes are shaped by imports. This has made adoption of an "import-reproducing" rather than an "import-substituting" industrialisation strategy more likely. These terms need clarification. Import reproduction strategies, such as those implemented in many of the least developed countries at various points in time, take "product" as their point of departure, rather than "purpose". They thus ignore the extent to which products incorporate concepts of functionality, cost, quality and aesthetics that correspond to the producer's principal market of sale. An import substitution strategy reverses this process by first identifying the purposes embodied in imported products, then determining their relevance to the local environments, and lastly by designing a product that conforms most closely to domestically available inputs, material and non-material.

Within the context of an import substitution strategy there is no automatic assumption that given product categories will necessarily slavishly reproduce the characteristics of imported goods. The reduction of imports as a share of domestic consumption thus occurs without requiring that the locally made product be identical to the formerly imported one. The economics of import reproduction strategies, on the other hand, are such that the tendency is not only to reproduce the characteristics of imported products but to do so with technology similar if not identical to that used in the original market of manufacture. Not only is the practice of import substitution different from that of import reproduction but its teleology differs as well. In a process of import substitution, production is part of a wider process of technological learning (see Chapters 1 and 2) that encompasses product specification and design, process choice and change and the social

organisation of production. Technological mastery becomes part of the process of industrial deepening and the qualitative changes in the structure of production that this implies. In contrast, import reproduction strategies neglect these vital dynamic learning effects with negative long-term consequences for technology acquisition, utilisation, absorption and generation[26]. The stagnation and indeed the decline of manufacturing output in the least developed countries (see Table 60) can be interpreted in this light.

Towards a user-oriented strategy for technology diffusion

A focus on technology application

For the vast majority of developing countries, the main issue relating to new technologies is how to apply them in order to eliminate critical bottlenecks that slow growth in productivity and thus in per capita income. If the problems of developing countries go well beyond technological issues, technology can nonetheless make a fundamental contribution, not only to industrial development but also to an enhancement of agricultural productivity. In most cases, applying traditional technologies to quite mundane tasks in agriculture and infrastructure may be the more immediate concern. Yet, for many applications, new technologies are far superior to older ones, often by an order of magnitude. Many potential applications to the specific problems of developing countries still need to be discovered, but this requires an active exploration based on sophisticated technology assessment capabilities. So far, the search for new applications has remained focused nearly exclusively on the requirements of OECD countries. While there are many fruitful avenues of research into the applications of biotechnology, new materials, and new energy technologies in developing countries, in keeping with the rest of this chapter the focus here is on the new ITs.

Traditional activities

Two fundamental categories associated with traditional activities concern *i)* activities of the subsistence sector, e.g. peasant agriculture and handicraft production; and *ii)* activities associated with the export of traditional agricultural commodities and unprocessed, or semi-processed, natural resources (e.g. forestry and mineral products). The subsistence agricultural sector is the source of supply of the labour force for industry. In many countries, it must also supply an investible surplus, since there are few other income-generating activities. How effective it is in providing both an industrial labour force and an investible surplus will depend critically on the productivity of the resources employed in it. In order to free up workers for employment in industry, agriculture must be able to feed an ever growing number of people not labouring on the land. In other words, the size of the marketable surplus must be steadily increased. Technological innovation is a very important means of raising the share of the marketable surplus in agricultural output. Organisational change, including change in land tenure arrangements, may be equally important. ITs may open up new possibilities for relatively small peasant cultivators to manage their own lands more efficiently, assuming they can achieve a certain basic level of literacy and numeracy. ITs also hold out the prospect of improving marketing and distribution systems, e.g. of managing the operations of marketing cooperatives. Post-harvest processing and storage are also amenable to efficiency gains (and wastage reduction) through utilisation of ITs. Such applications should contribute to raising the size of the marketable agricultural surplus, thereby facilitating the accumulation of capital for industrial investment.

In countries with an export crop sector (or other primary commodity exports), that sector can help to generate the investible surplus. It is not likely to be a major labour reserve, since employment of workers in export agriculture is more strictly governed by profitability considerations. It can, however, be a valuable source of the foreign exchange needed to import capital equipment and intermediate inputs required for the growth of industry, so long as the country does not need to import food. The commodity export sector is likely to be a more significant user of sophisticated technologies than the subsistence sector. Frequently, it is largely dominated by foreign corporations which have access to the latest available technologies. If productivity gains are to become available as investible surplus within the economy, some measures are needed to ensure that they are not immediately and exclusively repatriated by foreign subsidiaries and also that the use of IT in this sector diffuses to other parts of the economy.

Basic tradeoffs involved in a user-oriented strategy

A user-oriented strategy for IT diffusion in developing countries would place primary importance on making ITs as accessible as possible to the firms and industries that can make most effective use of them to improve productivity, product quality, market responsiveness, and other determinants of international competitiveness[27]. A local supplier sector of ITs would be promoted only to the extent necessary to improve the competitive prospects of users. Otherwise, users would be free to source IT needs from the most competitive supplier anywhere in the world, subject presumably to

the same foreign exchange constraints faced by all other firms in a particular country. If the strategy is effective, then the application of ITs to user industries should result in gains in international competitiveness that help to relax the foreign exchange constraint in the future.

Many developing countries may only have limited options for implementing such a strategy. Only one or two sectors may have the prospect of becoming internationally competitive. Thus, a user-oriented strategy would mean fostering IT diffusion in those sectors. In any event, given limited local expertise in the application of the ITs, it may be difficult to support the requirements of more than a few groups of users. Much depends on how specialised the applications for a particular user or group of users are. If the important

---

Box 55. Social conditions for the use of technology

Frontier technologies such as biotechnology or micro-electronics undoubtedly involve a more complex mix of skills, knowledge and productive capabilities than many established technologies. However, this alone does not provide the grounds for rejecting their use and even their production in the least developed countries. What appears to matter most is the selectivity with which the choice of techniques is made, and the degree of consistency with which a conscious process of building the indigenous technological capabilities required for their mastery is undertaken. In addition, the ability to assess the opportunities and constraints posed by technological change is crucial to the design of new investment programmes and rehabilitation projects in the least developed countries. There is clearly a need for a mixture of techniques in the least developed countries, but the absence of sufficient indigenous engineering design, prototype development and manufacturing capabilities limits the ability of actors in the local environment to identify opportunities for the recombination of technological capabilities and to act on that perception. The lack of assessment, selection, acquisition and negotiation skills, weak operating and organisational capabilities, and limited problem-identification and problem-solving capabilities further reduce the ability of least developed countries to deal with today's changing technological and competitive environment. Building up these capabilities and learning to manage the aid process, where foreign aid continues to be a major source of funds for technical assistance and capital goods imports, are thus central elements in the broader process of development (Mytelka, 1990).

applications are "horizontal" in nature – i.e. span a number of sectors, like accounting software packages for microcomputers – the amount of specialised local support needed by users is likely to be small. A user-oriented strategy would then do best to allow relatively free importation of microcomputer hardware and the low-cost diffusion of the relevant software. Where the important applications are "vertical" in nature – i.e. specific to a particular industry, e.g. computer process controls for a certain type of plant – or where a country's users have specific requirements distinct from those in other countries – e.g. banking automation systems – then local user support is likely to prove more critical. In this case, a user-oriented strategy would need to encourage the development of a sector capable of providing IT applications knowledge.

Due to the limited market size in most developing countries, the mere provision of products and/or services needed by users may not be sufficient to make supplier firms financially viable. If externalities and/or economies of scale exist, the government would need to provide some form of subsidy or protection to ensure the emergence of a viable industry. A government subsidy, however, may be unaffordable in many developing countries where fiscal constraints are binding. Protection of the industry in question imposes the burden on users and, by raising the costs of the product or service, undoubtedly reduces the extent of use. This may be a highly undesirable outcome, given that the purpose of the policy is to foster more widespread diffu-sion of the ITs. Thus, if there is a feasible alternative that would not penalise local users, it is preferable. One possibility is to provide suppliers with protection in a related business so that they can effectively cross-subsidise their supply of the critical product or service (Perez, 1990). Ideally, the "cash generator" line of business should bear some technical resemblance to the targeted product or service, so that the firms are able to reap economies of scope. That in turn would reduce the extent of protection needed to make the related product market profitable. Of course, protecting a related business rather than subsidising the business of interest is a "second best" policy.

Learning to apply ITs to a particular industry or activity is a time-consuming and costly process, as is the acquisition of any other technological capabilities. To be used effectively, ITs are more likely than older technologies to require complementary changes in organisational structures and managerial routines, placing additional demands for learning upon developing country firms. Some of that learning must be done by the users, some by specialised suppliers, and some can only result from intensive interaction between the two. Today, participation in the process of technological choice is itself a central factor in building indigenous technological capabilities and in ensuring a commitment to technological change (see Box 55). This will also require greater participation by users – women, villagers, artisans – in the planning of technological change.

5. SITUATION AND SPECIAL NEEDS OF LEAST DEVELOPED COUNTRIES

The situation of the developing countries, in particular the least developed countries, received special attention from DAC in the last phase of the TEP. This section draws heavily on the orientations which emerged and are addressed in turn to the developing countries themselves, to advanced donor countries and to agencies responsible for regional and international co-operation (OECD, 1991k).

Regarding the developing countries, the orientations and efforts suggested are built around two central ideas:

– Science and technology programmes and priorities must be derived from a national dialogue among the stake-holders on needs and opportunities, rather than be driven by the "suppliers" of science and technology, whether domestic or external. There must be a much enhanced partnership between economic and social policy-makers and science and technology professionals, but the defining and co-ordinating role should be assigned to the former, not to the latter.

– Systems of consultation and organisation have to be created which generate involvement and commitment to change as a multi-faceted, interactive endeavour across the whole spectrum of the community and its institutions, including non-governmental actors.

Consequently, the action orientations proposed for developing countries note that each developing country should, through a process of national dialogue, identify a prioritised list of missions in the major domains of development – agriculture, industry, services, education, health, energy, population, environment – which integrate science and technology programmes with development objectives and strategies. Such an approach should produce multi-disciplinary missions with strong interlinkages, making full use of the social sciences, and it should bring clarity and purpose to the national development agenda. A major object of these technological development missions should be to harness the innovative capacities of the whole society. The private sector should be involved as much or more than the government sector.

In order to organise and underpin the priority-setting and resource allocation process, developing country governments should establish a central unit for strategic thinking, with both a creative and resource management role. It should be close to the centre of economic decision-making, for example in the office of the Economy Minister, but able to reach out into key sections of the community, including at the grassroots level.

When science and technology needs have been redefined, educational priorities and management will need to ensure that the pattern of educational "output" broadly matches requirements. For example, if the development priority for raising incomes and output is in the agricultural sector, education systems must produce graduates able to disseminate more productive technologies throughout the rural areas of the economy.

With respect to the advanced countries, the action orientations stress that donor agencies should make it their primary objective in the least developed countries to help them to create a national capacity to manage technological change. This objective cannot be confined to their efforts in the field of science and technology, narrowly defined. Indeed donors have to recognise that the character and content of their entire aid programmes is involved. There is a strong tendency in the aid system to supply a wide range of capital equipment on a highly-subsidised basis while at the same time failing to ensure that the recurrent expenditure or human resource capacities needed to sustain the effective use of this capital are available. Particularly in least developed countries this propensity has a major negative impact on the national capacity to manage technological change.

More assiduous regard for comprehensive project appraisal principles and more effective and co-ordinated approaches to technical co-operation are essential to relieve this basic problem. The OECD work in these areas has been designed to help improve donor practices. However, the problem is also systemic in nature and requires systemic changes in the way aid agencies operate. To make these changes, donors need to have as a basis the prioritised set of national missions established by each developing country. These priorities need to be respected as parameters for the effective allocation of scarce resources (particularly human resources) provided by the developing country. Only then can the strong tendency for science and technology choices to be "donor-driven" and to lack overall coherence be averted.

Aid agencies will also need to provide opportunities for dialogue between science and technology experts and key aid managers to generate information and ideas that can be brought to bear on the shape and content of aid programmes. Aid agencies should seek ways to enhance dialogue with the wider science and technology community in their countries and to include non-governmental organisations and the private sector in this endeavour. In particular, donors should consider long-term "linkage grants", joining together non-governmental organisations, universities, and the private sector in their countries with counterparts in developing countries. In this way, the transfer of technology becomes a more spontaneous aspect of a broader web of long-term contacts.

Regarding the need for regional and international co-operation in science and technology, the orientations point out that while the capacity to manage technological change has to be, by definition, a broadly-based national capacity, most of the least developed countries will not be able, solely on a national basis, to create and sustain flourishing science and technology communities of the size and quality needed to cater to their needs. Furthermore, knowledge creation has, today more than ever, an international dimension and, as much as the development of technology applications, professional life is also increasingly being internationalised.

There is a sufficient degree of commonalty in the development problems faced by the least developed countries, particularly in regional contexts, to make co-operative problem-solving endeavours and arrangements for sharing experience an obvious course to follow. Such arrangements can provide a mechanism for bringing the financial and technological resources of the wider world community to bear on the development needs of the least developed countries. Particularly in those areas of the developing world where there are many small countries unable to achieve a critical mass of their own expertise, a few regional centres of excellence could provide the most realistic option for developing a vital science and technology community.

NOTES

1. EANIEs = East Asian Newly Industrialising Economies, defined to include Hong Kong, Singapore, South Korea and Taiwan.

2. See the UNCTAD Secretariat reports, UNCTAD (1986 and 1989*b*).

3. These new information technologies include a variety of computer-based technologies for generating, processing, storing and communicating information used in the design and production of goods and services and in the co-ordination of related support services, such as R&D, procurement and marketing.

4. Other countries that have been classified as NIEs or near-NIEs are not included, largely due to limited published information, especially with regard to the development of technological capabilities.

5. Typical examples are: Dahlman *et al.* (1987); Thailand, TDRI (1989); and Westphal *et al.* (1990).

6. Sound empirical case studies on the role of organisational structure and finance for latecomer strategies of technological capability formation are rare. Some further evidence will be presented here. Much of the debate is still very much on the level of conceptualisation. One of the few attempts to go beyond this stage to analyse country-specific evidence of organisational restructuring requirements can be found in Perez (1990).

7. For recent findings on the Brazilian electronics industry see Ernst and O'Connor (forthcoming). Amsden (1989) provides the results of case studies on the subcontractors to Hyundai Motor Company, which highlight some of the important features of the principal-subcontractor relationship.

8. For a definition see Vickery and Campbell (1989, pp. 108-109).

9. As evidenced by inclusion in the *Fortune 500* list, only a handful of Korean *chaebol* and a few previously state-owned firms in Brazil, and some other NIEs are in the same league as OECD-based corporations.

10. See the papers prepared for the Helsinki TEP Conference on Technological Change as a Social Process, December 1989, particularly those of Pries and Trinczek, Adler, Watanabe, and Edquist, and the synthesis report by Boyer (OECD*d*, forthcoming).

11. See the proceedings of the Stockholm TEP Conference on Technology and Investment, Deiaco, *et al.* (1990) and Chapter 5 of this report.

12. Making that information available to a dozen firms is no more costly than making it available to one.

13. The OECD Development Centre project on the electronics sector is one attempt to fill in some of the micro-economic details. See Ernst and O'Connor (forthcoming); Perez (1990).

14. For example, Korea's lathe exports were roughly the same as Sweden's in 1987. For details, see the case study on the Brazilian machine tool industry in Ernst (forthcoming).

15. For details, see Ernst and O'Connor (forthcoming); Perez (1990); Ernst (forthcoming).

16. See the case study on the Taiwanese electronics industry, in Ernst (forthcoming).

17. They are addressed in more detail in Ernst and O'Connor (1989).

18. Ministry of Trade and Industry estimate, in *Electronics Korea,* February 1990.

19. On Japan, see *J.P. Morgan World Financial Markets* (1989); on outward-bound FDI by EANIEs, see *Nomura Asia Focus* (February 1990).

20. The conditions under which customers would have the incentive to help local firms acquire the necessary technical know-how and technological capabilities to supply their more sophisticated parts and components needs need to be better understood. The issue will be researched in detail in the Development Centre's new project on "Technology, Globalisation and Regionalisation – the Case of East Asia".

21. One Malaysian firm, Accent Technology, has entered the international market for personal computers. It concluded a $240 million deal with the ex-Soviet Union to supply microcomputer systems over the next five years. The agreement includes a joint venture, technology transfer, and training. Interestingly, the Malaysian firm was not a major player in the international PC market before this agreement *(Electronic World News,* 19 February 1990).

22. For data on Malaysia, Philippines and Thailand, see Ernst and O'Connor (forthcoming).

23. For an analysis of the growing "information poverty" gap in developing countries, see Hanna (1990).

24. Rada (1980). For an early refutation of the "relocation back to the North" hypothesis, see Ernst (1983).

25. The nature of skill requirements in the specific context of developing countries is still a very much under-researched topic. See Bessant (1989, pp. 47-52).

26. This is borne out by individual case study material examined by Mytelka (1990).

27. For details on the concept of a "user-oriented strategy" in the electronics industry, see Perez (1990).

CONSEQUENCES OF THE TEP FOR THE DEVELOPMENT OF INDICATORS

1. Introduction

The aim of the Technology/Economy Programme is to identify some of the policy implications of the present state of understanding of the relationships between science and technology and the medium- to long-term growth and welfare performance of national economies. From the start of the exercise it was recognised that the availability of relevant statistics and indicators would affect the understanding of these relationships and that the implications of the TEP for statistical policies should be dealt with explicitly. In consequence, work was carried out by the Secretariat culminating in a Conference in July 1990 at the OECD to assess how far the indicators which were currently or potentially available met the new needs revealed by the other TEP conferences, studies etc. and to review the options and priorities for either improvement or development, taking into consideration the resources required at the OECD and in national capitals.

This annex presents these new needs and options. While some consideration is given to the resources needed, the main stress is on the demand for new and improved statistics and indicators.

It begins with some general explanations about statistical work and then discusses the need for new and improved indicators in the light of a number of unifying TEP themes which have implications for a wide range of statistics. A set of fact sheets is provided at the end of the annex to assist readers who are not familiar with the main types of data discussed[1].

2. Some initial terminology

Indicators can be defined as series of data designed to answer questions about a given phenomenon or system. Statistics are the material (atoms) from which indicators (molecules) are constructed. Statistics themselves are built up on the basis of the underlying source information available about the phenomenon under review. The questions which indicators should answer concern those aspects of wider issues which can be examined using quantitative techniques.

Thus science and technology indicators are traditionally defined as "series of data designed to answer questions about the Science and Technology System (STS), its internal structure, its relation with the economy, the environment and society and the degree to which it is meeting the goals of those who manage it, work within it or are otherwise affected by its impacts"[2].

However the issues dealt with by the TEP go beyond the S&T system and straddle the boundaries between traditionally separate areas of statistics and indicators development such as economic statistics (national accounts, macro-economic data, industrial and trade data, etc.), science and technology statistics (R&D, patents, technology, balance-of-payments, bibliometrics, etc.), human resource statistics (education, training, etc.) and environment indicators. The challenge is to improve and combine the statistics from these traditions in order to build indicators to answer new questions as well as to develop entirely new categories of indicators and data, which is a long process (Box 56).

Much of the data quoted in this TEP background report are taken from case studies. While these can be a very valuable source of information, particularly in the early stages of understanding a phenomenon or for specific industries or technologies, the present annex concentrates on indicators derived from standard statistics, i.e. data which are based on general concepts and which are collected and published on a regular basis.

3. General aspects of indicators work

The users and producers of statistics form a complex network which adjusts relatively slowly to changes in the economic and social system and our understanding of how it works. There are three roles in statistical work: users, with needs for statistics and indicators; producers, who collect the source information and process it into statistics and indicators; and the observed, who are involved in the activity under review and are usually called on to provide the source information. Both public and private bodies are involved. Box 57 shows the involvement of the private sector in the production and use of statistics and indicators. It is necessary to understand the roles of these actors and of their differing interests and attitudes when assessing the desirability and feasibility of future work on indicators at national and international levels.

Box 56. Developing new indicators

New indicators and the underlying statistics usually take two or more decades to reach general acceptance and regular collection and publication.

Blue sky

At the origin of a new class of indicators there is generally an enterprising academic who designs entirely new indicators in order to analyse the phenomenon on which he or she is working or who identifies a new use for existing data. In short they "innovate". In this phase the new approach will be known to only a very restricted network of those who specialise in the area concerned.

Experimental stage

As knowledge of the new class of indicators is diffused, a larger network of specialists comes into existence. Most will still be academics though there may be some support from government in the form of grants. The main means of exchange will be academic papers and conferences. There will be *ad hoc* data collection, testing various ways of obtaining and analysing the indicators concerned. At the end of this stage it becomes clear that some official co-ordination would be useful.

Development stage

During the development stage a first preliminary statistical framework is set up and tested by experimental surveys. As the utility of the framework and the resulting data has been illustrated, the official statisticians move in and take the place of the academics. The first official surveys are launched at national and at international levels.

Established statistics

Regular surveys are undertaken and results are then published in established formats (publications, diskettes etc.) and used in regular policy and analytical reports. Time series are long enough to permit trend analysis. Established statistics need to be reviewed from to time to see whether they are still useful and how they can be improved in order to meet new needs.

Box 57. The role of the private sector in the production and use of indicators

Individual firms maintain and use the data regularly needed to manage their own resources and to fulfil legal requirements. These are mainly based on company accounting procedures.

They also need information on the economy at large and on their industry in order to plan their future operations. Large firms may use an extensive and complex battery of indicators for policy-making. Unsophisticated firms within the system have simpler needs.

Indicators to support planning may be general or specific to their industry or products. For general economic indicators, enterprises count on government statistics. For very up-to-date or highly specific indicators, they sometimes find the very expensive services offered by commercial data bases worthwhile. In some branches, industrial federations collect and supply specific indicators. Commercial services offered by university units are also beginning to compete in this market.

Enterprises are called upon to fill in many questionnaires from government departments and from offices of statistics. These questions are based on economic or special approaches which may or may not match the companies' own data bases. Response may be mandatory or voluntary.

In large firms, the users of indicators published by government statistical services are not necessarily the respondents to government statistical surveys. In SMEs the same people are involved but they have less perceived use for the results of government surveys. In consequence, industry is generally resistant to new government surveys unless those at a sufficiently high level in the individual firms or in the industrial federations can be persuaded of the utility of the results.

Resistance will be less if the concepts in the new survey correspond to the way enterprises organise or think about their activities and if the expenditure categories bear some relation to those in company accounts.

Users and their needs

The demand for new and improved indicators comes from a number of different types of users. Each has special areas of interest and needs in terms of topics, detail, timeliness, etc. A method of presenting information may be more appropriate to one type than to another. The OECD's obligation to provide information to the categories of users also varies.

Using indicators in decision-making

Both public and private bodies use statistics and indicators to help reduce uncertainty in decision-making. For this they need series which tell them:

- what is happening to their own resources and activities (internal indicators);
- how far these activities are meeting their goals (goal indicators);
- what is happening in the area of the economy/society/technology related to their activities (a scoreboard);
- what is likely to happen in this area in the future (predictive indicators).

Decision-makers are essentially interested in the present and the near future. They need the most up-to-date information available and are often willing to pay a price, in terms of cost and sometimes even of data quality, in order to get it early. They operate, broadly speaking, at two levels. There are policymakers whose decisions can actually change not only their own activities but also the system itself, and there are those who can act only within the system.

On the government side, the policymakers are ministers of science, technology, industry, education, employment, the environment, etc. Their internal indicators (i.e. the budget and associated information) are on such a scale as to be significant for others, and their scoreboards have to be established at a national or even international level.

Of course, ministers and their very senior staff are rarely direct consumers of indicators because their decisions are based on qualitative political considerations. Most of the statistics and indicators collected and issued by national statistical offices and by the OECD are, in fact, designed for use by the staff who prepare decisions for ministers and monitor the results. In general, ministers and senior staff only retain perhaps a half dozen key figures for their area of responsibility such as Gross Domestic Expenditure on R&D (GERD) as a percentage of GDP. A point discussed during the TEP was whether these half-dozen figures should be a selection of "strategic" indicators for key points in the STS and industry or whether they should be composite indicators based on some kind of weighted average of other indicators.

As the TEP has stressed, many of the most important international policy decisions are taken in the private sector. For example, General Motors commits as much money to R&D as the Italian government (OECD, 1989e, Table 8). Such private decision-makers will probably keep their own key internal indicators in terms of company accounting systems, but they also have an interest in more general "scoreboards" prepared and interpreted for them by their supporting staff. The TEP offered a welcome opportunity to find out what kind of science, technology and industry (STI) indicators such industrialists would find interesting, and a number of their supporting staff attended the TEP Indicators Conference.

Decision-makers working within the system are persons actually carrying out R&D, technological and related industrial activities in the public and private sectors. They want indicators which measure how the S&T system is meeting their particular goals. Some are looking for information about their specific areas in order to identify recent trends or to situate their projects in a slightly wider context. Such users generally need data for individual fields of science, technology or industry, often in more detail than the OECD system can supply. Others are seeking indicators in order to justify their projects and the need for more resources.

Using indicators for research and analysis

Researchers need indicators in order to test models about how the S&T system actually works and how it fits into the larger universe of interest to the specialists concerned. For the TEP, much attention was initially given to the needs of economists to test various key hypotheses. However, economics was not the only frame of reference. Indicators are used by political scientists interested in examining how government S&T policy is made and discovering the sociological barriers to its successful implementation; they are also used by the "science of science" specialists who track the emergence and diffusion of new technologies at a more abstract level. The analytical framework for environment indicators is different again.

Once the researchers' hypotheses have been thoroughly tested and are generally accepted, they become part of the explicit or implicit conceptual framework for standard indicators work, underlying, for example, statistical frameworks, "scoreboards" and predictive indicators.

Researchers are usually academics and/or members of government policy teams. They generally prefer to build their own indicators from a stock of statistics and need long time series for modelling. The pioneers work at the cutting edge of S&T indicators development and often have to invent new methods and make imaginative use of new sources and types of statistics in order to advance ("blue sky work"). However, most so-called analytical work in policy offices or at the OECD involves redeploying existing or improved statistics to devise indicators to test current hypotheses on a larger number of cases, industries or countries. Chapter 8 on "Technology and Economic Growth" is a typical example.

Informing the general public

Public attitudes towards science and technology are formed by a number of factors, including articles in the press and television and radio programmes. Supply of accurate and attractively presented S&T indicators to the media could contribute to promoting a social climate which is receptive to technological change, meaningful public discussion on technology issues, and public participation in the assessment of technological choice. Data on public attitudes to science and technology can assist with planning, notably for S&T education and, at the other end of the scale, for marketing new science-based goods and services. Environment indicators can also be valuable in communicating with the public[3].

The main categories of producers

The main public sources of STI indicators are government statistical services, the bodies whose primary task is to collect and issue government scientific, technological or industrial statistics. Most of these are part of central statistical offices, though some are special units in S&T ministries (notably those preparing "science budgets") or in industry ministries. Some, but not all, also design and use standard indicators. The statistical services of inter-governmental agencies such as the OECD[4], the United Nations, the EC (Eurostat) or UNESCO collect data from the national government statistical services. At the OECD the link between the national and international level is made by statistical working parties such as the Group of National Experts on Science and Technology Indicators (NESTI), Working Party N° 9 of the Industry Committee or the Group of National Experts on Training Statistics.

In general these official bodies, whether at the national or international level, work under a number of constraints which make it difficult for them to react rapidly to new needs. For example, they may not introduce new surveys, especially of industry, without lengthy consultations, and they may not publish for individual firms data which have not undergone rigorous quality control. Furthermore, their work is cumulative: changes in the economic and social system and in the understanding of how it works are rarely so great as to make entire categories of statistics redundant, hence the need to retain and improve existing series as well as to develop entirely new ones[5].

There are other public bodies which collect and issue data intended for other purposes but which are also used to construct S&T indicators. Notable examples are patent offices and central banks (for Technology Balance of Payments – TBP – data). In the case of patents, there are at least two international agencies, the World Intellectual Property Organisation (WIPO) in Geneva and the European Patent Office in Munich.

Box 58.   The United Nations System of National Accounts & statistical classifications

The System of National Accounts (SNA) provides a comprehensive and detailed framework for the systematic and integrated recording of the flows and stocks of the economy for the use of national statistical authorities and in the international reporting of comparable national accounting data. The general framework provides the basis for detailed definitions and classifications and for the standard accounts and tables; its gradual elaboration provides guidance for countries wishing to extend their system of national accounts (Afriat, 1990; and at greater length, Caspar and Afriat, 1988).

International Standard Classification of all Economic Activities (ISIC) (UN, 1990)

The third revision was approved by the Statistical Commission in 1989.

Provisional Central Product Classification (CPC) (UN, 1991)

This is a general purpose classification intended to provide a framework for the international comparison of various kinds of statistics on both goods and services. It has been approved by the Commission in order to gain experience in obtaining international comparability for product data pending finalisation.

Standard International Trade Classification (SITC) (UN, 1986*b*)

The third revision was approved by the Statistical Commission in 1985 together with the request that data in respect of 1988 be based on it.

Harmonised Commodity Description and Coding System (HS) (Customs Co-operation Council, 1986)

This is an update of the Customs Co-operation Council Nomenclature which entered into force in 1988. It is the basis of the parts relating to goods of all the classifications listed above.

International Standard Classification of Occupations (ISCO) (ILO, 1990)

The 1988 version was endorsed by a resolution adopted by the International Conference of Labour Statisticians in October 1987.

International Standard Classification of Education (ISCED) (UNESCO, 1976)

The UN-ECE expert groups are currently working on the international comparability problems of ISCED, mainly concerning levels rather than classifications. This may lead to minor revisions in ISCED but not before the middle of the 1990s.

International Classification of the Status in Employment (ICSE)

This classification is currently being revised.

Classification of the Functions of Government (COFOG) (UN, 1980)

The current version was established in 1980.

The main private data sources are companies or consultants who collect and issue statistics or who produce indicators as a commercial service. In some cases they actually collect basic data themselves, in others they repackage official statistics in a more accessible form. In most cases their main preoccupation is timeliness. They are more flexible than official statistical offices as they are not affected by the same constraints, but their services are usually more expensive. They may be particularly useful for one-off surveys which cannot readily be undertaken by official statistical offices. Such commercial suppliers dominate the market in detailed and up-to-date statistics on the information and communication technology field and, at the other end of the spectrum, for bibliometrics data, where it is usual to have to pay both for the use of the original American database and the construction of indicators by a consultancy group.

Where a new need for specialist indicators for a given industry arises, industrial federations may play an important role. This has been the case for data on robotics and on the use of micro-electronics.

University units are increasingly entering the market in order to earn a commercial return on databases set up in connection with their research projects, e.g. on the number and type of innovations. Sometimes small consultancy firms are set up at universities for this purpose.

The main activities of producers

i) Setting up and improving statistical frameworks

A statistical framework is a coherent set of concepts and classifications to be used when collecting and analysing a given type of data.

In theory, a statistical framework should be strictly based on a single explicit conceptual model of the system to be surveyed. In practice, this is not always so, for a number of reasons. Setting up a new survey is generally an expensive business and may only be worthwhile if the results serve the

---

Box 59. OECD methodological manuals

Type of data	Title
R & D	*Proposed Standard Practice for Surveys of Research and Experimental Development* ("Frascati Manual") (OECD, 1981*b* and 1989*i*)
Technology Balance of Payments	"Manual for the Measurement and Interpretation of Technology Balance of Payments Data" (OECD, 1990*k*)
Innovation[a]	"OECD Proposed Guidelines for Collecting and Interpreting Innovation Data" ("Oslo Manual") (OECD*a*, forthcoming)
Information Technology Statistics[a, c]	"Interim Manual for the Statistical Measurement of Information, Computer and Communications Products and Sectors"[d]
Patents[a, c]	"Practical Guidelines for the Use and Interpretation of Patent Data as Science and Technology Indicators"
High-tech[b, c]	"Measurement of High- Medium- and Low-Technology Products and Sectors"
Bibliometrics[b, c]	"Recommendations for the Use of Indicators Derived from Statistical Studies of Scientific and Technical Literature" ("Bibliometrics")
Intangible Investment[b]	"Statistical Framework for the Measurement of Intangible Investment"
S&E Personnel[b]	"Recommendations for the Development of Indicators of Human Potential in Science and Engineering"
Education Statistics	"Methods and Statistical Needs for Educational Planning"
Education indicators[a]	"Handbook for International Educational Indicators" (OECD*b*, forthcoming)
Training Statistics[b]	"Proposed Best Practice for Surveys of Training"

a) In preparation.
b) Planned.
c) Dealing mainly with problems of classifying and interpreting existing information.
d) "Interim Manual for Information, Computer and Communications Statistics" will contain initial recommendations for a common framework and measurement methodology on which related indicators can be based. The principal objective of this manual is to achieve improved international comparability of data without requiring new surveys but by adding a limited number of questions to existing surveys.

needs of several types of users. If these users are decision-makers, then the concepts underlying the questions will probably be implicit rather than explicit. If the data to be collected by a survey are to be compatible with established statistics, it is useful to adopt existing standard classifications or categories, but in doing so implicit concepts may be imported from the framework concerned. Even if a survey is originally constructed for a single type of user and based on an agreed conceptual model, theories change over time and so do users' interests. In consequence, the original framework may be extended and adjusted to meet their new needs: for example, the current efforts to incorporate environmental issues in a wide variety of statistical systems. Last but not least, a statistical framework must find a compromise between the requirements of the underlying conceptual model(s) and the practical possibility of actually obtaining the data from the units surveyed.

The most widely known and applied framework for economic and social statistics is the United Nations System of National Accounts (SNA) (UN, 1968), and the associated UN family of classifications (Box 58). A similar set of EC classifications exists[6]. The System of National Accounts (SNA) was originally largely based on Keynesian concepts but has been extended to deal with many other economic preoccupations; both the system and its classifications are currently under revision. When setting out to establish a statistical framework for a new set of social or economic data, one of the first issues to be settled is the degree to which it should be linked to the SNA. It may be developed: entirely within its framework; as a "satellite account"; by adopting SNA concepts and classifications where useful and others where not (the case for most S&T statistics); or wholly separately from the SNA.

Beginning with the first edition of the Standard Practice for Surveys of R&D ("The Frascati Manual") (OECD,1981b and 1989i) issued in the 1960s[7], the OECD has been the lead organisation for the development of international frameworks for the measurement of scientific and technological activities. Statistical frameworks have been developed for the Technology Balance of Payments (OECD, 1990k) and for surveys of innovation (OECDa, forthcoming); and others are planned (see Box 59). Some are pure statistical frameworks: others deal more with the problems of use and interpretation of existing data categories. The Frascati Manual is currently under revision and will take into consideration both the results of the TEP and the changes in international systems and classifications.

Hence, over the next few years national statistical offices will be working on international standards and classification systems produced at round-table discussions at the United Nations, the European Community and the OECD. This overhauling of statistical systems offers an opportunity to introduce new categories and concepts incorporating or revealing technology. Even revised international classification systems, however, may not be satisfactory because:

– The classifications and other systems adopted by worldwide institutions such as the United Nations are based on a consensus between a very wide range of countries, the results of which may not necessarily be appropriate for the advanced countries of the OECD. However, in the future, the OECD may play a more important role in establishing statistical standards and classifications for the "industrial-

ised countries" – a group somewhat wider than the present membership of the OECD.

– Reaching agreement on a new classification system is a long, drawn-out process, whereas technical change proceeds rapidly. There is a danger therefore that a new classification may already be obsolete by the time it is introduced.

ii) Data collection and diffusion

Most of the regular statistics analysed in the TEP exercise were obtained by sending questionnaires to the units involved in the activities under review. A few (e.g. patents, TBP), as noted above, were obtained as by-products of other government information gathering exercises.

It is noticeable that most of the statistics quoted in this report are several years out of date. In the case of official surveys, the amount of work and the time needed to carry them out, first at the national then at the international level, are often underestimated, particularly the need for quality control. The steps involved are shown in Box 60. This explains why, despite the advances in data-processing, it takes so long for international indicators to become available from even standard surveys, especially if national and international systems are incompatible to some degree.

The need for a combined approach

As seen above, the preoccupations of the various categories of users and producers vary in terms of flexibility, timeliness, detail and accuracy. Effective development of new indicators involves strengthening the links in the network in order to reach consensus.

Users and producers

A close association between users and producers can be extremely fruitful. For example, the OECD Group of National Experts on Science and Technology Indicators (NESTI) contains both users (policy staff and some academics) and producers.

In addition, close co-operation is essential in order to deal with problems of data interpretation. The TEP has stressed the need for making the underlying conceptual frameworks explicit and well understood.

It is usual for the first versions of manuals of international statistical standards to also discuss the purpose of the resulting indicators and the main pitfalls attending their interpretation. Even in the case of some well-known S&T indicators such as patents or bibliometric data, international manuals are necessary to help new users to understand the strengths and weaknesses of the various types of data available for different types of analysis and interpretation.

It is only practical use of the indicators and the underlying statistics which shows whether they are accurate and detailed enough to meet decision-makers' and analysts' needs. However, the needs of users are often rather narrow; they concern an individual industry or technology or test a specific hypothesis. Although such use will identify some

Box 60. Timetable for a statistical survey of industry

Year T.

The firm
- carries out the activity to be measured by the statistics.

The national survey agency
- designs a questionnaire on the basis of the relevant national statistical framework
- identifies the target list of units to be surveyed (for example from business registers)
- where full coverage is not possible, designs the sample frame.

Year T+1

Early

The firm
- completes its own internal accounts and records for year T.

The national survey agency
- sends out the questionnaire.

Then

The firm
- must spend time assembling the necessary source information and filling in the questionnaire and send it back to the survey agency.

The national survey agency
- follows up non-respondents
- assembles the responses
- checks and enters the results into a database
- makes estimates for non-response
- grosses up data where sampling techniques have been used
- undertakes overall quality control and documents problems
- prepares statistical publications, etc.

Meanwhile

The international organisation
- designs a questionnaire on basis of relevant international statistical framework
- sends the questionnaire to national survey agencies.

The national survey agency
- rearranges national data to fit the international questionnaire
- documents differences between national and international specifications
- returns the questionnaire to the international organisation.

Year T+2

The international organisation
- follows up non-respondents
- assembles the responses
- checks and enters the results in a database
- undertakes overall quality control and documents problems
- prepares international statistical publications
- makes further adjustments to data for use in analytical publications and studies (for example, estimating data for missing countries).

---

deficient or implausible statistics, it tests only part of the data base concerned. It is also worthwhile to make systematic studies drawing on the whole data base at regular intervals. Though such studies may be more descriptive than strictly analytical, they often turn up trends or phenomena which might not otherwise have been noticed as well as problems of comparability etc., and if undertaken by producer agencies will assist them with quality control.

Co-operation between producers and original sources

The producer developing a new survey has to balance the needs of the users and the capacity and willingness of the original sources to provide the basic information. This is complicated by the fact that most of the producers are in the public sector whereas many of the new indicators emerging from the TEP require obtaining more information from and about industry (Box 56). Co-operation between producers and original sources is also essential in the case of surveys of the activities of the public sector.

Such co-operation is also required to deal with problems of "subjective" response. While most survey questions have no links to the subsequent finances of the unit surveyed, such cases may arise. For example, where fiscal incentives are offered for R&D, industrial respondents are liable to report more activity than before. Similarly, in the public sector respondents may be tempted to give "politically desirable" replies rather than "true" ones to questions requiring some qualitative or even subjective judgement, for example, concerning the breakdown between basic and applied research.

4. Unifying TEP issues

There are a number of analytical themes running through the TEP reports which have implications for a wide range of statistics. They are presented already translated into methodological issues. They imply both extending and improving existing types of statistics and setting up new frameworks and surveys.

Integrating Science and Technology

One of the main lessons of the TEP is that science and technology are not exogenous and independent but develop in line with a wide range of economic factors, social values and arbitrations. Furthermore, they are essentially cumulative. This has both theoretical and practical consequences for statistical work.

On the theoretical side it is necessary to find a model which allows science and technology to be endogenised with the corresponding social and economic data. From a statistical point of view, of those examined in the reports, that of "intangible investment" (Box 61) described in Chapter 5 seems particularly promising. Therefore, a statistical framework based on this concept must be designed.

The obvious way to do so would be within the SNA. During the current revision of SNA, the desirability of treating the acquisition of software and the performance of industrial R&D as investment activities was discussed. At the time of writing, it seems that the development and acquisition of software will be considered as a capital asset in the revised SNA but R&D will not. However, a new functional classification in the SNA will be developed which will reveal a number of intangible investment activities.

In the meantime, the OECD intends to develop its own statistical framework for intangible investment and to encourage countries to begin collecting data for those components not covered by existing surveys. The main challenge will be to establish internationally agreed ways of treating expenditures on these "long-term" activities so that they can be incorporated as investments into economic models. It is one thing to recognise implicitly that R&D, or training, is an investment; it is another to define explicitly and put values on the assets concerned, how they are traded, built up into stocks, depreciated, etc.

Box 61. Intangible investment

There is a need to measure the level and structure of investment in intangible assets, to compare them with investment in fixed assets and to assess their contribution to trends in industrial production.

No standard international methodology for collecting and comparing intangible investment data has been established, although manuals do exist or are under development for some of the component types of data. French experts have worked actively in recent years on the conceptual framework for studying intangible investment (Afriat, 1990; and Caspar and Afriat, 1988). The OECD plans to prepare a standard international statistical framework in the early 1990s.

There is as yet on no internationally agreed definition of intangible investment which covers all long-term outlays by firms aimed at improving their performance other than by the acquisition of fixed assets. According to the TEP analysis (see Chapter 5), intangible investment covers expenditure on *software* and on *technology* (the performance of R&D or the acquisition of its results), on *enabling* activities (training, management, information, organisation) and on some forms of *marketing*.

Future measurement work will have to take into consideration work both by statisticians and by accountants who are developing their own concepts of intangible investments (for example, a round table was held in the OECD in 1991 on Intangible Assets – Accounting and Disclosures Issues).

In order to include intangible investment in economic models it will be necessary to develop measures of stocks of the corresponding assets; this implies establishing constant price series and depreciation rates.

At present, there is no regular source of internationally comparable data on intangible investment. Only one or two OECD countries have undertaken specific surveys of intangible investment by enterprises (Finland, Central Statistical Office, 1989). Most studies have compiled data on individual components from various sources with differing methodology. The OECD currently issues or is developing series on the technology components (Fact Sheets 1, 3 and 6), on software and on training (Fact Sheet 10).

Future improvements in statistics will also be necessary to incorporate spillover resulting from cumulativeness and complementarity in the form of returns to scale required by the new economic models.

At a practical level, integrating scientific, technological, industrial and social statistics means ensuring comparability in coverage and classification. In the past this was not so. For example, R&D statistics, industrial statistics and education statistics were collected via different surveys. As a first contribution to the TEP analysis an effort was made at the OECD to combine S&T and industrial statistics in one international data base (the STructural ANalysis data base or STAN) (Box 62).

A similar data base (an academic STAN) might be established for the higher education sector with a standard classification by field which would include statistics on expenditures (R&D and teaching), staff (R&D and teaching), numbers of students and graduates at different levels and possibly bibliometric indicators, etc. (Box 63). These statistics could be used to construct R&D indicators, education indicators and also S&T and research personnel indicators.

A wider view of technology

For many years, R&D was often considered as the only significant technology development activity, and the gap between the end of the R&D process and the economic application of the resulting new technology was largely ignored. In consequence, R&D data was treated as a reasonable proxy of the overall technological level of an industry or country and was related directly to measures of competitiveness or growth. However, such analyses proved to have only modest explanatory value.

A first problem is that R&D data measure only the resources put into developing new technology ("inputs"). Because R&D is an uncertain activity, it is necessary to develop measures of its "output". A number of output indicators such as patents (Fact Sheet 2), the technology balance of payments (Fact Sheet 3) and bibliometrics (Fact Sheet 4) were developed before the TEP. Most were adapted from existing data sources, are themselves proxies and give only partial measures of R&D output (Freeman, 1987c).

In addition, R&D statistics measure only organised activities undertaken on a regular basis in established laboratories, whereas new technology may be generated elsewhere, e.g. by "shop floor engineering" in the machine tool industry. It was suggested in the TEP that, given the key role of software and design, they should be included with R&D. The question arose whether the concept of R&D should be extended to cover them or whether new categories should be developed. The experts working on the revision of the "Frascati Manual" decided that organised R&D remains a coherent and significant activity which merits continued monitoring and that its basic definitions should not be changed.

A wider view of technology takes account both of intangible investment and, more specifically, the conceptual framework of innovation discussed in Chapter 1. Hence the interest of the innovation surveys undertaken in a number of OECD countries (see Chapter 5) and for which guidelines are being developed (see Fact Sheet 6).

The TEP exercise has placed particular stress on the diffusion of technology and the application of an innovation after its initial development. This is because most of the economic benefits from new technologies are realised when they are widely adopted in the economy rather than when they are first introduced by the innovating firms (Chapter 7). In consequence it is very important to measure the flows of

---

Box 62. The OECD's structural analysis data base (STAN)

In 1988 the OECD Scientific, Technological and Industrial Indicators Division launched the Structural Analysis (STAN) programme. The aim is to establish a comprehensive, disaggregated and internationally comparable analytical database linking R&D, industrial and import/export data at the individual industry level. It will serve as the basis for studies of structural analysis in OECD countries over the last 25 years.

The database covers some 46 manufacturing industries and 11 service industries from 1967 to 1989. The unifying classification of the data stored in STAN is ISIC, Rev. 2. At present, it combines industrial and technological data for selected OECD countries. The major variables are production, value added, investment (gross fixed capital formation and investment in machinery and equipment), employment (employees and number engaged), wages and salaries, trade (imports and exports) and business enterprise expenditure on R&D.

Work is in progress to link STAN to internationally comparable input-output tables for several major countries from the early 1970s to the mid-1980s.

The sectoral disaggregation for most analytical work is limited by the less fine disaggregation of international R&D and input/output data and by the intermittent availability of the latter.

Further improvements and refinements are under way but the STAN database already is one of the largest and most complex in the OECD. When complete, it will represent the most comprehensive international database of industrial, trade and R&D data available.

technology among industries, to examine all levels of innovation rather than "world-wide firsts" and to identify the degree of use of "key technologies", for example, using manufacturing technology surveys[8] (Fact Sheet 7).

An alternative method is to trace the transfer of technology incorporated in inputs of intermediate and investment goods using input-output tables constructed on the basis of National Accounts data.

This methodology allows:

- the construction of "technology flow matrices" that trace the flows of embodied R&D from one industry to another;
- the calculation of indicators of "total technology intensity" and "technological level" at the sectoral or the national level that combine an industry's own R&D expenditures with the R&D expenditures of industries upstream that flow to the user industry through its purchase of inputs;
- the estimation of the value of the technology embodied in exports or in other components of final demand. By using separate input-output matrices for the imports of intermediate and investment goods, it is possible to examine *international* as well as *inter-sectoral* technology transfer through the R&D embodied in products.

The level of analysis

Many current statistical systems were designed to build up totals for the nation-state defined either in terms of activities on national territory (Gross Domestic Product, Gross Domestic Expenditure on R&D) or the world-wide activities and resources of citizens/residents of the country (Gross National Product, Gross National Expenditure on R&D).

Their principal use was for macro-economic analysis of relatively few key parameters. For example, the original version of the SNA had only six basic tables. The key indicators known to policymakers were at national level (growth in GDP, GERD as a percentage of GDP). Such systems generally provided for analysis of activities and flows of funds between a small number of transactors (e.g. industry, government, private non-profit sector). The TEP exercise has confirmed what had already become obvious during the earlier structural analysis exercise (OECD, 1987): this macro-economic analysis does not have sufficient explanatory value, notably when relating technology and economic growth (see Chapter 8).

The industrial structure of each country, the policy aims of the government and the structure of S&T activities carried out in the government, higher education and private non-profit sectors must be taken into consideration. Classifications and statistical procedures already exist for providing data for such "meso" studies, and particular efforts were made during the TEP exercise to improve the respective OECD data bases, especially in connection with the STAN project. Data which distinguishes large from small and medium enterprises, both for industrial and for science and technology indicators, are also needed.

The emergence of techno-globalism poses a much greater challenge as existing statistical frameworks and surveys are not geared to describe an economic system composed of world-wide networks. Based as they are on the concept of the nation-state, they pay relatively little attention to the international dimension or the "rest of the world" as it is described in the SNA. For example, R&D surveys based on the "Frascati Manual" currently lump all international flows of funds into the single category "abroad". This has consequences for the conception, construction and interpretation of new indicators and the methods by which they are collected.

The three main issues are:
- how to measure the degree and spread of globalisation and its impacts;
- how to define and measure "national" efforts in a globalised world in ways which are meaningful for policymaking; and
- how to trace the activities of individual MNEs.

Measuring the progress of globalisation involves obtaining a wide range of information about international movements of technology, investment, human resources, goods and services and their characteristics. In some cases, new indicators can be constructed using existing statistics (e.g. Technology Balance of Payments or patents), or questions can be added to existing surveys or to new ones being developed for other purposes (innovation surveys, data on scientists and engineers, etc.). In other cases, completely new surveys may be required (Box 64).

Unfortunately the process of globalisation is accompanied by a progressive loss of some of the traditional sources of data on international flows of goods, funds and human resources, as the administrative regulations concerned are lifted. For example, commodity trade data are currently a by-product of customs controls; they will disappear in Europe from 1992 onwards and elsewhere as customs unions are introduced. Other ways of collecting this data must be established.

It is clear that the "domestic" concept currently used at the OECD for most economic and science and technology statistics no longer meets the needs of policymakers. New measures are needed which allow for the fact that company decision-making, R&D activities and actual investment and production no longer occur on the same territory. Whatever new definitions are chosen – and these would probably have to differ according to the policy area – it will be essential to be able to distinguish subsidiaries of multinational enterprises throughout standard statistical surveys (see, for example, Fact Sheet 14). Work on standard international definitions is in progress at the OECD in connection with the revision both of the SNA and of the International Monetary Fund (IMF) handbook.

MNEs are such important actors in the global economy that indicators are needed to trace their activities worldwide. This poses a particular problem to official statistical offices, because the rules of confidentiality under which they operate do not permit them to release any data for named firms, as this might reveal sensitive information to competitors. However, the activities of many MNEs are so complex and the overall resources involved so large that it should be possible to design a set of indicators for individual companies which would be useful for economic analysis but would not reveal information which could be of value to competitors. Much of the information can already be assembled from published sources (such as annual company reports, trade

---

Box 64. Restructuring and industrial performance at global level

In order to measure and analyse world level trends in industrial restructuring it will be necessary to collect and develop new indicators, of which the following may be the most significant:

- The three categories of data linked to foreign direct investment (see Fact Sheet 14)
- Production by subsidiaries of MNEs abroad, broken down by industry and by country of installation
- Intra-firm flows, i.e. commodity trade, flows of funds and flows of technology among subsidiaries of the same group or holding companies situated in different countries
- Imports by national firms from their foreign subsidiaries, broken down by industry
- Mergers and acquisitions involving firms from different countries, broken down by industry and by geographical zone
- Inter-firm agreements or alliances: their number and monetary value, the industry concerned and the country of origin of the firms involved
- Identification, in conceptual and empirical terms, of the strategic sectors of an economy
- Costs and losses, broken down by industry and country, due to non-respect of intellectual property rights
- Measures of industrial concentration in terms of production, employment, number and size of firms
- Degree of diversification by industry; measures of vertical integration
- Structure of SME-SMI by sector and by country.

---

journals, etc.) and, given the confidentiality constraints on official statistical bodies, commercial firms and university units are stepping in to fill the gap[9].

New statistics and indicators are also needed to measure the growth in research co-operation in connection with the growing role of the European Community funding of and plans for "big science". For example, regular R&D statistics currently exclude activities undertaken on "international" territory, e.g. at CERN or the European Space Agency.

The TEP also revealed the need for indicators at the regional and site level. Countries with federal systems of government regularly collect and analyse data at the level of the province/state, and the European Commission has made significant progress in developing regional indicators of all kinds. Much less work has been done at the level of the site[10].

The TEP has confirmed the view that the role of government is to establish an appropriate environment for technological development by firms. However, more stress is now placed on the desirability of establishing the "rules of the game" for when and how governments should intervene. In consequence, it is necessary to reorient and extend existing indicators on government funding of scientific, technological and industrial activities.

Given the importance of government's role in supporting long-term research and facilitating the transfer of its results to industry, there is a need to re-examine the existing statistical concepts (basic research, applied research and experimental development) and to include new areas, notably the "transfer sciences" described in Chapter 1 (Blume, 1990). For the latter, it would be necessary to develop an internationally accepted detailed classification by field of science and technology to be used as a basis for an overall scoreboard of indicators for the higher education sector, which should also include information on direct and indirect government support.

A further set of improvements and changes are needed for the measurement of long-term research. The data currently available for many OECD countries do not permit evaluation of whether the real amount of resources available is growing or declining, let alone whether it is sufficient. This stems from the general difficulty of distinguishing the financial and human resources devoted to R&D in the higher education sector from those devoted to teaching, to administration, etc., particularly in the case of block grants from Ministries of Education. Considerable methodological work has been done on this issue and a special supplement to the "Frascati Manual" was issued (OECD, 1989*i*). Though some countries do collect R&D data for the higher education sector by means of special surveys, most estimate the amounts involved by applying sets of sometimes very arbitrary or out-of-date "R&D percentages" (coefficients) to the total expenditure and personnel data kept by or for Ministries of Education. At best, these estimates are made at the level of the main fields of science and technology and the method of calculation is clearly explained. At worst, a global estimate is made on the basis of some unexplained "rule of thumb". In all cases the R&D data are based on estimates of what they are actually intended to reveal. The widespread poor quality and lack of international comparability of R&D data for the higher education sector has been clearly documented by authors outside the OECD (Irvine *et al.*, 1990). Until governments are willing to do something serious about this problem, establishing utilisable data for an old category such as basic research or for a new one such as the transfer sciences remains wishful thinking.

In order to meet the fear that government support for technology may confer unfair competitive advantage, it is necessary to review and compare policies in areas such as innovation, public procurement, public ownership of companies, taxation, competition, financial and foreign investment, with a view to limiting the scope for international friction. Such a review needs to be based at least in part on internationally comparable indicators (see Fact Sheet 13).

Most of these policies involve some sort of direct or indirect transfer of funds between government and industry.

Measuring these flows, whether for R&D, innovation or new equipment, is notoriously difficult. It is necessary to find a statistical framework which can effectively bridge:
– the budget categories used in different countries;
– the budget categories used by governments and the company accounts used by the firms receiving the support.

It is particularly difficult to match payments and receipts in the case of indirect support, for example, via fiscal incentives. Nevertheless, systematic efforts need to be made to measure the level and influence of government support in all relevant surveys addressed to industry in addition to the data collected in connection with policy reviews. The role of governments in the provision of infrastructures is changing. Indicators are needed to map the availability of the new advanced telecommunications infrastructures (OECD, 1990*l*) and to assess the pace and impact of government standard setting for this and other areas (OECD, 1990*m*). Some measures of how intellectual property regulations affect the innovative behaviour of firms would also be valuable (see Levin *et al.*, 1987).

Quantity, quality and complexity

Most of the discussion so far has been of purely quantitative data. In stressing the role of more "social" activities, such as management or marketing, the TEP has shown the need for more qualitative indicators. Furthermore, it has reinforced the observation that analysis of the S&T system, and particularly the public sector part, also requires information about the associated institutional structures and procedures, to allow for the fact that each OECD country has its own history and traditions.

However, the TEP has not only highlighted the need for information on organisation but has extended the analysis to the industrial sector. There is, thus, a need to find ways of identifying and constructing indicators which describe corporate culture. This is only one of the areas where the TEP points to a need for systematic analysis and comparison of social and qualitative information.

The TEP has also stressed the complexity of many relationships which were previously thought to be relatively straightforward and has added some new phenomena to be measured. Perhaps the most challenging to the statistician is that of networks. As TEP ends, this concept is poised to move from the "blue sky" to the "experimental" stage (Box 56) but considerable case-study work (Callon, 1990) is needed before a regular statistical framework could be envisaged.

Human resource development

Human resources constitute the crucial link for translating technological progress into economic growth and social well-being. In consequence, it is essential that personnel issues should be stressed both when improving existing surveys and data bases (R&D, industrial structures, short-term industrial indicators, etc.) and when developing new ones (innovation, intangible investment, manufacturing technology). This covers both straight quantitative data (numbers of researchers, employers' expenditure on training, etc.) but also qualitative information about perceptions of the availability of skilled personnel and its impact on the activity under examination (innovation, introduction of manufacturing technology, etc.). Here the focus is on scientists and engineers and the general skill levels of the labour force[11].

Scientists and engineers are the lifeblood of the S&T system. They assure the generation of knowledge and participate in its transfer to institutions, sectors and countries and its further development into new products and processes and their diffusion in industries. As teachers, they prepare the next generation, and, whether in R&D, in production or, "beyond technology", in management, they are the stuff of which networks are made.

The number and distribution of scientists and engineers were recognised to be important indicators of a nation's S&T potential when the first S&T statistics were being designed in the 1960s (Freeman and Young, 1965). However, only the United States set up and has systematically maintained a coherent system for monitoring stocks and flows of scientists and engineers. Other countries have generally expressed a need of internationally comparable data only in the context of short-term policy issues such as the "brain drain" or "ageing". Thus, Chapter 6 is largely based on data for the United States and concentrates on a single issue, the possible mismatch between supply and demand resulting from demographic factors.

Establishing complete databases on scientists and engineers is a costly business, requiring, in the United States for example, about the same amount of resources as R&D statistics. There are difficult problems of international comparability, but the return to such an exercise at both national and international levels could be significant, especially if data were designed from the outset to fit into structural data bases such as STAN (Box 62) and Academic STAN (Box 63).

An adequate supply of appropriately educated and trained human resources is a critical factor for successful industrial innovation. To assist with education and training policies, information is needed on the quality of the labour force and on investments in its improvement. Information on basic education can be derived from standard education statistics (see Fact Sheet 9). As life-long education and training are needed to equip better the labour force, information on educational attainment of the labour force was published in the *1989 Employment Outlook* (OECD, 1989j) to adapt to and profit from technical change. The OECD is currently working to develop training statistics (Fact Sheet 10) which will also contribute to the work in intangible investment (Box 61).

Special areas

There are two areas discussed in the TEP which actually or potentially require special sets of indicators. They are technology and the environment and technology and developing countries. Both cut across the other main categories of S&T and industrial statistics.

The OECD already has a substantial programme to develop and collect environment statistics (Fact Sheet 15). Indicator work is needed on integrating environmental issues into economic accounting and measuring the environmental effects of specific industries. Here it is necessary to inject environmental concerns into existing surveys. For example, there is currently no satisfactory international standard for measuring environmental R&D. Similarly, the techniques developed for surveys of manufacturing technology might be extended to cover pollution-related industrial processes.

The TEP did not provide any opportunity for discussing how far the sets of scientific, technical and industrial indicators designed for OECD countries were also relevant to countries at a much lower stage of development. The UN (Madeuf, 1991; and UNESCO, 1978) has already done significant work in this area.

5. Fact sheets

The fact sheets which follow summarise fifteen areas of statistical work relevant to TEP preoccupations (OECD, 1990j). They are provided for information and reference.

Fact Sheet 1. **Research and development statistics**

Aim

To measure the investment in the creation of new scientific and technological knowledge by the performance of research and experimental development (R&D). To assess the scale, structure and direction of such R&D activities in various countries, sectors of the economy, industries, fields of science and technology, etc., by measuring the amount of financial and human resources (inputs) involved.

Methodology and coverage

The *Proposed Standard Practice for Surveys of Research and Experimental Development,* otherwise known as "the Frascati Manual" was the first in the OECD series of Manuals on "The Measurement of Scientific and Technical Activities" (OECD, 1981*b* and 1989*i*). First issued in 1963, it is the generally acknowledged international standard for such surveys. The draft of the fifth edition was discussed at a conference in Rome in Autumn 1991 and should be formally adopted in 1992.

Research and experimental development covers creative work undertaken on a systematic basis in order to increase the stock of knowledge, including knowledge of man, culture and society, and the use of this knowledge to devise new applications. R&D is a term covering three activities: basic research, applied research and experimental development.

In principle, R&D statistics are collected from the bodies who carry out the projects rather than those who finance them (except in the case of budget-based series). Financial resources include both current and capital expenditure with depreciation payments excluded. Human resources are collected in terms of "full-time equivalent" (FTE) on R&D and may be broken down either by occupation or by level of qualification.

R&D is an innovation activity (Fact Sheet 6) and also an important component of intangible investment (Box 61).

Data collection

The OECD maintains a substantial data base of R&D statistics with time series back to the late 1960s (OECD, 1990*e*). It is updated in detail by biennial benchmark surveys addressed to Member governments (known as ISY surveys) and by bi-annual surveys of the main series for the *Main Science and Technology Indicators* publication, also available on diskette. More detailed statistics are available in a publication entitled *Basic Science and Technology Statistics* and on tape and on diskettes.

Fact Sheet 2. **Patent statistics**

Aim

To use data collected by national and international patent agencies to construct indicators for the level, structure and evolution of inventive activities in countries, industries, companies and technologies by mapping changes in technology dependency, diffusion and penetration.

Methodology and coverage

The growing role of international patents organisations is contributing to greater comparability between the patent data available for individual countries, although these are still affected by special characteristics of patents. At the moment, there are no international guidelines for the use of patent statistics as science and technology indicators. The OECD plans to produce such a volume (for reviews of current literature see Basberg, 1987 and Griliches, 1990). A guideline for the use and interpretation of patent data as an indicator of S&T is being prepared. A first draft will be ready next spring.

A patent is a right granted by a government to an inventor in exchange for the publication of the invention; it entitles the inventor to prevent any third party from using the invention in any way, for an agreed period.

Patent data cover applications and grants classified by field of technology. International applications series distinguish four sub-categories: *i)* patents taken out by residents of a country in that country; *ii)* patents taken out in a country by non-residents of that country; *iii)* total patents registered in the country or naming it; *iv)* patents taken out outside a country by its residents. Data on patents granted only distinguish between patents awarded to residents and to non-residents.

Patent descriptions also contain much technological information unavailable elsewhere and therefore constitute a significant complement to the traditional sources of information for measuring diffusion of technological/scientific information (see Fact Sheet 4 on bibliometrics).

Data collection

The raw data are available from national and international patent offices. The OECD assembles, stocks (1990*e*) and publishes total applications data for its Member countries for the four categories identified above in *Main Science and Technology Indicators* and *Basic Science and Technology Statistics* and the associated diskettes and tapes. It also holds a base of patents applied for in the United States broken down by the country of residence of applicants and by industry.

Fact Sheet 3. The technology balance of payments (TBP)

Aim

To measure the international diffusion of disembodied technology by reporting all intangible transactions relating to trade in technical knowledge and in services with a technology content between partners in different countries.

Methodology and coverage

The OECD issued the *Proposed Standard Method of Compiling and Interpreting Technology Balance of Payments Data - TBP Manual* in 1990 (OECD, 1990*k*). It is the second in the series of OECD Manuals on Science and Technology Indicators.

The following operations should be *included* in the TBP: patents (purchase, sales); licenses for patents; know-how (not patented); models and designs; trade-marks (including franchising); technical services; finance of industrial R&D outside national territory.

The following should be *excluded:* commercial, financial, managerial and legal assistance; advertising, insurance, transport; films, recordings, material covered by copyright; design; software.

The international comparability of national TBP indicators is only improving progressively as national practices are changed to match the guidelines of the new Manual.

Data collection

National TBP data may be collected by means of special surveys but more often are assembled from existing records kept by Central Banks, exchange control authorities, etc.

The OECD has assembled a data base of "macro" TBP data for most of its Member countries covering total transactions (receipts and payments) by partner-country back to 1970 (OECD, 1990*e*). Data for recent years have been published in *Main Science and Technology Indicators* and *Basic Science and Technology Statistics* and the associated diskettes and tapes. It has recently created a new international data base for detailed TBP series (broken down by industry, type of operation and geographical area) starting with Japan, Germany, Italy and Sweden, and a special international survey was launched in Summer 1991. In parallel, detailed data based on national practices and classifications have been assembled and updated for about ten countries.

Fact Sheet 4.   **Bibliometrics**

Aim

To use data on the number and authors of scientific publications, articles and the citations therein (and in patents), to measure the "output" of research teams, institutions and countries, to identify national and international networks and to map the development of new (multi-disciplinary) fields of science and technology.

Methodology and coverage

Bibliometric methods have essentially been developed by university groups and by private consultancy firms. As yet there are no official international guidelines for the collection of such data or for their use as science and technology indicators. During the TEP, the OECD commissioned a report on the "state of the art" in bibliometrics (van Raan and Tijssen, 1990) which constitutes a basis for an OECD manual on the use and interpretation of bibliometric indicators. It may be prepared and issued in co-operation with the European Commission (European Network on S&T Indicators of the MONITOR-SPEAR Program).

Bibliometrics is the generic term for data about publications. Originally, work was limited to collecting data on numbers of scientific articles and publications, classified by authors and/or by institutions, fields of science, country, etc.

in order to construct simple "productivity" indicators for academic research. Subsequently, more sophisticated and multidimensional techniques based on citations in articles (and more recently also in patents) were developed. The resulting citation indexes and co-citation analyses are used, both to obtain more sensitive measures of research quality and to trace the development of fields of science and of networks.

Data collection

Most bibliometrics data come from commercial companies or professional societies. The main general source is the Science Citation Index (SCI) set of data bases created by the Institute for Scientific Information (United States) on which are based several major bases of science indicators developed by Computer Horizons Inc. (for the National Science Foundation). Other specialised bases are Medline (United States) and Excerpta Medica (Netherlands) for medical bibliometrics, and Chemical Abstracts (United States).

A number of other international and/or national databases, frequently interlinked, are currently being developed. The OECD currently has neither plans, resources nor competence to undertake basic data collection, although bibliometric data are regularly used in its analytical reports.

Fact Sheet 5. **High-technology products and industry**

Aim

To construct indicators to measure the technology content of the goods produced and exported by a given industry and country with a view to explaining their competitive and trade performance in "high-tech" markets. These markets are characterised by rapid growth in world demand and oligopolistic structures, they offer higher than average trade returns, and they affect the evolution of the whole structure of industry.

Methodology and coverage

Indicators on trade in high-tech products/industries were originally designed as measures of the "output" or "impact" of R&D; they are now seen as having a wider use in the analysis of competitiveness and globalisation.

There are yet no officially approved international standards for identifying high-tech industries and products. Two main approaches have been used to date, by *industry* where OECD work (drawing on earlier studies by the US Department of Commerce) has been the basis for most exercises in individual countries and by *product*. The OECD plans to produce a volume of recommendations covering both approaches.

a) In the *industry* approach, used by the OECD, the main criterion used in the past has been R&D expenditures as a percentage of the production, turnover or value added of the industry concerned. Industries were divided into three categories, "high", "medium" and "low" R&D

intensity (OECD, 1986c). Further work will allow industries to be divided up according to their "technology content", taking into consideration not only direct investment in R&D but the indirect acquisition of its domestic results incorporated in *i)* intermediate consumption, *ii)* capital goods, and *iii)* results of foreign R&D-incorporated in imported goods. All these technology inputs must be estimated econometrically using input-output matrices.

b) The *product* approach has the advantage of allowing more detailed analysis and identification of the technology content of products and hence a weeding out of mature products manufactured by otherwise R&D-intensive industries. This approach requires the use of detailed R&D data by product field.

Data collection

To date the OECD has favoured the industry approach. Using an OECD trade data base classified by ISIC, a series of import-export ratios for the main R&D-intensive industries has been set up and published twice a year in *Main Science and Technology Indicators* and the associated diskette. Series for trade by high, medium and low R&D-intensive industries are analysed in the annual *Review of Industrial Policy in OECD Countries* and summarised in *OECD in Figures*. In addition to the other improvements mentioned above, a new trade base by product that will offer greater analytical possibilities will be established in 1992.

Fact Sheet 6. Innovation statistics

Aim

To measure aspects of the industrial innovation process and the resources devoted to innovation activities. To provide qualitative and quantitative information on the factors enhancing or hampering innovation, on the impact of innovation, on the performance of the enterprise and on the diffusion of innovation.

Methodology and Coverage

The *OECD Proposed Guidelines for Collecting and Interpreting Innovation Data – Oslo Manual,* prepared jointly by the OECD and the Nordic Fund for Industrial Development (Nordisk Industrifond, Oslo) in 1990, is currently under revision. It will then be officially adopted by the OECD as the third in the "Frascati" family of manuals and should be issued late in 1991 or early in 1992.

Technological innovations comprise new products and processes and significant technological changes of products and processes. An innovation has been *implemented* if it has been introduced on the market (product innovation).

- *Major product innovation* describes a product whose intended use, performance characteristics, attributes, design properties or use of materials and components differ significantly compared with previously manufactured products. Such innovations can involve radically new technologies or can be based on combining existing technologies in new uses.
- *Incremental product innovation* concerns an existing product whose performance has been significantly enhanced or upgraded. This again can take two forms. A simple product may be improved (in terms of improved performance or lower cost) through use of higher performance components or materials, or a complex product which consists of a number of integrated technical subsystems may be improved by partial changes to one of the subsystems.
- *Process innovation* is the adoption of new significantly improved production methods. These methods may involve changes in equipment or production organisation or both. The methods may be intended to produce new or improved products which cannnot be produced using conventional plants or production methods or to increase the production efficiency of existing products.

Data collection

National data on innovation activities are generally collected by means of surveys addressed to industrial firms. Over half of the OECD Member countries have organised such surveys, and it is on their experience that the Manual is based (OECD, 1990d).

It is also possible to collect data on the number and nature of actual innovations. Such information can be obtained by special surveys or assembled from other sources such as the technical press.

So far, the only internationally comparable series of data are those collected under the auspices of the Nordic Industrial Fund. The OECD is planning to set up an international survey of innovation activities, working in close cooperation with the Nordic Fund and with Eurostat.

Fact Sheet 7. Measuring the use of advanced manufacturing technology

Aim

To measure the extent of use of different kinds of manufacturing technology, including the patterns of diffusion and the effects of use (disadvantages, difficulties, constraints and barriers to wider use) as well as skills and training and employment issues.

Methodology and coverage

A list of "Key Survey Questions" was published in *Government Policies and the Diffusion of Micro-electronics* (OECD, 1989*a*). These questions covered the applications of micro-electronics in processes where they are used for monitoring and controlling purposes as well as in products. The OECD has been playing a clearing-house role in this area, regularly reviewing and exchanging information on surveys that have been carried out or are under way, and promoting greater comparability between national surveys. The diffusion and use of manufacturing technology was reviewed in *Managing Manpower for Advanced Manufacturing Technology* (OECD, 1991*a*).

Advanced manufacturing technology is defined as computer-controlled or micro-electronics based equipment used in the design, manufacture or handling of a product. Typical applications include computer-aided design (CAD), computer-aided engineering (CAE), flexible machining centres, robots, automated guided vehicles, and automated storage and retrieval systems. These may be linked by communications systems (factory local area networks) into integrated flexible manufacturing systems (FMS) and ultimately into an overall automated factory or computer-integrated manufacturing system (CIM).

Data collection

National data have been collected through special surveys of manufacturing firms. About half of the OECD Member countries have carried out surveys, and their comparability has improved due to the use of common survey questions.

So far detailed international comparisons of the use of AMT have been made in France, Germany and the United Kingdom, and subsequently in Canada and the United States. Other countries have made more limited comparisons. The OECD has followed and encouraged these comparisons and is planning to carry out further work on comparisons and data collection in 1992.

Fact Sheet 8. **Science and engineering personnel**

Aim

To establish coherent data bases on the current and possible future supply, use and demand (at home and abroad) for science and engineering (S&E) personnel with a view to evaluating the consequences for future research and industrial performance, planning education and training, measuring the diffusion of knowledge incorporated in human resources and assessing the roles of women (and minorities) in science and technology activities.

Methodology and coverage

There are, as yet, no international standards for measuring stocks and flows of S&E personnel. The main exponent of this type of work is the National Science Foundation (United States). The OECD is undertaking a review of the current state of studies and data in its Member countries prior to preparing some initial guidelines.

S&E personnel may be defined in terms of their qualifications or their current employment. In the former case, the appropriate classification is the International Standard Classification of Education (ISCED) and, in the latter, the International Standard Classification of Occupations (ISCO). They may cover only persons with university qualifications/ professional occupations or also include those with other post-secondary qualifications and technical jobs. A combination of both criteria and levels is needed if supply and demand issues are to be analysed correctly.

A typical system should cover total national stocks of S&E personnel at a given point in time, broken down by employment status and by sector and type of employment, and the intervening inflows (mainly educational output and immigration) and outflows (mainly retirement and emigration). Both stocks and flows should be broken down by field of science and technology, age and sex and possibly also national or ethnic origins.

Data collection

While a few very small OECD countries are able to maintain complete nominal registers of all S&E graduates and their whereabouts, data bases on S&E personnel have to be built up in most countries from several sources, notably education statistics (numbers of teachers and graduates), employment statistics, and population censuses supplemented by special surveys. UNESCO collects and publishes data annually on total national stocks of scientists and engineers (UNESCO, 1990). The OECD hopes to build a more sophisticated set of indicators. The establishment of a consolidated data base for the higher education sector (see Academic STAN) could contribute to supplying this data.

Fact Sheet 9. **Education statistics and indicators**

Aim

To provide comprehensive and reliable international education statistics and indicators which take full account of the complexity of providing adequate education in modern societies in order to assist with policy formulation and debate.

Methodology

The underlying OECD methodology for education statistics is still basically that suggested in *Methods and Statistical Needs for Educational Planning* (OECD, 1967), although educational structures and international classification systems have since evolved considerably. UNESCO's International Standard Classification of Education (UNESCO, 1976) gives guidelines on fields and levels of education which have been adopted by the OECD.

The definition of "education" is conventional (but increasingly problematic given the growing diversity of provision), as it concerns only education provided in schools, universities and other establishments considered as part of the overall education system, with access to each level normally possible only on completion of the preceding level. All training activities for adults or specific target groups such as young people seeking jobs, the unemployed, etc. are thus excluded.

The OECD is currently working to develop the methodology for sets of more refined, internationally comparable education indicators of greater interpretative utility. This is still in a development phase but will feed into the work on educational statistics from 1992 in the form of a handbook that considers methodology, definitions and measurement issues.

Data collection

The OECD currently collects education statistics using a set of questionnaires common to the OECD, UNESCO and Eurostat. These cover, first, data on enrolments, qualifications, and teachers in pre-primary, primary, and secondary education; second, similar breakdowns for further and higher (third level) education; third, educational expenditure and financing. This data is stored in a data base and published in *Education in OECD Countries: A Compendium of Statistical Information* (OECD, 1989m and 1990n).

After a stocktaking, consolidation and review exercise, the surveys and data base will be overhauled; both the new needs identified and modern methods of data transmittal and access will be taken into consideration.

Fact Sheet 10. **Training statistics**

Aim

To collect data on the amount, costs and beneficiaries of "structured" training provided and sponsored by employers. Subsequently, to identify what determines employer spending, the costs and benefits to employers and employees, and the success of different types of training. The information will contribute to the design of policies for dealing with loss of competitiveness, an ageing work force and the challenges posed by new technologies. Data on training expenditure will contribute to measuring "intangible investment".

Methodology and coverage

There is as yet no international statistical framework for the collection and analysis of training statistics. However a group of OECD experts has reviewed the available national statistical information on training activities for about a dozen participating Member countries in order to propose useful and appropriate definitions and methodologies for measuring training for policy purposes and to allow meaningful inter-country comparisons.

The first results are given in a restricted report now being considered by the OECD Education Committee and Employment, Labour and Social Affairs Committee. It covered enterprise-based training, i.e. education and training activities organised or sponsored by enterprises, as described in surveys of employers and employees. The next task will be to develop a set of definitions for the most commonly used training terms and to work towards comparable surveys of employers and employees that are more transparent with respect to the definition of the components of training; and the associated classifications, reference periods and practical survey methods.

Data collection

All the participating countries conducted surveys of individuals to discover who receives training, although in varying degrees of detail. Most countries also conducted surveys of employers on employer-based training but only a few covered expenditure on training. Only one or two countries had time series. National surveys were undertaken using different methodologies, definitions and reference periods and asking different questions. Some approaches worked better than others.

The actual data are not given in the methodological report cited above. There has, however, been a parallel and closely related exercise assessing the scale and pattern of enterprise training whose results were published in the 1991 OECD *Employment Outlook* (1991j).

Fact Sheet 11. Structural statistics and indicators

Aim

To make available annual series over a long period for the main industrial variables (see below) broken down by industry; they can be used either alone or in connection with science and technology statistics and trade statistics, to construct derived indicators, to measure structural change and to assess government adjustment policies.

Methodology and coverage

Structural statistics are among the oldest series collected by OECD countries. The variables concerned are: production, value added, employment, total investment (of which investment in machinery and equipment), the number of firms or establishments, labour costs and hours worked.

There are two main types of data: a) responses to surveys of industrial firms; b) series derived from National Accounts. The former are based on methodology established by the United Nations (UN, 1986c) which can differ significantly from that in the SNA. The latter include a wide range of service industries; the former deal almost exclusively with manufacturing, but provide more industrial detail than the SNA (3 or 4 digits in ISIC).

Data collection

The OECD organises an annual survey on structural statistics and the resulting data go back up to twenty years depending on the variables concerned. They are published annually in *Industrial Structure Statistics*. Other data bases with similar contents have been established by the United Nations (United Nations Statistical Office at New York, United Nations Industrial Development Organisation in Vienna) and by Eurostat. All these are assembled and fed into the OECD structural analysis data base (STAN) (see Box 62).

Fact Sheet 12. **Short-term industrial indicators**

Aim

To supply information on trends in the volume of activities in the main industrial sectors (with a lag of at most one-quarter) in order to assess the current industrial climate and to update the indicators based on structural statistics (Fact Sheet 11) in order to give timely support for industrial policy and analysis.

Methodology

There are no established world-wide methodological guidelines for the collection and comparison of short-term industrial indicators other than those associated with the OECD quarterly survey (see below).

Short-term indicators are collected and presented on a monthly and quarterly basis. They are generally expressed as indices and are systematically corrected for seasonal variations.

The indicators concerned are: production, deliveries, production prices, new orders and unfilled orders and employment of operatives. Other qualitative indicators are collected, notably: judgements on stocks of finished goods, on order books, on total order inflow, on production prospects, on capacity utilisation and on labour force expectations.

The basic classification is by industry defined in terms of ISIC. At the OECD the quantitative data are available for about 25 industries (levels 2, 3 and 4 of ISIC) whereas the qualitative series are available only at level 2 of ISIC.

Data collection

National data are collected by means of monthly and quarterly surveys. The OECD has been holding a quarterly international survey since 1975. It covers monthly, (seasonally adjusted) quarterly, and annual series. The resulting data base is the only international source of up-to-date statistics by industry. In future it is intended to extend the data base to include selected high-tech industries and additional variables, notably for investment and certain financial indicators. The data are published quarterly in *Indicators of Industrial Activity* and are used in the *Annual Review of Industrial Policy in OECD Countries.*

Fact Sheet 13. **The OECD's subsidies and structural adjustment project**

Aim

To fulfil a Ministerial mandate by quantifying the scope of government programmes to support industry and the trends therein, to improve international transparency and to evaluate the subsidy element of these programmes and their impact, notably on competitiveness.

Methodology and coverage

A standard international methodology is being developed as the project progresses. In the present phase, the Net Cost to Government (NCTG) measurement has been added to the Gross Government Budget Expenditure (GGBE); it gives a better approximation of the subsidy equivalent of government support for industry. Two categories of support are identified: financial instruments and fiscal aid.

The main financial instruments included are: direct subsidies (grants, non-repayable advances, repayable advances, and interest rate subsidies), loans (included in the budget or accorded via financial intermediaries), loan guarantees and the provision of equality capital.

Fiscal aid covers tax exemption, tax allowances, tax credits, special rate reliefs, tax deferrals and accelerated depreciation. Civil and military procurement are also taken into consideration.

The different types of budgetary expenditure are broken down into eight policy areas: sectoral measures, enterprises in difficulty, research and development, regional policy, investment aid, SMEs, employment and training, and exports and trade-related assistance.

In the next phase, support programmes will be classified into more homogeneous categories, so as to be able to identify their direct economic impacts.

Data collection

There have been two surveys undertaken for the exercise. The results of the second one, covering the period 1986-89, are held in a data base at the OECD. The results are not simple statistics but give a quantitative and qualitative description of some 800 industrial support programmes which can be used to establish new classifications and measure the impact on competitiveness. Given the nature of this information and certain problems of confidentiality, it is stocked and analysed in local mode.

Fact Sheet 14. **Foreign direct investment**

Aim

Within the framework of the globalisation of industrial activities, to collect annual data on stocks and flows of FDI, broken down by industry, which are comparable with other industrial and technological statistics, in order to quantify the impact of flows of foreign investment on trade, technology transfer and the industrial structure and competitiveness of the investor and host countries.

Methodology and coverage

The basic world-wide definitions are those developed by the International Monetary Fund. The OECD is preparing a revised Detailed Benchmark Definition of Foreign Direct Investment which is compatible with the forthcoming revised versions of the "IMF Balance of Payments Manual" (fifth edition to be published at latest in 1992) and of the SNA.

A foreign direct investor is an individual, incorporated or unincorporated enterprise which possesses a subsidiary, associate or branch operating in a country other than the country or countries of residence of the investor. This should involve ownership of at least 10 per cent of the ordinary shares in the voting stock of the enterprise in which the investment is made. Depending on the degree of control "foreign direct investment enterprises" may be:

- Subsidiaries, i.e., where the parent company owns more than 50 per cent of the voting stock of an incorporated enterprise (or has the right to appoint or dismiss the majority of its directors).
- Associated corporations, i.e., where the parent company owns 10-50 per cent of the voting stock of an incorporated enterprise which enables it to influence its management.

- Branches, i.e., non-incorporated enterprises (treated as a quasi-corporation in the SNA) set up by the parent company with others, involving the purchase or rent of land or buildings and the acquisition of other fixed assets for use in a significant level of production.

Data collection

Data on foreign direct investment are collected in most OECD countries by Ministries of Finance, Central Banks and Ministries of Industry and Trade. The three main categories of data are:

-the amounts of flows of investment within and out of each country by sector (industry and services) and by geographic zone;

-the value of stocks of investment (or failing this, the cumulative sums over a long period), within and outside the country by industry and zone;

-the characteristics of firms whose capital is more than 50 per cent owned by non-residents, broken down by industry. The variables concerned are the number of firms, employment, gross production, value added, gross capital formation, gross operating surplus, gros capital stock, investments in R&D, exports, imports and technology transfers.

The OECD has surveyed the third category of FDI, and data are available for 12 countries and for various years during the period 1972-87. Information on the first and second categories is available from national sources only.

Fact Sheet 15. **Environmental indicators**

Aim

To develop indicators to be used for the integration of environmental and economic decision-making at the national, international and global levels.

To inform the ongoing process of policy dialogue among countries and to lay the basis for international co-operation and agreements. To respond to the public's "right to know" about basic trends in air and water quality and other aspects of their immediate environment affecting health and well-being.

Methodology and coverage

There is no universal set of environmental indicators; rather, there are sets of indicators responding to specific conceptual frameworks and policy purposes.

Three broad families of indicators are involved:

i) Indicators for reporting environmental conditions and trends and broadly measuring environmental performance where the existing OECD statistical framework is currently being improved and extended.
ii) Indicators for the integration of environmental concerns in sectoral policies, i.e., the development of sectoral indicators showing environmental efficiency and the link-

ages between economic policies and trends in key sectors (e.g. agriculture, energy, transport) on the one hand, and the environment on the other.

iii) Indicators for the integration of environmental concerns in economic policies more generally through environmental accounting, particularly at the macro-economic level. Priority is being given to two aspects: the development of satellite accounts to the system of national accounts, and work on natural resource accounts (e.g., pilot accounts on forest resources).

All the above methodological work is undertaken in close co-operation with other international organisations, notably Eurostat.

Data collection and publication

The OECD maintains a core set of statistics on the state of the environment in Member countries in its SIREN data base, and data are published regularly in *Compendia of Environmental Data* (OECD, 1991*i*). The results of this and other work on environment indicators are presented in the series of reports on the state of the environment prepared about every five years for Ministerial meetings at the OECD. A special publication *Environment Indicators: A Preliminary Set* was issued for the 1991 meeting (OECD, 1991*e*).

NOTES

1. These cover only the main categories. For a fuller description see OECD (1990*j*).

2. The general concept of science and technology indicators was addressed in the early 1960s but was not raised again seriously until the publication in 1973 of the first of the US National Science Board reports on Science Indicators (later Science and Engineering Indicators). The OECD definition quoted here is developed from ideas in these reports and first appeared in the general conceptual paper prepared by the Secretariat for the major OECD Science and Technology Indicators Conference held in 1980.

3. On S&T indicators see, for example, United States, National Science Board (1989) and for environmental indicators see "Public Opinion" in OECD (1991*e*).

4. Unlike the EC or the United Nations, the OECD does not have a separate Statistical Office or Directorate. Statistical divisions, units or individual staff are situated within the Directorates concerned. The work covered by this paper is carried out in the Directorate for Science Technology and Industry (Scientific and Technological and Industrial Indicators Division), the Directorate for Social Affairs, Manpower and Education, the Centre for Education Research and Innovation, the Economics and Statistics Department (Economic Statistics and National Accounts Division) and the Environment Directorate.

5. Such a major change in the economic and social system is currently in process in Central and Eastern Europe where official statistical agencies are having to adopt completely new statistical frameworks and methods.

6. EC (1990), *General Industrial Classification of Economic Activities within the European Communities* (NACE, Rev.1). NACE-70 will continue to be used up to 1993, at which time this new version of the classification will be applied.

EC (draft), Central Product Classification of the European Communities (CPC-COM).

EC (1987), *Combined Nomenclature* (CN). It has replaced the Common Customs Tariff and the Nomenclature of Goods for the External Trade Statistics of the Community and Statistics of Trade between Member States (NIMEXE).

7. These basic definitions have also been adopted by UNESCO for its world-wide statistics. See UNESCO (1978).

8. A forthcoming number of the OECD *STI Review* will be devoted to presenting the results of selected innovation and manufacturing technology surveys.

9. See, for example, the MERIT data quoted in Chapter 10.

10. In the TEP framework, "sites" are large agglomerations such as Silicon Valley or Sophia-Antipolis in which are concentrated high-level R&D laboratories, prestige universities and high-technology firms organised in networks. These sites provide significant positive externalities for the industrial development of the regions in which they are situated and act as poles of attraction for highly qualified manpower, high-tech industries and foreign investors.

11. It has been suggested, rather late in the TEP, that statistics should also be collected on technicians. In fact, a data base covering persons with at least university level qualifications in science and technology would already include most higher level technicians. Obtaining information on "shop floor" technicians with lower and more practical qualifications or in-house training would be best associated with general surveys of the quality of the labour force.

BIBLIOGRAPHY

Abernathy, W.J. (1978), *The Productivity Dilemma: Roadblock to Innovation in the Automobile Industry,* Johns Hopkins University Press, Baltimore.

Abernathy, W.J. and J.M. Utterback (1975), "A Dynamic Model of Product and Process Innovation", *Omega,* Vol. 3, No. 6.

Abramovitz, M. (1991), "The Postwar Productivity Spurt and Slowdown – Factors of Potential and Realisation", in OECD (1991c).

Ackerman, B. and W. Hassler (1981), *Clean Coal/Dirty Air,* Yale University Press, New Haven.

Afriat, C. (1990), "Measuring Intellectual Investment: Problems and First Results", paper presented to the Paris TEP Indicators Conference, July.

Aglietta, M., A. Brender and V. Coudert (1990), *Globalisation financière: l'aventure obligée,* CEPII-Economica, Paris.

Amable, B. (1990), "National Learning Effects, International Specialisation and Growth Trajectories", paper presented to the Paris TEP Technology and Competitiveness Conference, June.

Amable, B. and M. Mouhoud (1990), "Changement technique et compétitivité industrielle: une comparaison des six grands pays industriels", IRES Working Paper No. 90-01.

Amsden, A. (1989), *Asia's Next Giant: South Korea and Late Industrialisation,* Oxford University Press, Oxford and New York.

Andersen, E.S. and B.A. Lundvall (1988), "Small National Systems of Innovation", in Freeman and Lundvall.

Antonelli, C. (1982), "Technology Payments and the Economic Performance of Italian Manufacturing Industry – An Empirical Analysis", unpublished document, OECD, Paris.

Antonelli, C. (1984), *Cambiamento tecnologico e imprese multinazionale: il ruolo delle reti telematiche nelle strategie globali,* Franco Angeli, Milano; résumé in English, Antonelli (1986).

Antonelli, C. (1986), "TDF and the Structure and Strategies of TNCs", *The CTC Reporter,* No. 22, United Nations, New York.

Antonelli, C., ed., (1988a), *New Information Technology and Industrial Change: The Italian Case,* Kluwer Academic Publishers, Dordrecht.

Antonelli, C. (1988b), "The Emergence of the Network Firm", in Antonelli (1988a).

Antonelli, C. (1991a), *The Diffusion of Advanced Telecommunications in Developing Countries,* OECD Development Centre Studies, Paris.

Antonelli, C. (1991b), *The Economics of Information Networks,* North-Holland, Amsterdam.

Arcangeli, F., P. David and G. Dosi (1991), *Frontiers in Technology Diffusion,* Oxford University Press.

Arena, R., J. de Bandt and L. Benzoni, eds. (1988), *Traité d'Economie Industrielle,* Economica, Paris.

Arthur, B. (1988), "Competing Technologies, Increasing Returns and 'Lock-in' by Small Historical Events", *Economic Journal,* March.

Aschauer, D.A. (1989), "Is Public Expenditure Productive?", *Journal of Monetary Economies,* March.

Athukorala, P. (1989), "Export, Performance of 'New Exporting Countries': How Valid is the Optimism?", *Development and Change,* Vol. 20, No. 2.

Atkinson, P. and J.C. Chouraqui (1985), "The Origins of High Real Interest Rates", *OECD Economic Studies,* No. 5, Autumn.

Atkinson, P. and J. Stiglitz (1969), "A New View of Technological Change", *Economic Journal,* September.

Australia, Bureau of of Industry Economics (1991), "Networks: A Third Form of Organisation", Discussion Paper 14, Canberra.

Axelrod, R. (1984), *The Evolution of Co-operation,* Basic Books, New York.

Bagnasco, A. (1977), *Tre Italie: La problematica territoriale dello sviluppo italiano,* Il Mulino, Bologna.

Bain, J.S. (1968), *Industrial Organisation,* John Wiley and Sons Inc., New York.

Balassa, B. (1977), "A 'Stages' Approach to Comparative Advantage", *World Bank Staff Working Paper,* No. 256, The World Bank, Washington D.C.

Baldwin, R. (1989), "On the Growth Effects of 1992", *Economic Policy,* Fall.

Bartel, R.D. (1980), "Dynamic Transformation of the World Economy: The US Policy Response", in Lawrence.

Basberg, B.L. (1987), "Patents and the Measurement of Technological Change: A Survey of the Literature", in Freeman (1987c).

Baumol, W. and K.S. Lee (1991), "Contestable Markets, Trade and Development", *The World Bank Observer,* January.

Becattini, G. (1987), *Mercato e forze locali: Il distretto industriale,* Il Mulino, Bologna.

Bell, J. and G. Vickery (1989), "International Perspectives on Technology and Free Trade", *Technology in Society,* Vol. 11.

Benbrook, C. (1990), "Agriculture's Next Technological Fix: A Sectoral Review", *Chemical Engineering News,* 5 March.

Bernal, J.D. (1971), *Science in History,* MIT Press, Cambridge, Mass. and Penguin Books, London.

Bernstein, J. (1989), "The Structure of Canadian Interindustry R&D Spillovers, and the Rates of Return to R&D", *Journal of Industrial Economics,* Vol. 37, No. 3.

Bernstein, J. and M.I Nadiri (1989), "Research and Development and Intra-industry Spillovers: An Empirical Application of Dynamic Duality", *Review of Economic Studies,* Vol. 56, pp. 249-269.

Bertrand, O. and T. Noyelle (1988), *Human Resources and Corporate Strategy: Technological Change in Banks and Insurance Companies,* OECD, Paris.

Bessant, J. (1989), *Microelectronics and Change at Work,* ILO, Geneva.

Bienaymé, A. (1988), "Technologie et nature de la firme", *Revue d'Economie Polique,* November-December.

Bizet, J.P. (1990), "Améliorer la commercialisation de la technologie", paper presented to the Paris TEP Technology and Competitiveness Conference, June.

Blair, J.M. (1972), *Economic Concentration: Structure, Behaviour and Public Policy,* Harcourt Brace Jovanovich, Inc., New York.

Blume, S. (1990), "Transfer Sciences: Their Conceptualisation, Functions and Assessment", paper to the Paris TEP Indicators Conference, July.

Booz-Allen Acquisition Services (1989), "Studies on Obstacles to Takeover Bids in the European Community", Executive Summary prepared for the Commission of the European Communities, PGXV-B-2, December.

Bourdieu, P. (1985), "Propositions pour l'enseignement de l'avenir", Collège de France.

Bowring, A. (1989), "Garments: Global Subsector Study", *World Bank Industry Series Paper,* No. 19, December.

Boyer, M. (1988), *The Search for Labour Market Flexibility,* Clarendon Press, Oxford.

Boyer, R. (1986), *La théorie de la régulation: une analyse critique,* La Découverte, Paris.

Boyer, R. (1988), "Technical Change and the Theory of Regulation", in Dosi *et al.*

Boyer, R. (forthcoming), "New Directions in Management Practices and Work Organisation. General Principles and National Trajectories", in OECDd (forthcoming).

Bradford, C. (1986), "East Asian Models: Myths and Lessons", in J.P. Lewis, ed., *Development Strategies Reconsidered,* Transaction Books, New Brunswick.

Bradford, C. (1987), "Trade and Structural Change: NICs and Next-Tier NICs as Transitional Economies", *World Development,* Vol. 15, No. 3.

Brainard, R. (1990), "Towards Technoglobalism: A Summary of the Tokyo TEP Conference", unpublished document, OECD, Paris.

Braunling, G. (1990), "Public Policies Supporting Technology Appropriation", paper presented to the Paris TEP Technology and Competitiveness Conference, June.

Bresnahan, T. (1986), "Measuring the Spillovers from Technical Advance: Mainframe Computers in Financial Services", *American Economic Review,* Vol. 76, No. 4.

Bressand, A. (1990*a*), "From Trade to Networking: A Services and Information Perspective on Post-interdependence Globalisation Patterns", paper presented to the Tokyo TEP Technoglobalism Conference, February.

Bressand, A. (1990*b*), "The Producer-Consumer Link: From Competitive Advantage to Network Advantage", paper presented to the Paris TEP Technology and Competitiveness Conference, June.

Bressand, A. and C. Distler (1989), "Network Driven Interconnection", Project Prométhée Perspectives, January.

Bressand, A., and K. Nicolaidis, eds., (1989), *Strategic Trends in Services, An Inquiry into the Global Service Economy,* Harper and Row, New York.

Brusco, S. (1982), "The Emilian Model: Productive Decentralisation and Social Integration", *Cambridge Journal of Economics,* Vol. 6, No 2.

Buckley, P.J. and M.C. Casson (1985), *Economic Theory of the Multinational Enterprise,* Macmillan, London.

Burstall, M.L., J.H. Dunning and A. Lake (1981),'International Investment in Innovation', in N. Wells, ed., *Pharmaceuticals among the Sunrise Industries,* Croom Helm, London.

Caicarna, G.C., M.G. Colombo and S. Mariotti (1989), *Tecnologie dell'informazione e accordi tra imprese,* Edizioni Comunità, Milano.

Caicarna, G.C., M.G. Colombo and S. Mariotti (1990), "Firm Size and the Adoption of Flexible Automation", *Small Business Economics,* Vol. 12, No. 2.

Callon, M., ed., (1989), *La science et ses réseaux: genèse et circulation des faits scientifiques,* La Découverte, Paris.

Callon, M. (1990), "Les réseaux techno-économiques et leurs indicateurs", paper presented to the Paris TEP Indicators Conference, July.

Callon, M., P. Laredo and V. Rabeharisoa (1990), "The Management and Evaluation of Technological Programs and the Dynamics of Techno-economic Networks: The Case of the Agence Française de la Maîtrise de l'Energie (AFME)", Centre de Sociologie de l'Innovation, Ecole des Mines de Paris.

Canada, Statistics Canada (1990), "Software Research and Development (R&D) in Canadian Industry, 1988", *Science Statistics,* Vol. 14, No. 5.

Cantwell, J. (1989), *Technological Innovation and Multinational Corporations,* Basil Blackwell, Oxford.

Caspar, P. and C. Afriat (1988), *L'investissement intellectuel: Essai sur l'économie de l'immatériel,* CPE/Economica, Paris.

Casson, M., ed., (1991), *Global Research Strategy and International Competitiveness,* Department of Economics, Reading University.

Caves, R.E. (1974), "Industrial Corporations: The Industrial Economics of Foreign Investment", *Economica,* August.

Caves, R.E. (1982), *Multinational Enterprise and Economic Analysis,* Cambridge University Press.

Cecchini, P. (1989), *The European Challenge 1992,* Gower Publishing Company, London.

Chandler, A.D. (1977), *The Visible Hand: The Managerial Revolution in American Business,* Harvard University Press, Cambridge, Mass.

Chenery, H. (1986), "Growth and Transformation" in H. Chenery, R. Robinson, and M. Syrquin, *Industrialisation and Growth,* Oxford University Press.

Chesnais, F. (1981), "The Notion of International Competitiveness: A Discussion Paper", unpublished document, OECD, Paris.

Chesnais, F. (1982), "Schumpeterian Recovery and the Schumpeterian Perspective – Some Unsettled Issues and Alternative Interpretation", in H. Giersch, ed., *Proceedings of the 1981 Kiel Symposium on Emerging Technology: Consequences for Economic Growth, Structural Change and Employment in Advanced Open Economies,* J.C.B. Mohr, Tübingen.

Chesnais F. (1986), "Science, Technology and Competitiveness", *STI Review,* No. 1, OECD, Paris .

Chesnais, F. (1988a), "Multinational Enterprises and the International Diffusion of Technology", in Dosi *et al.*

Chesnais, F. (1988b), "Technical Co-operation Agreements between Firms", *STI Review,* No. 4, OECD, Paris.

Chesnais, F. (1989), "Petrochemicals" in Oman *et al.*

Chesnais, F. (1990a), "Present International Patterns of Foreign Direct Investment – Underlying Causes and Some Policy Implications for Brazil", *A Inserção do Brasil Nos Anos 90,* Vol. II, SEADE and CERI-UNICAMP, University of Campinos, March.

Chesnais, F., ed., (1990b), *Compétitivité internationale et dépenses militaires,* CPE/Economica, Paris.

Chisman, F. (1989a), "Jump Start: the Federal Role in Adult Literacy", Commission on Workforce Quality and Labor Market Efficiency, US Department of Labor, Washington D.C.

Chisman, F. (1989b), "Investing in People, A Strategy to Address America's Workforce Crisis", Commission on Workforce Quality and Labor Market Efficiency, US Department of Labor, Washington D.C.

Choffel, P., P. Cuneo and F. Kramarz (1988), "Des trajectoires marquées par la structure de l'entreprise", *Economie et Statistique,* No. 213, September.

Chung-Hua Institution for Economic Research (1987), *1987 Joint Conference on the Industrial Policies of the Republic of China and the Republic of Korea,* Conference Series No. 6, Taipei.

Ciborra, C. (1991), "Alliances as Learning Experiments: Cooperation, Competition and Change in High-tech Industries", in Mytelka (1991b).

Clark, B., B. Chen, and T. Fujimoto (1987), "Product Development in the World Auto Industry: Strategy, Organization and Performance", Brookings Macroeconomics Conference, December.

Clark, J. and F. Field (1990), "Recycling: Boon or Bane of Advanced Materials Technologies? Automotive Materials Substitutes", paper presented to the World Resource Institute Conference, "Towards 2000: Environment, Technology and the New Century", June.

Cline, W.R. (1991), "Scientific Basis for the Greenhouse Effect", *Economic Journal,* Vol. 101, No. 407.

Coase, R. (1937), "The Nature of the Firm", *Economica,* Vol. 4, No. 3.

Cohen, W.M. and D.A. Levinthal (1989), "Innovation and Learning: The Two Faces of R&D", *Economic Journal,* September.

Cohendet, P., M.J. Ledoux and E. Zuscovitch (1991), "The Evolution of New Materials: A New Dynamic for Growth", in OECD (1991c).

Collins, H.W. (1974), "The TEA Set: Tacit Knowledge and Scientific Networks", *Science Studies,* Vol. 4, No. 2.

Contractor, F.R. and P. Lorange (1985), "Why Should Firms Cooperate? The Strategy and Economic Basis for Co-operative Ventures", in F.R. Contractor and P. Lorange, eds., *Cooperative Strategies in International Business,* Lexington Books, Lexington, Mass.

Cooke, P. and P. Wells (undated), "Strategic Alliances in ICT: Learning by Interaction", Regional Industrial Research Report, No 4, Regional Industrial Research, Cardiff.

Cooley, M., ed., (1989), *European Competitiveness in the 21st Century,* Brussels, FAST Programme, Commission of the European Communities, Brussels.

Coriat, B. (1989), "Information Technologies, Productivity and New Job Content – Skill as a Competitive Issue", paper presented to the OECD International Seminar on Science, Technology and Economic Growth, Paris, June.

Coriat, B. (1990), *L'atelier et le robot: Essai sur le fordisme et la production de masse à l'âge de l'électronique,* Christian Bourgois, Paris.

Cotta, A. (1978), *La France et l'impératif mondial,* Presses Universitaires de France, Paris.

Cowan, R. (1991), "Technological Variety and Competition: Issues of Diffusion and Intervention", in OECD (1991c).

Crane, D. (1972), *Invisible Colleges,* Chicago University Press, Chicago.

Cumings, B. (1987), "The Origins and Development of the Northeast Asian Political Economy: Industrial Sector, Product Cycles and Political Consequences", in Deyo.

Customs Co-operation Council (1986), *Harmonised Commodity Description and Coding System,* New York.

Dahlman, C. (1989), "Impact of Technological Change on Industrial Prospects for the LDCs", *World Bank Working Papers,* No. 12, World Bank, Washington D.C., June.

Dahlman, C., B. Ross-Larson and L.E. Westphal (1987), "Managing Technological Development: Lessons from the Newly Industrialising Countries", *World Development,* Vol. 15, No. 6.

Dasgupta, P. and P. David (1988), "Priority, Secrecy, Patents and the Socio-Economics of Science and Technology", Stanford University, Center for Economic Policy, Research Paper No. 127, March.

Dasgupta, P. and P. Stoneman, eds., (1987), *Economic Policy and Technological Performance,* Cambridge University Press.

David, P. (1986), "Understanding the Economics of QWERTY: The Necessity of History", in W.N. Parker, ed. *Economic History and the Modern Economist,* Basil Blackwell, Oxford.

David, P. (1987), "New Standards for the Economics of Standardization" in Dasgupta and Stoneman.

David, P., D. Mowery and W.E. Steinmueller (1988), "The Economic Analysis of Pay-offs from Basic Research – An Examination of the Case of Particle Physics Research", The Center for Economic Policy Research, Stanford University.

Davies, S. (1979), *The Diffusion of Process Innovations,* Cambridge University Press.

Davis, L. (1988), "Technology Intensity of US, Canadian and Japanese Manufactures Output and Exports", US Department of Commerce, International Trade Administration, June.

De Bresson, C. (1986), "Les pôles de développement technique: vers un concept opérationnel", paper presented to the French Canadian Association for the Advancement of Science, May.

Deiaco, E., E. Hornell and G. Vickery (1990), *Technology and Investment: Crucial Issues for the 1990s,* Pinter Publishers, London.

De Laubier, D. (1986), "Les firmes européennes et l'internationalisation des services commerciaux", *Economie Prospective Internationale,* No. 28.

Denmark, Institute of Production (IKE) (1981), *Technical Innovation and Economic Performance,* Industrial Development Research Monographs, No. 15, Aalborg.

Dertouzos, M.L., R.K. Lester and R.M. Solow, eds., (1989), *Made in America: Report of the MIT Commission on US Industrial Productivity,* The MIT Press, Cambridge, Mass.

Deyo, I.C., ed., (1987), *The Political Economy of the New Asian Industrialism,* Cornell University Press, Ithaca and London.

Distler, C. (1989), "Beyond Free Trade", Projet Prométhée Perspectives, No. 8, January.

Dollar, D. and E. Wolff (1988), "The Convergence of Industry Labour Productivity Among Advanced Economies", *The Review of Economics and Statistics,* November.

Dosi, G. (1981), "Technology, Industrial Structures and International Economic Performance", unpublished document, OECD, Paris.

Dosi, G. (1984a), *Technical Change and Industrial Performance,* Macmillan, London.

Dosi, G. (1984b), "Technological Paradigms and Technological Trajectories. The Determinants and Directions of Technological Change and the Transformation of the Economy", in Freeman.

Dosi, G. (1988), "Sources, Procedures and Microeconomic Effects of Innovation", *Journal of Economic Literature,* Vol. 26, September.

Dosi, G., C. Freeman, R. Nelson, G. Silverberg and L. Soete (1988), *Technical Change and Economic Theory,* Pinter Publishers, London.

Dosi, G., K. Pavitt and L. Soete (1990), *The Economics of Technical Change and International Trade,* Harvester Wheatsheaf, Hertfordshire.

Dosi, G., and L. Soete (1988), "Technical Change and International Change", in Dosi *et al.*

Doyle, D. (1989), "Why Johnny's Teacher Can't Do Algebra", *Across the Board,* Vol. 26, No. 6.

Drucker, P. (1983), "Quality Education – The New Growth Era", *Wall Street Journal,* 19 July.

Du Granrut, C. (1990), "Ampleur et conséquences des investissements directs internationaux", *Futuribles,* No. 150, December.

Dunning, J.H. (1981), *International Production and the Multinational Enterprise,* George Allen and Unwin, London.

Dunning, J.H. (1984), "Domestic and National Competitiveness, International Technology Transfer, and Multinational Enterprises", unpublished document, OECD, Paris.

Dunning, J.H., (1988a), *Multinationals, Technology and Competitiveness,* Unwin Hyman, London.

Dunning, J.H., ed., (1988b), *Explaining International Production,* Unwin Hyman, London.

Dunning, J.H. and J. Cantwell (1991), "MNEs, Technology and Competitiveness of European Industries", *Aussenwirtschaft,* Vol. 1, No. 46.

Durand, M. and C. Giorno (1987), "Indicators of International Competitiveness: Conceptual Aspects and Evaluation", in *OECD Economic Studies,* No. 9.

Easterly, W.R. and D.L. Wetzel (1989), "Policy Determinants of Growth", *World Bank Working Papers,* WPS 343, December.

Edquist, C. and S. Jacobson (1988), *Flexible Automation: The Global Diffusion of New Technology in the Engineering Industry,* Basil Blackwell, Oxford.

Eliasson, G. and P. Ryan (1986), "The Human Factor in Economic and Technological Change", *OECD Educational Monograph,* No. 3, November.

Englander, S. (1991), "Tests for Measurement of Service Sector Productivity", in OECD (1991c).

Englander, S., R. Evenson and M. Hanazaki (1988), "R&D, Innovation, and the Total Factor Productivity Slowdown", *OECD Economic Studies,* No. 11.

Englander, S. and A. Mittelstadt (1988), "Total Factor Productivity: Macroeconomic and Structural Aspects of the Slowdown", *OECD Economic Studies,* No. 10.

Erdilek, A., ed., (1985), *Multinationals as Mutual Invaders: Intra-Industry Direct Foreign Investment,* Croom Helm, London.

Ergas, H. (1984), "Why Do Some Countries Innovate More than Others", Centre for European Policy Studies, CEPS Paper No. 5, Brussels.

Ergas, H. (1987a), "The Importance of Technology Policy", in Dasgupta and Stoneman.

Ergas, H. (1987b), "Does Technology Policy Matter", in B.R. Guile and H. Brooks, eds., *Technology and Global Industry,* National Academy Press, Washington D.C.

Ernst, D. (1980), "International Transfer of Technology, Technological Dependence and Underdevelopment: Key Issues", in D. Ernst, ed., *The New International Division of Labour, Technology and Underdevelopment,* Campus, Frankfurt and New York.

Ernst, D. (1983), *The Global Race in Microelectronics,* Campus, Frankfurt and New York.

Ernst, D. (1989), "Innovation, International Diffusion of Technologies and the World Economy – Comments", in F. Arcangeli, P. David and G. Dosi, eds. *The Diffusion of*

Innovation, Oxford University Press, New York and Oxford.

Ernst, D. (1990), "Global Competition, New Information Technologies and International Technology Diffusion – Implications for Industrial Latecomers", paper presented to the Paris TEP Technology and Competitiveness Conference, June.

Ernst, D. and D. O'Connor (1989), *Technology and Global Competition: The Challenge for Newly Industrialising Economies,* OECD Development Centre Studies, Paris.

Ernst, D. and D. O'Connor (forthcoming), *Competing in the Electronics Industry – The Experience of Newly Industrialising Economies,* OECD Development Centre Studies, Paris.

European Commission (1989), *Partnership between Small and Large Firms,* Proceedings of the Conference Organised by the Commission, Graham and Trotman, London.

European Management Forum (1980), *Report on the Competitiveness of European Industry,* European Management Forum, Davos.

Ezra, E. (1990), "The Importance of Technology in Regional Development", paper presented to the Lisbon Seminar on European Inter-regional Co-operation in the Field of Innovation and Transfer of Technology, October.

Fagerberg, J. (1988a), "International Competitiveness", *The Economic Journal,* Vol. 98, No. 391.

Fagerberg, J. (1988b), "Why Growth Rates Differ", in Dosi *et al.*

Far Eastern Economic Review (1989), "Part Exchange", 21st September.

Favereau, O. (1989), "Organisation et marché", *Revue Française d'Economie,* No. 1.

Fechter, A. (1991), "Science and Engineering: Human Resource Needs in the New Three Decades", in D.L. Zinberg, ed., *The Changing University: How Increased Demand for Scientists and Engineers is Transforming Academic Institutions Internationally,* Kluwer Academic Publishers, Dordrecht.

Felix, D. (1989), "Import Substitution and Late Industrialisation: Latin America and Asia Compared", *World Development,* Vol. 17, No. 9.

Ferguson, C.H. (1990), "Computers and the Coming of the US Keiretsu", *Harvard Business Review,* Vol. 90, No. 4.

Ferguson, C.H. (1991), "Macroeconomic Variables, Sectoral Evidence and New Models of Industrial Performance", in OECD (1991c).

Finland, Central Statistical Office (1989), *The Intangible Investment of Industry in Finland 1987,* Helsinki.

Finland, Central Statistical Office (1990), *Innovation Activity of Finnish Industry,* 1990:3, Helsinki.

Flamm, K. (1990), "Seminconductors" in G.C. Hufbauer, ed., *Europe 1992: An American Perspective,* The Brookings Institution, Washington D.C.

Foray, D. (1989), "Les modèles de compétition technologique", *Revue d'Economie Industrielle,* No. 28.

Foray, D. (1990a), "L'économie des rendements croissants et l'économie de la firme innovatrice", paper presented to the Paris TEP Technology and Competitiveness Conference, June.

Foray, D. (1990b), "Exploitations des externalités de réseau vs. innovation de normalisation", *Revue d'Economie Industrielle,* No. 51.

Foray, D. and D. Mowery (1990), "L'intégration de la R-D industrielle", *Revue Economique,* No. 3.

Ford, R., P. Poret and G. Nicoletti (1990), "Business Investment in the OECD Economies", OECD Working Paper, Paris.

Fouquin, M. (1990), "France: performances technologiques et faiblesse industrielle, éléments d'une explication", in Chesnais (1990c).

France, Centre d'Etudes et de Recherches sur les Qualifications (1989), "Statistiques de la formation professionnelle continue financée par les entreprises", December.

Fransman, M. (1986), *Machines and Economic Development,* MacMillan, London.

Freeman, C. (1963), "The Plastics Industry: A Comparative Study of Research and Innovation", *National Institute Economic Review,* No. 26.

Freeman, C. (1965), "Research and Development in Electronic Capital Goods", *National Institute Economic Review,* No. 34.

Freeman, C. (1981), *Technical Innovation and Economic Performance,* IKE, Industrial Development Research Monographs, No. 15, Aalborg.

Freeman, C. (1982), *The Economics of Industrial Innovation,* Pinter Publishers, London.

Freeman, C. (1984), *Long Waves in the World Economy,* Pinter Publishers, London.

Freeman, C. (1987a), *Technology Policy and Economic Performance: Lessons from Japan,* Pinter Publishers, London.

Freeman, C. (1987b), "The Challenge of New Technologies", in *Interdependence and Co-operation in Tomorrow's World,* OECD, Paris.

Freeman, C., ed., (1987c), *Output Measurement in Science and Technology: Essays in Honour of Yvan Fabian,* North-Holland, Amsterdam.

Freeman, C. (1989), "Induced Innovation, Diffusion of Innovations, and Business Cycles," in B. Elliott, ed., *Technology and Social Process,* Edinburgh University Press.

Freeman, C. (1991), "The Nature of Innovation and the Evolution of the Production System", in OECD (1991c).

Freeman, C. and B.A. Lundvall, eds., (1988), *Small Countries Facing the Technological Revolution,* Pinter Publishers, London.

Freeman, C. and L. Soete (1985), "Information Technology and Employment: An Assessment", Science Policy Research Unit, University of Sussex.

Freeman, C. and L. Soete, eds., (1987), *Technical Change and Full Employment,* Basil Blackwell, Oxford.

Freeman, C. and L. Soete, eds., (1990), *New Explorations in the Economics of Technical Change,* Pinter Publishers, London.

Freeman, C. and A. Young (1965), *Research and Development in Western Europe, North America and the Soviet Union: An Experimental International Comparison of Research Expenditures and Manpower in 1962,* OECD, Paris.

Fusfeld, H.I. (1986), *The Technical Enterprise: Present and Future Patterns,* Ballinger, Cambridge, Mass.

Fusfeld, H.I. (1990), "Enterprise Strategies and Technical Networks", paper presented to the Tokyo TEP Technoglobalism Conference, February.

Gaffard, J.L. (1990a), *Economie industrielle et de l'innovation,* Dalloz, Paris.

Gaffard, J.L. (1990b), "Sunk Costs and the Creation of Technology", paper presented to the Paris TEP Technology and Competitiveness Conference, June.

Gaffard, J.L. and E. Zuscovitch (1988), "Mutations technologiques et choix stratégiques des entreprises", in Arena *et al.*

Gagey, F. (1990), "Competitiveness of Firms and Economic Policies", paper presented to the Paris TEP Technology and Competitiveness Conference, June.

Galbraith, J.K., (1967), *The New Industrial State,* NAL/Dutton, New York.

Gardner, H. (1985), *Frames of Mind: The Theory of Multiple Intelligence,* Basic Books, Inc., Cambridge, Mass.

GATT (1989), *International Trade 1988-89,* Vol. 1, Geneva.

Geiser, K., K. Fisher and N. Beecher (1986), *Foreign Practices in Hazardous Waste Management: A Report to the United States Environment Protection Agency,* August.

Gervasi, S. (1991), "Changes in the Pattern of Demand, Consumer Learning and Firm Strategies. An Examination of Postwar Economic Growth in the United States and France", in OECD (1991c).

Giarini, O. and H. Loubergé (1978), *La civilisation technicienne à la dérive,* Dunod, Paris.

Giget, M. (1990), "Key Factors of Industrial Co-operation and Competition in an Age of Interdependence", paper presented to the Paris TEP Technology and Competitiveness Conference, June.

Glytsos, N.P. (forthcoming), "Anticipated Graduate Job Mismatches and Higher Education Inadequacies in Greece", *Higher Education.*

Gold, B. (1989), "Technological Diffusion in Industry: Research Needs and Shortcomings", *Journal of Industrial Economics,* March.

Golding, A.M. (1972), "The Semi-Conductor Industry in Britain and the United States: A Case Study in Innovation Growth and the Diffusion of Technology", D.Phil. thesis, University of Sussex.

Gordon R.J. and M.N. Baily (1991), "Measurement Issues and the Productivity Slowdown in Five Major Industrial Countries", in OECD (1991c).

Goto, A. (1982), "Business Groups in a Market Economy", *European Economic Review,* Vol. 19, No. 2.

Goto, A. and K. Suzuki (1989), "R&D Capital, Rate of Return on R&D Investment and Spillover of R&D in Japanese Manufacturing Industries", *The Review of Economics and Statistics,* Vol. 71, No. 4.

Granstrand, O., C. Oskarsson, N. Sjöberg and S. Sjölander (1990), "Business Strategies For New Technologies", in Deiaco *et al.*

Graves, A. (1987), *Comparative Trends in Automotive Research and Development,* MIT International Vehicle Program, MIT, Cambridge, Mass., August.

Griliches, Z. (1979), "Issues in Assessing the Contribution of Research and Development to Productivity Growth", *Bell Journal of Economics,* Vol. 10, No. 1.

Griliches, Z. (1989), "Productivity Puzzles and R&D: Another Non-explanation", *The Journal of Economic Perspectives,* Vol. 2, No. 4, Fall.

Griliches, Z. (1990), "Patent Statistics as Economic Indicators: A Survey", *Journal of Economic Literature,* Vol. XXVIII, December.

Griliches, Z. and F. Lichtenberg (1984), "Interindustry Technology Flows and Productivity Growth: A Reexamination", *The Review of Economics and Statistics,* Vol. 65.

Grimwade, N. (1989), *International Trade: New Patterns of Trade, Production and Investment,* Routledge, London.

Guellec, D. and P. Ralle (1990), "Product Innovation and Non-Price Competitiveness", paper presented to the Paris TEP Technology and Competitiveness Conference, June.

Gummet, P., ed., (1991), "Future Relations Between Defence and Civil Science and Technology", Report for the Parliamentary Office of Science and Technology, Houses of Parliament, Westminster.

Hagedoorn, J. and J. Schakenraad (1990a), "Leading Companies and the Structure of Strategic Alliances in Core Technologies", MERIT Working Paper, Maastricht.

Hagedoorn, J. and J. Schakenraad (1990b), "Interfirm Partnerships and Co-operative Strategies in Core Technologies" in Freeman and Soete.

Hahn, R. (1989), "Economic Prescriptions for Environmental Problems", *Journal of Economical Prescriptions,* Vol. 3, No. 2, Spring.

Hakanson, L. (1981), "Organization and Evolution of Foreign R&D in Swedish Multinationals", *Geografiska annaler,* Ser. B 63, 47-56.

Hakansson, H. (1989), *Corporate Technological Behaviour: Cooperation and Networks,* Routledge, London.

Hamel, G., Y. Doz and C.K. Prahalad (1989), "Collaborate with Your Competitors – and Win", *Harvard Business Review,* No. 1.

Hanel, P., J.F. Angers and M. Cloutier (1986), *L'effet des dépenses en R-D sur la croissance de la productivité,* Ministère de l'Enseignement Supérieur et de la Science, Direction de la Maîtrise du Développement Scientifique et Technologique, Québec.

Hanna, N. (1990), "The Information Technology Revolution and Economic Development", background paper prepared for the Task Force on Information Technology for Development, The World Bank, Washington D.C., May.

Harvard Business School (1985), "'Daewoo Group' Case Study 9-385-014", *Harvard Business School Case Services,* Boston, Mass.

Hatzichronoglou, T. (1991), "Indicators of Industrial Competitiveness: Results and Limitations", in J. Niosi, ed., *Technology and National Competitiveness,* McGill-Queen's University Press, Montreal.

Haveman, R. (1989), "Thoughts on the Sustainable Development Concept and the Environmental Effects of Economic Policy", paper presented to an OECD Seminar on Sustainable Development, October.

Hayes, R.H. and R. Jaikumar (1988), "Manufacturing's Crisis: New Technologies, Obsolete Organisations", *Harvard Business Review,* Vol. 66, No. 5.

Heaton, G.R. (1990), "Regulation and Technological Change", paper presented to the World Resources Institute Conference "Towards 2000: Environment, Technology and the New Century", June.

Heaton, G. and J. Maxwell (1984), "Patterns of Automobile Regulation: An International Comparison", *Journal of Environmental Policy,* March.

Helleiner, G.K. (1979), "Transnational Corporations and Trade Structure: The Role of Intra-firm Trade", in H. Giersch, ed., *On the Economics of Intra-Industry Trade,* J.C.B. Mohr, Tübingen.

Helleiner, G.K. (1981), *Intra-Firm Trade and the Developing Countries,* MacMillan, London.

Helpman, E. and P. Krugman (1985), *Trade Policy and Market Structure,* The MIT Press, Cambridge, Mass.

Hippel, E. von (1987), "Co-operation between Rivals: Informal Know-how Trading", *Research Policy,* Vol. 16, No. 6.

Hippel, E. von (1988), *The Sources of Innovation,* Oxford University Press, Oxford and New York.

Hirsch, S. (1965), "The US Electronics Industry in International Trade", *National Institute Economic Review,* No. 34.

Hirschhorn, J.S. (1990), "The Technological Potential: Industrial Pollution Prevention", paper presented to the World Resources Institute Conference, "Towards 2000: Environment, Technology and the New Century", June.

Hirschman, A.O. (1986), "The Political Economy of Latin American Development: Seven Exercises in Retrospective", paper presented to the Thirteenth International Congress of the Latin American Studies Association, Boston, October.

Hoeller, P., A. Dean and I. Nicolaisen (1991), "Macroeconomic Implications of Producing Greenhouse Gas Emissions: A Survey of Empirical Studies", *OECD Economic Studies,* No. 16.

Hollingsworth, R. (1990), "Variation among Nations in the Logic of Manufacturing Sectors and International Competitiveness", paper presented to the Paris TEP Technology and Competitiveness Conference, June.

Hong, W. (1987), "Export-Oriented Growth and Trade Patterns of Korea", in C. Bradford and W. Branson, eds., *Trade and Structural Change in Pacific Asia,* University of Chicago Press, Chicago.

Hooker, S. (1990), "Use of Training at Motorola to Support the Development, Acquisition and Application of New Technologies in Six Sigma Quality Culture", paper presented to the Stockholm TEP Technology and Investment Conference, January.

Horn, E.J. (1981), "Technology and International Competitiveness: the German Evidence and Some Overall Comment", unpublished document, OECD, Paris.

Howells, J. (1990), "The Location and Organization of Research and Development: New Horizons", *Research Policy,* Vol. 19, No. 2.

Hufbauer, G. (1966), *Synthetic Materials and the Theory of International Trade,* Harvard University Press, Cambridge, Mass.

Hufbauer, G. (1989), "Beyond GATT", *Foreign Policy,* No. 77, Winter 1989-90.

Hymer, S. (1960), *The International Operations of National Firms,* The MIT Press, Cambridge, Mass.

IEEP (Institute for European Environmental Policy) (1990), "Environmental Problems in the Member States", May.

ILO (International Labour Office) (1990), *ISCO-88 International Standard Classification of Occupations,* Geneva.

Imai, K.J. (1988a), "Technological Change in the Information Industry and Implications for the Pacific Region", paper presented to the 17th Pacific Trade and Development Conference, Bali, Indonesia, July.

Imai, K.J. (1988b), "Patterns of Innovation and Entrepreneurship in Japan", paper presented to the Second Congress of the International Schumpeter Society, Siena.

Imai, K.J. (1988c), "International Corporate Networks: A Japanese Perspective", Project Prométhée Perspectives, June.

Imai, K.J. (1990), "Japan's Business Groups and Keiretsu in Relation to the Structural Impediments Initiative", paper presented to "The US/Japan Relationships Entering a New Phase: Issues for Structural Impediments" Symposium, Washington D.C., April.

Imai, K.J. (1991), "Globalization and Cross-border Networks of Japanese Firms", paper for the conference "Japan in a Global Economy" at the Stockholm School of Economics, September.

Imai, K.J. and Y. Baba (1991), "Systemic Innovation and Cross-Border Networks, Transcending Markets and Hierarchies to Create a New Techno-Economic System", in OECD (1991c).

Imai, K.J. and H. Itami (1984), "Mutual Infiltration of Organisation and Market – Japan's Firm and Market in Comparison with the US", *International Journal of Industrial Organisation,* Vol. 1, No. 2.

Institute of Manpower Studies (1990), *Scientific Research Manpower: A Review of Supply and Demand Trends,* Report No. 169, University of Sussex.

IPCC (Intergovernmental Panel on Climate Change) (1990), "Climate Change: The IPCC Scientific Assessment", Cambridge University Press.

Irvine, J., B. Martin and P.A. Isard (1990), *Investing in the Future: An International Comparison of Government Funding of Academic and Related,* Edward Elgar, Aldershot.

Jacobs, D. (1990), "The Policy Relevance of Diffusion", report for *Policy Studies on Technology and Economy,* Ministry of Economic Affairs of the Netherlands, The Hague.

Jacoby, H. and J. Steinbrunner (1973), "Salvaging the Federal Attempt to Control Auto Pollution", *Public Policy,* Vol. 21, No. 1.

Jaffe, A. (1986), "Technological Opportunity and Spillovers in R&D: Evidence from Firms' Patents, Profits and Market Value", *American Economic Review,* Vol. 76, No. 5.

Jaffe, D. and R. Stavins (1990), "Evaluating the Efficiency of Economic Incentives and Direct Regulation for Environmental Protection: Impacts on the Diffusion of Technology", paper presented to the World Resource Institute Conference, "Towards 2000: Environment, Technology and the New Century", June.

Japan, Chemical Fibre Association (1988), *Annual Report 1988,* quoted in *The Japan Economic Journal,* 12 April.

Japan, Environment Agency (1988), *Quality of the Environment in Japan,* Tokyo.

Jeelof, G. (1987), "Keynote Speech at the Financial Times' World Electronics Conference", *Financial Times* (1988), Speakers' Papers, June, London.

Johnson, B., "Institutional Learning", in Lundvall, (forthcoming).

Jones, D.T. (1988), "Structural Adjustment in the Automobile Industry", *STI Review,* No. 3, OECD, Paris.

Julius, D. (1990), *Global Companies and Public Policy: The Growing Challenge of Foreign Direct Investment,* Royal Institute of Foreign Affairs, London.

Kaldor, N. (1978), "The Effect of Devaluations on Trade in Manufactures", in *Further Essays on Applied Economics,* Duckworth, London.

Kaldor, N. (1981), "The Role of Increasing Returns, Technical Progress and Cumulative Causation in the Theory of International Trade and Economic Growth", *Economie Appliquée* (ISMEA), Vol. 34, No. 4.

Kamien, M. and N. Schwartz (1982), *Market Structure and Innovation,* Cambridge University Press.

Kanawaty, G., G.A. Gladstone, J. Prokopenko and G. Rodgers (1989), "Adjustment at the Micro Level", *International Labour Review,* Vol. 128, No. 3.

Kaneko, I. and K.J. Imai (1987), "A Network View of the Firm", paper presented at the first Hitotsubashi University/ Stanford University Joint Conference, March.

Kaplan, M.C. (1987), "Intangible Investment: An Essay at International Comparison", unpublished document, OECD, Paris.

Kaysen, C. and D.F. Turner (1971), *Anti-Trust Policy,* Harvard University Press, Cambridge, Mass.

Kelley, M.R. and H. Brooks (1988), *The State of Computerised Automation in US Manufacturing,* Harvard University Press, Cambridge, Mass.

Kellman, M. (1983), "Relative Prices and International Competitiveness: an Empirical Investigation", *Empirical Economics,* Vol. 8, Nos. 3-4.

Kemp, R. and L. Soete (1990), "Inside the 'Green Box': on the Economics of Technical Change and the Environment", in Freeman and Soete.

Kendrick, J. (1984), *The Formation and Stocks of Total Capital,* National Bureau of Economic Research, New York.

Kendrick, J. (1988), "The Problem in Measuring Service Productivity", *The Service Economy,* Vol. 2, No. 2.

Kendrick, J. (1991), "Total Factor Productivity: What It Measures and What It Does Not Measure", in OECD (1991c).

Kimbel, D. (1987), "Information Technology: Increasingly the Engine of OECD Economies", *The OECD Observer,* No. 147, August/September.

Kimbel, D. (1988), "Innovation and Growth Opportunities through IT-related Networking", paper presented to the Workshop on Information Technology and New Growth Opportunities for the 1990s, Tokyo, September.

Kirst, M. (1988), "Recent State Education Reforms in the US; Looking Backward and Forward", *Education Administration Quarterly,* Vol. 24, No. 3.

Klein, B.H. (1979), "The Slowdown in Productivity Advances: A Dynamic Explanation", in C.T. Hill and J.M. Utterback, eds., *Technological Innovation for a Dynamic Economy,* New York.

Kline, S.J. and N. Rosenberg (1986), "An Overview of Innovation", in National Academy of Engineering, *The Positive Sum Strategy: Harnessing Technology for Economic Growth,* The National Academy Press, Washington D.C.

Knickerbocker, F.T. (1973), *Oligopolistic Reaction and Multinational Enterprises,* Harvard University Press, Boston, Mass.

Kodama, F. (1990), "Can Changes in the Techno-economic Paradigm be Identified through Empirical and Quantitative Study?", *STI Review,* No. 7, OECD, Paris.

Krugman, P. (1979), "Increasing Returns, Monopolistic Competition, and International Trade", *The Journal of International Economics,* Vol. 9, No. 4.

Krugman, P. (1984), "Import Protection as Export Promotion: International Competition in the Presence of Oligopoly and Economies of Scale", in H. Kierzkowski, ed., *Monopolistic Competition and International Trade,* Clarendon Press, Oxford.

Krugman, P., ed., (1987), *Strategic Trade Policy and the New International Economics,* The MIT Press, Cambridge, Mass.

Krugman, P. (1990), *Rethinking International Trade,* The MIT Press, Cambridge, Mass.

Kuhn, T. (1962), *The Structure of Scientific Revolutions,* Chicago University Press, Chicago.

Kurt Salmon Associates (1987), *Soft Goods Outlook,* New York.

Lall, S. (1979), "The International Allocation of Research Activities by US Multinationals", *Oxford Bulletin of Economics and Statistics,* Vol. 41, No. 3.

Lall, S. (1990), *Building Industrial Competitiveness in Developing Countries,* OECD Development Centre Studies, Paris.

Lanvin, B. (1990a), "Technology-based Competition: Globalisation vs. Fragmentation?", paper presented at the Tokyo TEP Technoglobalism Conference, February.

Lanvin, B. (1990b), "Technology and Competitiveness", report of the Paris TEP Technology and Competitiveness Conference, June.

LARA/GEST (1985), "Grappes technologiques et stratégies industrielles", Etude CPE No. 57, Centre de Prospective et d'Evaluation, Paris.

Lawrence, R.Z. (1980), "The United States Current Account: Trends and Prospects", in United States Congress, *Special Study on Economic Change: The International Economy: US Role in a World Market,* US Government Printing Office, Washington D.C.

Leaf, A. (1989), "Potential Health Effects of Global Climatic and Environmental Changes", *New England Journal of Medicine,* Vol. 321, No. 23.

Leborgne, D. (1987), "Equipement flexible et organisation productive: les relations industrielles au cœur de la modernisation: éléments de comparaison internationale'', *Couverture Orange,* CEPREMAP, Paris.

Leborgne, D. and A. Lipietz (1988), ''L'après-fordisme et son espace'', *Les Temps Modernes,* April; in English as ''New Technologies, New Mode of Regulation: Some Spatial Implications'', *Society and Space,* Vol. 6.

Leborgne, D. and A. Liepietz (1989), ''Deux stratégies sociales dans la production des nouveaux espaces économiques'', *Couverture Orange,* CEPREMAP, Paris.

Lee H. (1988), ''Review on the State-of-the-Art of Programmable Automation in Korea'', in Korea Advanced Institute of Science and Technology (KAIST), *Proceedings of the Korean/German Seminar on Automation Technology and Co-operation,* Seoul.

Lester, R.K. (1989), ''Determinants of US Productive Performance: Findings of the MIT Commission on Industrial Productivity'', paper presented to the OECD International Seminar on Science, Technology and Economic Growth, Paris, June.

Levin, R. (1988), ''Appropriability, R&D Spending and Technological Performance'', *American Economic Review,* Papers and Proceedings, May.

Levin, R., A. Klevorick, R. Nelson and S. Winter (1987), ''Appropriating the Returns from Industrial Research and Development'', *Brookings Papers on Economic Activity,* No. 3.

Levin, R. and P. Reiss (1984), ''Tests of a Schumpeterian Model of R&D and Market Structure'' in Z. Griliches, ed., *R&D, Patents and Productivity,* University of Chicago Press, Chicago.

Levin, R. and P. Reiss (1989), ''Cost-reducing and Demand-creating R&D with Spillovers'', *Rand Journal of Economics,* Vol. 19, No. 4.

Levy, J.D. and R.J. Samuels (1991), ''Research Collaboration as Technology Strategy'', in Mytelka (1991*b*).

Link, A.N. and L.L. Bauer (1989), *Cooperative Research in US Manufacturing: Assessing Policy Initiatives and Corporate Strategies,* Lexington Books, Lexington, Mass.

Little, G. (1987), ''Intra-firm Trade: An Update'', *New England Economic Review,* May/June.

Lucas, R.E. (1988), ''On the Mechanics of Economic Development'', *Journal of Monetary Economics,* Vol. 22, No. 2.

Lundvall, B.A. (1988), ''Innovation as an Interactive Process: From User-Producer Interaction to the National System of Innovation'', in Dosi *et al.*

Lundvall, B.A. (1991), ''Innovation, the Organised Market and the Productivity Slowdown'', in OECD (1991*c*).

Lundvall, B.A. (forthcoming), *National Systems of Innovation,* Pinter Publishers, London.

MacDonald, S. (1986), ''Controlling the Flow of High-Technology Information from the United States to the Soviet Union: A Labour of Sisyphus'', *Minerva,* Vol. XXIV, No. 1.

Maddison, A. (1989), ''Growth and Slowdown in Advanced Capitalist Economies: Techniques of Quantitative Assessment'', *Journal of Economic Literature,* June.

Madeuf, B. (1991), ''Technology Indicators and Developing Countries'', UNCTAD, ITP/TEC/19, Geneva.

Malmberg, A. (1990), ''Linkages, Labour and Location, Local Industrial Change in an Internationalised Economy'', paper presented to the Berlin Workshop on the Socio-Economics of Inter-firm Co-operation.

Mangano, S. (1990), ''Italian Survey on Technological Innovation'', Istituto Nazionale di Statistica, Rome.

Mansfield, E. (1968), *Industrial Research and Technological Innovation,* W.W. Norton, New York.

Mansfield, E. (1974), ''Technology and Technical Changes'', in J.H. Dunning, ed., *Economic Analysis and the Multinational Enterprise,* George Allen and Unwin, London.

Mansfield, E. (1985), ''How Rapidly Does New Industrial Technology Leak Out?'', *Journal of Industrial Economics,* December.

Mansfield, E., J. Rapoport, A. Romeo, S. Wagner and G. Beardsley (1977), ''Social and Private Rates of Return from Industrial Innovations'', *Quarterly Journal of Economics,* Vol. 77, No. 2.

Mariti, P. and R.H. Smiley (1983), ''Co-operative Agreements and the Organisation of Industry'', *The Journal of Industrial Economics,* Vol. 31, No 4.

Martinez-Alier, J. (1987), *Ecological Economics: Energy, Environment and Society,* Basil Blackwell, Oxford.

Mathis, J., J. Mazier and D. Rivaud-Danset (1988), *La compétitivité industrielle,* Publications de l'IRES, Dunod, Paris.

Meadows, D.H, D. Meadows, E. Zahn and P. Milling (1972), *Limits to Growth,* Universe Books, New York.

Mellman, S. (1983), *Profits without Production,* Knopf, New York.

Metcalfe, S. (1990), ''On Diffusion, Investment and the Process of Technological Change'', in Deiaco *et al.*

Meyer-zu-Schlochtern, F.G.M. (1988), ''A Sectoral Databank for Thirteen OECD Countries'', *OECD/ESD Working Paper Series,* No. 57, November.

Michalet, C.A. (1981), ''Competitiveness and Internationalisation'', unpublished document, OECD, Paris.

Michalet, C.A. (1985), *Le capitalisme mondial,* Presses Universitaires de France, Paris.

Michalet, C.A. (1991), ''Global Competition and its Implications for Firms'', in OECD (1991*c*).

Michalet, C.A. and M. Delapierre (1978), ''The Impact of Multinational Enterprises on National Scientific and Technical Capacities in the Computer Industry'', unpublished document, OECD, Paris.

Mistral, J. (1978), ''Formation de capital et compétitivité en longue période'', *Economie et Statistique,* No. 97.

Mistral, J. (1983), ''Competitiveness of the Productive System and International Specialisation'', unpublished document, OECD, Paris.

Mohnen, P. (1989), ''New Technologies and Inter-Industry Spillovers'', *STI Review,* No. 7, OECD, Paris.

Momigliano, F. and G. Dosi (1983), *Tecnologia e organizzazione industriale internazionale,* Il Mulino, Bologna.

Mordschelles-Regnier, G., G. Dahan and A. Reboul (1987), ''Le rôle des sociétés de recherche sous contrat vis-à-vis des PME'', *Annales des Mines,* July/August.

Mowery, D.C. and N. Rosenberg (1979), ''The Influence of Market Demand upon Innovation: A Critical Review of

Some Recent Empirical Studies'', *Research Policy*: Vol. 8, No. 2.

Mowery, D.C. and N. Rosenberg (1989), *Technology and the Pursuit of Economic Growth*, Cambridge University Press.

Mucchielli, J.L. (1985), *Les firmes internationales, mutations et nouvelles perspectives*, Economica, Paris.

Muldur, U. (1990), ''Fusions et acquisitions dans le secteur financier européen'', *Revue d'Economie Financière*, Spring-Summer.

Muller, P. (1990), *The Enlargement of the Concept of Gross Fixed Capital Formation and its Impact on National Accounts*, INSEE, Paris.

Mytelka, L.K. (1984), ''La gestion de la connaissance dans les entreprises multinationales: vers la formation d'oligopoles technologiques'', *Economie Prospective Internationale*, No. 20.

Mytelka, L.K. (1990), ''Transfer and Development of Technology in the Least Developed Countries: An Assessment of Major Policy Issues'', UNCTAD, Geneva.

Mytelka, L.K. (1991a), ''States, Strategic Alliances and International Oligopolies: The European ESPRIT Programme'', in Mytelka (1991b).

Mytelka, L.K., ed., (1991b), *Strategic Partnerships: States, Firms and International Competition*, Pinter Publishers, London.

Narin, F. and E. Noma (1985), ''Is Technology Becoming Science?'', *Scientometrics*, Vol. 7, Nos. 3-6.

Nature (1989), ''Students for Higher Education'', Vol. 342, 2nd November.

NAVF (Norwegian Research Council for Science and the Humanities) (1990), *Sustainable Development, Science and Policy*, NAVF, Oslo.

Nelson, K. (1989), ''Are There Any Energy Savings Left?'', *Chemical Processing*, January.

Nelson, R. (1980), ''Balancing Market Failure and Government Inadequacy: The Case of Policy Towards Industrial R&D'', Yale University Working Paper No. 840.

Nelson, R. (1981a), ''Research on Productivity Growth and Productivity Differences: Dead Ends and New Departures'', *The Journal of Economic Literature*, Vol. 19, No. 3.

Nelson, R. (1981b), ''Competition, Innovation, Productivity Growth, and Public Policy'' in H. Giersch, ed., *Proceedings of the 1980 Kiel Symposium on Towards an Explanation of Economic Growth*, J.C.B. Mohr (Paul Siebeck), Tübingen.

Nelson, R. (1986), ''Institutions Generating and Diffusing New Technology'', paper presented to the Conference on Innovation Diffusion, Venice.

Nelson, R. (1988), ''Institutions Supporting Technical Change in the United States'', in Dosi *et al.*

Nelson, R. (1991), ''Institutions Generating and Diffusing New Technology'', in Arcangeli *et al.*

Nelson, R. and N. Rosenberg (1990), ''Technical Advance and National Systems of Innovation'', Working paper for the Columbia University International Project on National Systems of Innovation, January.

Nelson, R. and S. Winter (1977), ''In Search of a Useful Theory of Innovation'', *Research Policy*, Vol. 6, No. 1.

NEPP (Netherlands Environment Policy Plan) (1989), ''To Choose or to Lose'', Second Chamber of the States General's Session 1988-89, The Hague.

Netherlands, Ministry of Education and Science (1989), ''Ontwerp Hoger Onderwijs en Onderzoeksplan (HOOP) 1990'', The Hague.

Nevens, T.M., G.L. Summe and B. Uttal (1990), ''Commercializing Technology: What the Best Companies Do'', *Harvard Business Review*, May-June.

Newfarmer, R.S. (1985), ''International Industrial Organisation and Development: A Survey'', in *Profits, Progress and Poverty: Case Studies of International Industries in Latin America*, University of Notre Dame Press, Indiana.

Nicolaisen, J., A. Dean and P. Hoeller (1991), ''Economics and the Environment: A Survey of Issues and Policy Options'', *OECD Economic Studies*, No. 16, Spring.

Noble, D.F. (1984), *Forces of Production: A Social History of Industrial Automation*, Alfred A. Knopf, New York.

Nordhaus, W.D. (1991), ''To Slow or Not to Slow: The Economics of the Greenhouse Effect'', *Economic Journal*, Vol. 101, No. 407.

O'Doherty, D., ed., (1990), *The Co-operation Phenomenon: Prospects for Small Firms and the Small Economy*, Graham and Trotman, London.

OECD (1967), *Methods and Statistical Needs for Educational Planning*, Paris.

OECD (1971), *The Conditions for Success in Technological Innovation*, Paris.

OECD (1973), *Recurrent Education: A Strategy for Lifelong Learning*, Paris.

OECD (1979a), *Technology on Trial*, Paris.

OECD (1979b), *Impact of Multinational Enterprises on National Scientific and Technical Capacities: The Food Industry*, Paris.

OECD (1979c), *Economies in Transition*, Paris.

OECD (1980), *Technical Change and Economic Policy*, Paris.

OECD (1981a), *The Future of University Research*, Paris.

OECD (1981b), *The Measurement of Scientific and Technical Activities, Proposed Standard Practice for Surveys of Research and Experimental Development, "Frascati Manual" 1980*, Paris.

OECD (1981c), *Science and Technology Policy for the 1980s*, Paris.

OECD (1984), *Industry and University: New Forms of Co-operation and Communication*, Paris.

OECD (1985a), *The Pharmaceutical Industry: Trade Related Issues*, Paris.

OECD (1985b), *The Semiconductor Industry: Trade Related Issues*, Paris.

OECD (1985c), *The Space Products Industry: Trade Related Issues*, Paris.

OECD (1985d), *The State of the Environment*, Paris.

OECD (1986a), *New Information Technologies, A Challenge for Education*, Paris.

OECD (1986b), *Competition Policy and Joint Ventures*, Paris.

OECD (1986c), *OECD Science and Technology Indicators No. 2: R&D, Invention and Competitiveness*, Paris.

OECD (1987), *Structural Adjustment and Economic Performance*, Paris.

OECD (1988a), *Science and Technology Policy Outlook*, Paris.

OECD (1988b), *New Technologies in the 1990s: A Socio-economic Strategy*, Paris.

OECD (1988c), *Internationalisation of Software and Computer Services*, ICCP Report No. 17, Paris.

OECD (1988d), "Trends in Patterns, Needs and Sources of Industrial Investment", Paris (free document).

OECD (1989a), *Government Policies and the Diffusion of Microelectronics*, Paris.

OECD (1989b), *IT and New Growth Opportunities*, ICCP Report No. 19, Paris.

OECD (1989c), *Industrial Policy in OECD Countries, Annual Review 1989*, Paris.

OECD (1989d), *Research Manpower – Managing Supply and Demand*, Paris.

OECD (1989e), *OECD Science Indicators: R&D, Production and Diffusion of Technology*, Report No. 3, Paris.

OECD (1989f), *Economic Outlook*, No. 46, Paris.

OECD (1989g), *Towards an Enterprising Culture*, Educational Monograph No. 4, Paris.

OECD (1989h), *Education and the Economy in a Changing Society*, Paris.

OECD (1989i), *The Measurement of Scientific and Technical Activities, R&D Statistics and Output Measurement in the Higher Education Sector, "Frascati Manual" Supplement*, Paris.

OECD (1989j), *OECD Employment Outlook 1989*, Paris.

OECD (1989k), *United States, OECD Economic Surveys*, Paris.

OECD (1989l), *Biotechnology: Economic and Wider Impacts*, Paris.

OECD (1989m), *Education in OECD Countries 1986-87: A Compendium of Statistical Information*, Paris.

OECD (1989n), *OECD Environmental Data: Compendium 1989*, Paris.

OECD (1989o), "Technological Change and Human Resource Development: The Service Sector. Main Trends and Issues", paper presented to the Utrecht TEP Technological Change and Human Resource Development Conference, November.

OECD (1989p), *Agricultural and Environmental Policies, Opportunities for Integration*, Paris.

OECD (1989q), "Education and Structural Change – A Statement by the Education Committee", Paris (free document).

OECD (1989r), *Economic Instruments for Environmental Projection*, Paris.

OECD (1989s), *Trends in Patterns, Needs and Sources of Industrial Investment: Industrial Investment in a Financial Economy*, Paris.

OECD (1989t), "Technological Change and Human Resources Development: The Service Sector: Educational Implications", paper presented to the Utrecht TEP Technological Change and Human Resource Development Conference, November.

OECD (1989u), *Renewable Natural Resources – Economic Incentives for Improved Management*, Paris.

OECD (1990a), *Advanced Manufacturing Technology and the Organisation of Work*, Paris.

OECD (1990b), "Pollution Control Expenditure in OECD Countries, A Statistical Compendium", *Environment Monographs*, No. 38, November.

OECD (1990c), *Industrial Policy in OECD Countries: Annual Review 1990*, Paris.

OECD (1990d), "Description of Innovation Surveys and Surveys of Technology Use Carries Out in OECD Member Countries", Paris (free document).

OECD (1990e), "Inventory of the S&T Indicators Data Base", paper presented to the Paris TEP Indicators Conference, July.

OECD (1990f), "Low Consumption/Low Emission Automobile", Summary Report of the OECD/IEA Expert Panel Meeting, February.

OECD (1990g), *Advanced Materials Policies and Technological Challenges*, Paris.

OECD (1990h), *Labour Market Policies for the 1990s*, Paris.

OECD (1990i), *Economic Outlook*, No. 48, Paris.

OECD (1990j), "Overview of TEP-related Statistical Work at OECD", paper presented to the Paris TEP Indicators Conference, July.

OECD (1990k), "The Measurement of Scientific and Technological Activities, Proposed Standard Method of Compiling and Interpreting Technology Balance of Payments Data, TBP Manual 1990", Paris, (free document).

OECD (1990l), *Information, Computer, Communications Policy: Communication Outlook 1990*, Paris.

OECD (1990m), "The Relevance of Standard-Related Indicators", paper presented to the Paris TEP Indicators Conference, July.

OECD (1990n), *Education in OECD Countries 1987-88: A Compendium of Statistical Information*, 1990 Special Edition, Paris.

OECD (1991a), *Managing Manpower for Advanced Manufacturing Technology*, Paris.

OECD (1991b), *Information Technology Standards: The Economic Dimension*, Paris.

OECD (1991c), *Technology and Productivity: The Challenges for Economic Policy*, Paris.

OECD (1991d), *The State of the Environment*, Paris.

OECD (1991e), *Environmental Indicators: A Preliminary Set*, Paris.

OECD (1991f), *Choosing Priorities in Science and Technology*, Paris.

OECD (1991g), *STI Review*, No. 8, Paris.

OECD (1991h), *Strategic Industries in a Global Economy: Policy Issues for the 1990s*, Paris.

OECD (1991i), *OECD Environmental Data. Compendium*, Paris.

OECD (1991j), *OECD Employment Outlook 1991*, Paris.

OECD (1991k), *Managing Technological Change in Less-Advanced Countries*, Paris.

OECDa (forthcoming), "OECD Proposed Guidelines for Collecting and Interpreting Innovation Data", Paris (free document).

OECDb (forthcoming), *Handbook for International Educational Indicators*, Paris.

OECDc (forthcoming), *Science and Technology Policy Outlook*, Paris.

OECDd (forthcoming), *Technological Change as a Social Process: Society, Enterprises and the Individual*, Paris.

Ohlin, B. (1967), *Inter-regional and International Trade*, Harvard University Press, Cambridge, Mass.

Ohmae, K. (1985), *Triad Power*, The Free Press, New York.

Oldenburg, K.U. and J.S. Hirschhorn (1987), "Waste Reduction from Policy to Commitment", *Hazardous Waste and Hazardous Materials*, Vol. 4, No. 1.

Olson, M. (1988), "The Productivity Slowdown, the Oil Shocks, and the Real Cycle", *The Journal of Economic Perspectives*, Vol. 2, No. 4.

Olson, M. (1989), "A Microeconomic Approach to Macroeconomic Policy", *The American Economic Review*, Vol. 79, No. 2.

Oman, C. (1984), *New Forms of International Investment in Developing Countries*, OECD Development Centre Studies, Paris.

Oman, C., F. Chesnais, J. Pelzman, and R. Rama (1989), *New Forms of Investment in Developing Country Industries: Mining, Petrochemicals, Automobiles, Textiles, Food*, OECD Development Centre Studies, Paris.

Osterman, P. (1990), "New Technology and Work Organisation", in Deiaco *et al.*

Ostry, S. (1990a), *Governments and Corporations in a Shrinking World: Trade and Innovation Policies in the United States, Europe and Japan*, Council on Foreign Relations, New York.

Ostry, S. (1990b), "The Impact of Globalisation: Convergence or Conflict", paper presented to the Tokyo TEP Technoglobalism Conference, February.

Ostry, S. (1991), "Beyond the Border: The New International Policy Arena", in OECD (1991h).

Ouchi, W.G. (1984), *The Reform Society*, Addison-Wesley, Reading, Mass.

Owen, R.F. (1989), "The Evolution in Japan's Relative Technological Competitiveness Since the 1960s: A Cross-Sectional, Time-Series Analysis", *Bank of Japan Monetary and Economic Studies*, November.

Pack, H. (1987), *Productivity, Technology and Industrial Development: A Case Study in Textiles*, Oxford University Press, Oxford and New York.

Papaconstantinou, G. (1991), "Research Spillovers, International Competition and Economic Performance", Ph.D. dissertation, London School of Economics.

Patel, P. and L. Soete (1987), "Technological Trends in the UK Manufacturing Industry", in Freeman and Soete.

Patel, P. and K. Pavitt (1990), "Do Large Firms Control the World's Technology?", in K. Pavitt, "What Makes Basic Research Economically Useful", paper presented to the Paris TEP Technology and Competitiveness Conference, June.

Pavitt, K. (1984), "Sectoral Patterns of Technical Change: Towards a Taxonomy and a Theory", *Research Policy*, Vol. 13, No. 6.

Pearce, D.W. (1989), "Economic Incentives and Renewable Natural Resource Management", in OECD (1989u).

Pearce, D.W. (1991), "The Role of Carbon Taxes in Adjusting to Global Warming", *Economic Journal*, Vol. 101, No. 407.

Pearce, D.W., A. Markandya and E. Barbier (1989), *Blueprint for a Green Economy*, Earthscan, London.

Pearce, R.D. (1989), *The Internationalisation of Research and Development by Multinational Enterprises*, Macmillan, London.

Pearson, R. (1990), "Scientific Research Manpower: A Review of Supply and Demand Trends", *Institute of Manpower Studies Report*, No. 169, University of Sussex.

Pearson, R., H. Connor and C. Pole (1988), *The IT Manpower Monitor 1988*, Institute of Manpower Studies, University of Sussex.

Perelman, L. (1984), "The Learning Enterprise: Adult Learning, Human Capital and Economic Development", The Council of State Planning Agencies, Washington D.C.

Perez, C. (1989), "Technical Change, Competitive Restructuring and Institutional Reform in Developing Countries", SPR Publications, Discussion Paper No. 4, December, The World Bank, Washington D.C.

Perez, C. (1990), "Electronics and Development in Venezuela – A User-Oriented Strategy and its Policy Implications", OECD Development Centre Technical Paper, No. 25, Paris.

Perez, C. and C. Freeman (1988), "Structural Crises of Adjustment, Business Cycles and Investment Behaviour", in Dosi *et al.*

Peters, L.S. (1989), "Academic Crossroads: The US Experience", *STI Review*, No. 5, OECD, Paris.

Petit, P. (1986), *Slow Growth and the Service Economy*, Pinter Publishers, London.

Peyrache, V. (1990), "L'influence du changement technique sur l'évolution économique et spatiale des régions métropolitaines", *La Revue de l'IRES*, No. 2, Winter.

Piasecki, B. and G. Davis (1987), *America's Future in Toxic Waste Management – Lessons From Europe*, Greenwood Publishing Group, Inc., Westport.

Picard, J.F. (1990), *La République des Savants*, Flammarion, Paris.

Piore, M. and C. Sabel (1984), *The Second Industrial Divide*, Basic Books, New York.

Polanyi, M. (1967), *The Tacit Dimension*, Doubleday Anchor, New York.

Porter, M.E. (1986), "Competition in Global Industries: A Conceptual Framework" in M.E. Porter, ed., *Competition in Global Industries*, Harvard Business School Press, Boston.

Porter, M.E. (1990a), "The Competitive Advantage of Nations", *Harvard Business Review*, No. 2.

Porter, M.E. (1990b), *The Competitive Advantage of Nations*, Macmillan, London.

Porter, M.E. and M.B. Fuller (1986), "Coalitions and Global Strategy", in M.E. Porter, ed., *Competition in Global Industries,* Harvard Business School Press, Boston.

Posner, M.V. (1961), "International Trade and Technical Change", *Oxford Economic Papers,* Vol. 13, No. 3.

Powell, W.W. (1990), "Neither Market nor Hierarchy: Network Forms of Organisation", in B.N. Straw and L.L. Cummings, eds., *Research in Organisational Behaviour,* Vol. 12, No. 2.

Press, F. (1990), "The Role of Education in Technological Competitiveness", *Siemens Review,* February.

Pugel, J.A. (1980), "The Changing Position of US Industries in the Global Pattern of Industrial Production" in the Joint Economic Committee of the US Congress, *Special Study on Economic Change: US Role in a World Market,* US Government Printing Office, Washington D.C.

Rada, J. (1980), *The Impact of Microelectronics. A Tentative Appraisal of Information Technology,* ILO, Geneva.

Ranta, J. (1989), "Tomorrow's Industries, Structural Changes and the Future of Work Organization", in Cooley.

Ray, G. (1984), *The Diffusion of Mature Technologies,* Cambridge University Press.

Ray, G. (1989), "Full Circle: The Diffusion of Technology", *Research Policy,* Vol. 18, No. 1.

Rehbinder, E. and R. Stewart (1985), *Environmental Protection Policy,* Walter de Gruyter, New York.

Reich, R.B. (1990), "But Now We're Global", review of Porter (1990) and Dertouzos *et al.* (1989), in the *Times Literary Supplement* (Economics Section), 31 August.

Reidel, J. (1980), "The Symptoms of Declining US International Competitiveness", in Joint Economic Committee, Congress of the United States, *Special Study on Economic Change: The International Economy: US Role in a World Market,* US Government Printing Office, Washington D.C.

Reinganum, J. (1984), "Practical Implications of Game-Theoretic Models of R&D", *American Economic Review,* Papers and Proceedings, Vol. 74, No. 2.

Reinganum, J. (1989), "The Timing of Innovation: Research, Development, and Diffusion", in R. Schmalensee and R. Willing, eds., *The Handbook of Industrial Organisation,* Vol. 1, North-Holland, Amsterdam.

Reppy, J. (1990), "Military Research and Economic Performance", paper presented to the Paris TEP Technology and Competitiveness Conference, June.

Resnick, D. (1987), *Education and Learning to Think,* National Academy Press, Washington D.C.

Richardson, G.B. (1972), "The Organisation of Industry", *Economic Journal,* Vol. 82, No. 327.

Robson, M., J. Townsend and K. Pavitt (1988), "Sectoral Patterns of Production and Use of Innovations", *Research Policy,* Vol. 17, February.

Rogers, E. (1982), "Information Exchange and Technological Innovation", in D. Sahal, ed., *The Transfer and Utilization of Technical Knowledge,* Lexington Books, Lexington, Mass.

Romer, P. (1987), "Crazy Explanations for the Productivity Slowdown", *Macroeconomics Annual,* National Bureau of Economic Research, Washington D.C.

Romer, P. (1988), "Comment on M.N. Baily and R.J. Gordon. – The Productivity Slowdown, Measurement Issues, and the Explosion of Computer Power", *Brookings Papers on Economic Activity,* Vol. 2.

Romer, P. (1989), "What Determines the Rate of Growth and Technological Change", The World Bank Working Papers, WPS 279, September.

Romer, P. (1990), "Capital, Labour and Productivity", *Brookings Papers on Economic Activity,* Microeconomics Issue.

Roos, D. (1991), "The Importance of Organisational Structure and Production System Design in the Development of New Technology", in OECD (1991*c*).

Rosenberg, N. (1976), *Perspectives on Technology,* Cambridge University Press.

Rosenberg, N. (1982), *Inside the Black Box: Technology and Economics,* Cambridge University Press.

Rosenberg, N. (1990), "Why Do Firms Do Basic Research (With Their Own Money)?", *Research Policy,* Vol. 19, No. 2.

Rosenberg, N. (1991), "Critical Issues in Science Policy Research", Opening Address to the SPRU 25th Anniversary Conference, in *Science and Public Polices,* Vol. 18, No. 6.

Rullani, E. and A. Zanfei (1988), "Networks Between Manufacturing and Demand: Cases from Textile and Clothing Industries", in Antonelli (1988*a*).

Rush, H. and J. Bessant (1990), "The Diffusion of Manufacturing Technology", *The OECD Observer,* No. 166, August/September.

Sahal, D. (1981), *Patterns of Technological Innovation,* Addison-Wesley, New York.

Salter, W.E.G. (1960), *Productivity and Technical Change,* Cambridge University Press.

Sauvant, K.P. (1986), *Trade and Foreign Direct Investment in Data Services,* Westview Press, London.

Sauviat, C. (1990), "Services et compétitivité: une relation équivoque", paper presented to the Paris TEP Technology and Competitiveness Conference, June.

Schelling, T. (1988), "Global Environmental Forces", Cambridge Energy and Environmental Policy Centre, John F. Kennedy School of Government, Harvard University, November.

Scherer, F.M. (1970), *Industrial Market Structure and Economic Performance,* Rand McNally College Publishing Company, Chicago.

Scherer, F.M. (1982*a*), "Interindustry Technology Flows in the United States", *Research Policy,* Vol. 11, No. 4.

Scherer, F.M. (1982*b*), "Interindustry Technology Flows and Productivity Growth", *Review of Economics and Statistics,* Vol. 64.

Scherer, F.M. (1984), "Using Linked Patent and R&D Data to Measure Interindustry Technology Flows", in Z. Griliches, ed., *R&D, Patents and Productivity,* Chicago University Press, Chicago.

Schmookler, J. (1966), *Invention and Economic Growth,* Harvard University Press, Cambridge, Mass.

Scholz, L. (1990), "Changing Structure of Investment in Different Industries", paper presented to the Stockholm TEP Technology and Investment Conference, January.

Scholz, L. (forthcoming), "Innovation Surveys in Germany", *STI Review*, OECD, Paris.

Schrader, S. (1991), "Information Technology Transfer between Firms: Co-operation through Information Trading", *Research Policy*, Vol. 20, No. 2.

Schumpeter, J.A. (1947), *Capitalism, Socialism and Democracy*, Harper and Row, New York.

Schware, R. (1989), "Trends in the World-wide Software Industry and Software Engineering: Opportunities and Constraints for the Newly-Industrialising Economies (NIEs)", unpublished document, OECD Development Centre, June.

Sciberras, E. (1980), "Technical Innovation and Industrial Competitiveness in the Television Industry", Occasional Paper, SPRU, University of Sussex.

Sciberras, E. and B. Payne (1985), *Technology and International Competitiveness: The Machine-Tool Industry*, Longman, London.

Scott, J. (1986), *Capitalist Property and Financial Power*, Wheatsheaf Books, Brighton.

Scott, M.F. (1989), *A New View of Economic Growth*, Clarendon Press, Oxford.

Sengenberger, W., G.W. Loveman and M.J.·Piore, eds., (1990), *The Re-emergence of Small Enterprises*, International Institute for Labour Studies, ILO, Geneva.

Shimada, H. (1991), "Humanware, Technology and Industrial Relations", in OECD (1991*c*).

Siebert, H. (1989), "The Single European Market: A Schumpeterian Event?", Kiel Discussion Papers, No. 157, November.

Sigurdson, J. (1986), *Industry and State Partnership in Japan: The VLSI Project*, Research Policy Institute, Lund.

Silverberg, G. (1990), "Adoption and Diffusion of Technology as a Collective Evolutionary Process", in Freeman and Soete.

Smith, C. (1990), "Comments Before the MIT Working Group on Business and the Environment", April.

Smith, K. (1990), "Interaction between Public and Private R&D Investment in Four Norwegian Industries", paper presented to the Stockholm TEP Technology and Investment Conference, January.

Smith, S. (1991), "Industrial Policy in Developing Countries: Reconsidering the Real Sources of Export-led Growth", Economic Policy Institute, Washington, D.C.

Soete, L. (1991), "National Support Policies for Strategic Industries: The International Implications", in OECD (1991*h*).

Solomon, R.J. (1990), "Broadband Communications as a Development Problem", in *STI Review*, No. 7, OECD, Paris.

Spence, M. (1984), "Cost Reduction, Competition and Industry Performance", *Econometrica*, January.

Speth, J.G. (1990), "Needed: An Environmental Revolution in Technology", paper prepared for the World Resources Institute Conference, Towards 2000: Environment Technology and the New Century, June.

SPRU (1984), *Innovation Survey*, University of Sussex, Brighton.

Steindl, G. (1952), *Maturity and Stagnation in American Capitalism*, Oxford University Press.

Stern, D. (1989), "Elements of Learning-Intensive Production", paper presented to the AERA Conference, United States.

Stewart, R. (1981), "Regulation, Innovation, and Administrative Law: A Conceptual Framework", *California Law Review*.

Stiglitz, J. (1987), "Learning to Learn, Localised Learning and Technological Progress", in Dasgupta and Stoneman.

Stopford, J.M. and J.R. Wells (1972), *Managing the Multinational Enterprise*, Basic Books, New York.

Stowsky, J. (1986), "Beating Our Plowshares into Double-Edged Swords: The Impact of Pentagon Policies of Commercialization of Advanced Technologies", BRIE Working Paper No. 17, University of California, Berkeley.

Summerville, P.A. (1990), "Regional Investment by Asian Nations Placing Area at Core of Global Economy", *The Japan Economic Journal*, 28 April.

Sweden, *Statistika Meddelanders*, F13 SM 8802, Stockholm.

Teece, D.J. (1982), "Towards an Economic Theory of the Multiproduct Firm", *Journal of Economic Behaviour and Organization*, Vol. 3.

Teece, D.J. (1986), "Profiting from Technological Innovation", *Research Policy*, Vol. 15, No. 6.

Teece, D.J. (1991), "Support Policies for Strategic Industries: Impact on Home Economies", in OECD (1991*h*).

Thailand Development Research Institute (TDRI) (1989), *The Development of Thailand's Technological Capability in Industry - Final Report*, 6 Vols., Bangkok.

Tilton, J. (1971), *International Diffusion of Technology: The Case of Semiconductors*, The Brookings Institution, Washington D.C.

Tirman, J. (1984), *The Militarization of High Technology*, Ballinger, Cambridge, Mass.

Toyne, B. J.S. Arpan, D.A. Ricks, T.A. Shimp and A. Barnett (1984), *The Global Textile Industry*, Allen & Unwin, London.

UN (United Nations) (1968), "A System of National Accounts", *Studies in Methods Series F*, No. 2, Rev. 3, Statistical Office of the United Nations, New York.

UN (1980), "Classification of the Functions of Government", *Statistical Papers Series M*, No. 70, Statistical Office of the United Nations, New York.

UN (1986*a*), *National Accounts Statistics*, New York.

UN (1986*b*), *Standard International Trade Classification*, Rev. 3, New York.

UN (1986*c*), "Handbook of National Accounting. Accounting for Production: Sources and Methods", *Studies in Methods, Series F*, No. 39, New York.

UN (1990), "International Standard Classification of All Economic Activities", *Statistical Papers Series M*, No. 4, Rev. 3, Statistical Office of the United Nations, New York.

UN (1991), "Provisional Central Product Classification", *Statistical Papers Series M*, No. 77, Statistical Office of the United Nations, New York.

UNCTAD (United Nations Conference on Trade and Development) (1986), "Impact of New and Emerging Technologies on Trade and Development", TD/B/C.6/136, Geneva.

UNCTAD (1987), *Trade and Development Report 1987*, Geneva.

UNCTAD (1988), *Trade and Development Report 1988*, Geneva.

UNCTAD (1989a), *Trade in Services. Sectoral Issues*, Geneva.

UNCTAD (1989b), "A Review of the Manufacturing Sector in the Least Developed Countries – the Implementation of SNPA in the Eighties and Proposals for Further Action", contribution by United Nation Industrial Development Organization, A/CONF/147/DR/3/Add.3, 10th March.

UNCTAD (1989c), *Trade and Development Report, 1989*, Geneva.

UNCTC (United Nations Center on Transnational Corporations) (1988), *Transnational Corporations in World Development: Trends and Prospects*, New York.

UNCTC (1990), "Recent Developments Related to Transnational Corporations and International Economic Relations", E/C-10/1990/2, New York.

UNCTC (1991), "World Investment Report 1991. The Triad in Foreign Direct Investment", United Nations, New York.

UN ECLAC (Economic Commission for Latin America and the Caribbean) (1990), *Changing Production Patterns with Social Equity*, Santiago de Chile.

UNESCO (1976), "International Standard Classification of Education (ISCED)", Paris.

UNESCO (1978), "Recommendations Concerning the International Standardization of Statistics on Science and Technology", 27th November, Paris.

UNESCO (1990), *Statistical Year Book*, Paris.

UNIDO (United Nations Industrial Development Organisation) (1985), *Global Report 1985*, New York.

United States Council on Environmental Quality (1981), *Environmental Trends*, US Government Printing Office, Washington D.C.

United States, National Academy of Engineering (1987), *Balancing the National Interest: US National Security Export Controls and Global Economic Competition*, National Academy Press, Washington D.C.

United States, National Academy of Engineering (1990), *National Interests in an Age of Global Technology*, Washington D.C.

United States, National Science Board (1989), *Science and Engineering Indicators – 1989*, US Government Printing Office, Washington D.C.

United States, NSF (National Science Foundation) (1986), "US Scientists and Engineers", US Government Printing Office, Washington D.C.

United States, NSF (1990), "Future Scarcities of Scientists and Engineers: Problems and Solutions", Private Communications, April.

United States, NSF (annual), *Science Indicators*, US Government Printing Office, Washington D.C.

United States, OTA (Office of Technology Assessment) (1986), "Serious Reduction of Hazardous Waste: For Pollution Prevention and Efficiency", OTA-ITE-317, US Government Printing Office, September, Washington D.C.

United States, President's Commission on Industrial Competitiveness (1985), "Global Competition: The New Reality", *Analytical Reports*, Vol. 2, US Government Printing Office, Washington D.C.

United States, The Conference Board (1988), "Beyond Business/Education Partnerships: The Business Experience Report from the Conference Board", Washington, D.C.

Van Raan, A.F.J. and R.J.W. Tijssen (1990), "An Overview of Quantitative Science and Technology Indicators Based on Bibliometric Methods", paper presented to the Paris TEP Indicators Conference, July.

Van Zon, A. (1991), "Vintage Capital and the Measurement of Technological Progress", in OECD (1991c).

Vernon, R. (1966), "International Investment and International Trade in the Product Cycle", *Quarterly Journal of Economics*, Vol. 80, No. 1.

Vernon, R. (1974), "The Location of Economic Activity", in J.H. Dunning, ed., *Economic Analysis and the Multinational Enterprise*, George Allen & Unwin, London.

Vickery, G. (1986), "International Flows of Technology", *STI Review*, No. 1, OECD, Paris.

Vickery, G. (1988) "A survey of International Technology Licensing", *STI Review*, No. 4, OECD, Paris.

Vickery, G. and D. Campbell (1989), "Advanced Manufacturing Technology and the Organisation of Work", *STI Review*, No. 6, OECD, Paris.

Walsh, V. (1987), "Technology, Competitiveness and the Special Problems of Small Countries", *STI Review*, No. 2, OECD, Paris.

Watanabe, S. (1983), *Market Structure, Industrial Organisation and Technological Development: The Case of the Japanese Electronics based NC-Machine Tool Industry*, ILO, Geneva.

Watanabe, S. (forthcoming), "The Diffusion of New Technologies, Management Styles and Work Organisation in Japan: A Survey of Empirical Studies", in OECDd.

WCED (World Commission on Environment and Development) (1987), *Our Common Future*, Oxford University Press.

Weidner, H. (1989), "Japanese Environmental Policy in an International Perspective: Lessons for a Preventive Approach," in S. Turo and H. Weidner, eds., *Environmental Policy in Japan*, Sigma, Berlin.

Weiermair, K. (1988), "The Labour Market and the Service Sector", The Frazer Institute, Canada.

Westphal, L., L. Kim and C. Dahlman (1985), "Reflections on the Republic of Korea's Acquisition of Technological Capability", in N. Rosenberg and C. Frischtak, eds., *International Transfer of Technology: Concepts, Measures and Comparisons*, Praeger, New York.

Westphal, L., K. Kritayakirana, K. Petchsuwan, H. Sutabutr and Y. Yuthavong (1990), "The Development of Technological Capability in Manufacturing: A Macroscopic Approach to Policy Research for Thailand", in R.E. Evenson and G. Ranis, eds., *Science and Technology Lessons for Development Policy*, Westview, Boulder.

Williams, S., S. Fenn and T. Clausen (1990), "Renewing Renewable Energy", *Issues in Science and Technology,* Spring.

Williamson, O.E. (1975), *Markets and Hierarchies: Analysis and Antitrust Implications: A Study in the Economies of International Organisation,* The Free Press, New York.

Williamson, O.E. (1985), *The Economic Institutions of Capitalism,* The Free Press, New York.

Willinger, M. and E. Zuscovitch (1988), "Towards the Economics of Information Intensive Production Systems: The Case of Advanced Materials", in Dosi *et al.*

Womack, J.P., D.T. Jones and D. Roos (1990), *The Machine that Changed the World,* Rawson Associates, New York.

World Bank (1989), *World Development Report 1989,* Washington D.C.

Wortmann (1990), "Multinationals and the Internationalization of R&D: New Developments in German Companies", *Research Policy,* Vol. 19, No. 2.

Yoshihara, K. (1988), *The Rise of Ersatz Capitalism in South-East Asia,* Oxford University Press.

Yoshimoto, M. (1990), "Economic Functions of Keiretsu in Japan", *Insights,* Institute for International Economics, Washington D.C.

Yoshitomi, M. (1991), "New Trends of Oligopolistic Competition in the Globalisation of High-Tech Industries: Interactions among Trade, Investment and Government", in OECD (1991*h*).

Yuan, R. (1987), "An Overview of Biotechnological Transfer in Our International Context", *Genetic Engineering News,* March.

Zinberg, D.L. (1991), "Contradictions and Complexity: International Comparisons in the Training of Foreign Scientists and Engineers" in Zinberg, D.L., ed., *The Changing University: How Increased Demand for Scientists and Engineers is Transforming Academic Institutions Internationally,* Kluwer Academic Publishers, Dordrecht.

Zysman, J. (1990), "Trade, Technology and National Competitiveness", in Deiaco *et al.*

WHERE TO OBTAIN OECD PUBLICATIONS – OÙ OBTENIR LES PUBLICATIONS DE L'OCDE

Argentina – Argentine
CARLOS HIRSCH S.R.L.
Galería Güemes, Florida 165, 4° Piso
1333 Buenos Aires   Tel. 30.7122, 331.1787 y 331.2391
Telegram: Hirsch-Baires
Telex: 21112 UAPE-AR. Ref. s/2901
Telefax:(1)331-1787

Australia – Australie
D.A. Book (Aust.) Pty. Ltd.
648 Whitehorse Road, P.O.B 163
Mitcham, Victoria 3132   Tel. (03)873.4411
Telefax: (03)873.5679

Austria – Autriche
OECD Publications and Information Centre
Schedestrasse 7
D-W 5300 Bonn 1 (Germany)   Tel. (49.228)21.60.45
Telefax: (49.228)26.11.04
Gerold & Co.
Graben 31
Wien I   Tel. (0222)533.50.14

Belgium – Belgique
Jean De Lannoy
Avenue du Roi 202
B-1060 Bruxelles   Tel. (02)538.51.69/538.08.41
Telex: 63220   Telefax: (02) 538.08.41

Canada
Renouf Publishing Company Ltd.
1294 Algoma Road
Ottawa, ON K1B 3W8   Tel. (613)741.4333
Telex: 053-4783   Telefax: (613)741.5439
Stores:
61 Sparks Street
Ottawa, ON K1P 5R1   Tel. (613)238.8985
211 Yonge Street
Toronto, ON M5B 1M4   Tel. (416)363.3171
Federal Publications
165 University Avenue
Toronto, ON M5H 3B8   Tel. (416)581.1552
Telefax: (416)581.1743
Les Publications Fédérales
1185 rue de l'Université
Montréal, PQ H3B 3A7   Tel.(514)954-1633
Les Éditions La Liberté Inc.
3020 Chemin Sainte-Foy
Sainte-Foy, PQ G1X 3V6   Tel. (418)658.3763
Telefax: (418)658.3763

Denmark – Danemark
Munksgaard Export and Subscription Service
35, Nørre Søgade, P.O. Box 2148
DK-1016 København K   Tel. (45 33)12.85.70
Telex: 19431 MUNKS DK   Telefax: (45 33)12.93.87

Finland – Finlande
Akateeminen Kirjakauppa
Keskuskatu 1, P.O. Box 128
00100 Helsinki   Tel. (358 0)12141
Telex: 125080   Telefax: (358 0)121.4441

France
OECD/OCDE
Mail Orders/Commandes par correspondance:
2, rue André-Pascal
75775 Paris Cédex 16   Tel. (33-1)45.24.82.00
Bookshop/Librairie:
33, rue Octave-Feuillet
75016 Paris   Tel. (33-1)45.24.81.67
     (33-1)45.24.81.81
Telex: 620 160 OCDE
Telefax: (33-1)45.24.85.00 (33-1)45.24.81.76
Librairie de l'Université
12a, rue Nazareth
13100 Aix-en-Provence   Tel. 42.26.18.08
Telefax : 42.26.63.26

Germany – Allemagne
OECD Publications and Information Centre
Schedestrasse 7
D-W 5300 Bonn 1   Tel. (0228)21.60.45
Telefax: (0228)26.11.04

Greece – Grèce
Librairie Kauffmann
28 rue du Stade
105 64 Athens   Tel. 322.21.60
Telex: 218187 LIKA Gr

Hong Kong
Swindon Book Co. Ltd.
13 - 15 Lock Road
Kowloon, Hong Kong   Tel. 366.80.31
Telex: 50 441 SWIN HX   Telefax: 739.49.75

Iceland – Islande
Mál Mog Menning
Laugavegi 18, Pósthólf 392
121 Reykjavik   Tel. 15199/24240

India – Inde
Oxford Book and Stationery Co.
Scindia House
New Delhi 110001   Tel. 331.5896/5308
Telex: 31 61990 AM IN
Telefax: (11)332.5993
17 Park Street
Calcutta 700016   Tel. 240832

Indonesia – Indonésie
Pdii-Lipi
P.O. Box 269/JKSMG/88
Jakarta 12790   Tel. 583467
Telex: 62 875

Ireland – Irlande
TDC Publishers – Library Suppliers
12 North Frederick Street
Dublin 1   Tel. 744835/749677
Telex: 33530 TDCP EI   Telefax: 748416

Italy – Italie
Libreria Commissionaria Sansoni
Via Benedetto Fortini, 120/10
Casella Post. 552
50125 Firenze   Tel. (055)64.54.15
Telex: 570466   Telefax: (055)64.12.57
Via Bartolini 29
20155 Milano   Tel. 36.50.83
La diffusione delle pubblicazioni OCSE viene assicurata
dalle principali librerie ed anche da:
Editrice e Libreria Herder
Piazza Montecitorio 120
00186 Roma   Tel. 679.46.28
Telex: NATEL I 621427
Libreria Hoepli
Via Hoepli 5
20121 Milano   Tel. 86.54.46
Telex: 31.33.95   Telefax: (02)805.28.86
Libreria Scientifica
Dott. Lucio de Biasio 'Aeiou'
Via Meravigli 16
20123 Milano   Tel. 805.68.98
Telefax: 800175

Japan – Japon
OECD Publications and Information Centre
Landic Akasaka Building
2-3-4 Akasaka, Minato-ku
Tokyo 107   Tel. (81.3)3586.2016
Telefax: (81.3)3584.7929

Korea – Corée
Kyobo Book Centre Co. Ltd.
P.O. Box 1658, Kwang Hwa Moon
Seoul   Tel. (REP)730.78.91
Telefax: 735.0030

Malaysia/Singapore – Malaisie/Singapour
Co-operative Bookshop Ltd.
University of Malaya
P.O. Box 1127, Jalan Pantai Baru
59700 Kuala Lumpur
Malaysia   Tel. 756.5000/756.5425
Telefax: 757.3661
Information Publications Pte. Ltd.
Pei-Fu Industrial Building
24 New Industrial Road No. 02-06
Singapore 1953   Tel. 283.1786/283.1798
Telefax: 284.8875

Netherlands – Pays-Bas
SDU Uitgeverij
Christoffel Plantijnstraat 2
Postbus 20014
2500 EA's-Gravenhage   Tel. (070 3)78.99.11
Voor bestellingen: Tel. (070 3)78.98.80
Telex: 32486 stdru   Tel. (070 3)47.63.51

New Zealand – Nouvelle-Zélande
GP Publications Ltd.
Customer Services
33 The Esplanade - P.O. Box 38-900
Petone, Wellington
Tel. (04)685-555   Telefax: (04)685-333

Norway – Norvège
Narvesen Info Center - NIC
Bertrand Narvesens vei 2
P.O. Box 6125 Etterstad
0602 Oslo 6   Tel. (02)57.33.00
Telex: 79668 NIC N   Telefax: (02)68.19.01

Pakistan
Mirza Book Agency
65 Shahrah Quaid-E-Azam
Lahore 3   Tel. 66839
Telex: 44886 UBL PK. Attn: MIRZA BK

Portugal
Livraria Portugal
Rua do Carmo 70-74, Apart. 2681
1117 Lisboa Codex   Tel.: 347.49.82/3/4/5
Telefax: (01) 347.02.64

Singapore/Malaysia – Singapour/Malaisie
See "Malaysia/Singapore" – Voir «Malaisie/Singapour»

Spain – Espagne
Mundi-Prensa Libros S.A.
Castelló 37, Apartado 1223
Madrid 28001   Tel. (91) 431.33.99
Telex: 49370 MPLI   Telefax: 575.39.98
Libreria Internacional AEDOS
Consejo de Ciento 391
08009 - Barcelona   Tel. (93) 301-86-15
     Telefax: (93) 317-01-41
Llibreria de la Generalitat
Palau Moja, Rambla dels Estudis, 118
08002 - Barcelona   Telefax: (93) 412.18.54
Tel. (93) 318.80.12 (Subscripcions)
(93) 302.67.23 (Publicacions)

Sri Lanka
Centre for Policy Research
c/o Mercantile Credit Ltd.
55, Janadhipathi Mawatha
Colombo 1   Tel. 438471-9, 440346
Telex: 21138 VAVALEX CE   Telefax: 94.1.448900

Sweden – Suède
Fritzes Fackboksföretaget
Box 16356, Regeringsgatan 12
103 27 Stockholm   Tel. (08)23.89.00
Telex: 12387   Telefax: (08)20.50.21
Subscription Agency/Abonnements:
Wennergren-Williams AB
Nordenflychtsvägen 74, Box 30004
104 25 Stockholm   Tel. (08)13.67.00
Telex: 19937   Telefax: (08)618.62.32

Switzerland – Suisse
OECD Publications and Information Centre
Schedestrasse 7
D-W 5300 Bonn 1 (Germany)   Tel. (49.228)21.60.45
Telefax: (49.228)26.11.04
Librairie Payot
6 rue Grenus
1211 Genève 11   Tel. (022)731.89.50
Telex: 28356
Subscription Agency – Service des Abonnements
Naville S.A.
7, rue Lévrier
1201 Genève   Tél.: (022) 732.24.00
Telefax: (022) 738.48.03
Maditec S.A.
Chemin des Palettes 4
1020 Renens/Lausanne   Tel. (021)635.08.65
Telex: (021)635.07.80
United Nations Bookshop/Librairie des Nations-Unies
Palais des Nations
1211 Genève 10   Tel. (022)734.14.73
Telex: 412962   Telefax: (022)740.09.31

Taiwan – Formose
Good Faith Worldwide Int'l. Co. Ltd.
9th Floor, No. 118, Sec. 2
Chung Hsiao E. Road
Taipei   Tel. 391.7396/391.7397
Telefax: (02) 394.9176

Thailand – Thaïlande
Suksit Siam Co. Ltd.
1715 Rama IV Road, Samyan
Bangkok 5   Tel. 251.1630

Turkey – Turquie
Kültur Yayinlari'Is-Türk Ltd. Sti.
Atatürk Bulvari No. 191/Kat. 21
Kavaklidere/Ankara   Tel. 25.07.60
Dolmabahce Cad. No. 29
Besiktas/Istanbul   Tel. 160.71.88
Telex: 43482B

United Kingdom – Royaume-Uni
HMSO
Gen. enquiries   Tel. (071) 873 0011
Postal orders only:
P.O. Box 276, London SW8 5DT
Personal Callers HMSO Bookshop
49 High Holborn, London WC1V 6HB
Telex: 297138   Telefax: 071 873 2000
Branches at: Belfast, Birmingham, Bristol, Edinburgh,
Manchester

United States – États-Unis
OECD Publications and Information Centre
2001 L Street N.W., Suite 700
Washington, D.C. 20036-4910   Tel. (202)785.6323
Telefax: (202)785.0350

Venezuela
Libreria del Este
Avda F. Miranda 52, Aptdo. 60337, Edificio Galipán
Caracas 106   Tel. 951.1705/951.2307/951.1297
Telegram: Libreste Caracas

Yugoslavia – Yougoslavie
Jugoslovenska Knjiga
Knez Mihajlova 2, P.O. Box 36
Beograd   Tel.: (011)621.992
Telex: 12466 jk bgd   Telefax: (011)625.970

Orders and inquiries from countries where Distributors
have not yet been appointed should be sent to: OECD
Publications Service, 2 rue André-Pascal, 75775 Paris
Cedex 16, France.

Les commandes provenant de pays où l'OCDE n'a pas
encore désigné de distributeur devraient être adressées à :
OCDE, Service des Publications, 2, rue André-Pascal,
75775 Paris Cédex 16, France.

75880–7/91

OECD PUBLICATIONS, 2 rue André-Pascal, 75775 PARIS CEDEX 16
PRINTED IN FRANCE
(92 92 02 1) ISBN 92-64-13622-3 — No. 45859 1992